MAGNETIC SPINELS-SYNTHESIS, PROPERTIES AND APPLICATIONS

Edited by **Mohindar Singh Seehra**

Magnetic Spinels- Synthesis, Properties and Applications
http://dx.doi.org/10.5772/63249
Edited by Mohindar Singh Seehra

Contributors

Jana Vejpravova, Daniel Niznansky, Barbara Pacakova, Simona Kubickova, Alice Reznickova, Tomohiro Iwasaki, Rajender Singh, Gadipelly Thirupathi, Gang Liu, Pengjun Ma, Qingfen Geng, Subhash Thota, Sobhit Kumar Singh, Zhen-Yu Tian, Patrick Mountapmbeme Kouotou, Guan-Fu Pan, Edilso Reguera, D Darminto, Malik Anjelh Baqiya, Mochamad Zainuri, Tri Wikantoro, Khuroti Ayun, Ahmad Taufiq, S Sunaryono, Suminar Pratapa, Sílvia Soreto, Manuel Pedro Graça, Manuel Valente, Luís Costa, Andréa Kalendová, Kateřina Nechvílová, Miroslav Kohl, Zhaorong Yang, Weiwei Xie, Oscar Perales-Perez, Yarilyn Cedeño-Mattei

Published by InTech
Janeza Trdine 9, 51000 Rijeka, Croatia

Publishing Process Manager Romina Rovan
Technical Editor SPi Global
Cover InTech Design team

First published March, 2017
Printed in Croatia
Legal deposit, Croatia: National and University Library in Zagreb

Additional hard copies can be obtained from orders@intechopen.com

Magnetic Spinels- Synthesis, Properties and Applications, Edited by Mohindar Singh Seehra
p. cm.
Print ISBN 978-953-51-2973-8
Online ISBN 978-953-51-2974-5

For EU product safety concerns:
IN TECH d.o.o., Prolaz Marije Krucifikse Kozulić 3, 51000 Rijeka, Croatia,
info@intechopen.com or visit our website at intechopen.com.

Meet the editor

Professor Mohindar Singh Seehra received his early education in India (BSc, Punjab University; MSc, Aligarh Muslim University). He joined West Virginia University as Assistant Professor of Physics in 1969 after receiving his PhD degree (University of Rochester, NY) and becoming Associate Professor in 1973, Professor in 1977, and Eberly Distinguished Professor in 1992. His honors in research include A. P. Sloan Foundation Research Fellow (1974–1976); Fellow of the American Physical Society (1984–present); Fellow of the Institute of Physics, UK (2001–present); and Outstanding Referee Award of the American Physical Society in 2010. He has guided the research of 65 graduate and postdoctoral students and authored over 300 publications on the magnetic, dielectric, optical, and catalytic properties of materials and nanoparticles. In 2016, he became Eberly Professor Emeritus.

Contents

Preface

The magnetism of lodestone (Fe_3O_4), now known to be a room temperature ferrimagnet with the spinel structure, was discovered many centuries ago. However, it was likely the research work of Snoek and others in the mid-twentieth century [1–3] that really kindled the research interest of the scientific community to this new class of magnetic materials whose structure is based on nonmagnetic mineral called spinel with the chemical formula $MgAl_2O_4$. These materials are magnetic spinels with the formula $AB_2X_4 = [A][B_2]X_4$, where A and B are generally divalent and trivalent cations, respectively, and X is usually oxygen although spinels with X = S, Se, Te, etc. have also been investigated. This spinel structure is cubic with cell length of about 0.84 nm, and it normally contains 8 formula units of the spinel. Ferrites are a special class of spinels with the general formula MFe_2O_4, M being a divalent metal ion.

The continued worldwide interest in these materials stems from two primary factors: (i) A variety of magnetic and nonmagnetic ions can be accommodated on the eight tetrahedral A sites and 16 octahedral B sites leading to new materials and a wealth of new physics and phenomena resulting from their fundamental investigations, and (ii) unlike metallic room temperature ferromagnets such as Fe, Co, and Ni, the magnetic spinels generally have very low electrical conductivity which leads to many applications of these spinels particularly in microwave communication devices and where reducing the eddy current losses is an important consideration. Recent developments in the use of magnetic spinels for microwave communication devices have been reviewed by Harris et al. [4]. The ferromagnetic-like magnetic moment observed in many of these spinels at room temperature is due to ferrimagnetism resulting from overall noncancelation of the magnetic moments aligned usually antiferromagnetically on the A and B sites. The theory of this ferrimagnetism was first advanced by Néel [5], and it is described in detail in the book by Morrish [6] where some of the other basic properties of ferrimagnets and magnetic spinels are also given.

The diversity of the properties of the magnetic spinels is best illustrated by comparing the nature of magnetism in the three simple spinels: $Mn_3O_4 = [Mn^{2+}][2Mn^{3+}]O_4$, $Fe_3O_4 = [Fe^{3+}][Fe^{2+}Fe^{3+}]O_4$, and $Co_3O_4 = [Co^{2+}][2Co^{3+}]O_4$. It is noted that for Mn_3O_4 and Co_3O_4, the A site is occupied by divalent cations as expected in *normal* spinels. However, Fe_3O_4 is an example of an *inverse* spinel because the positions of half of the Fe^{3+} ions on the A sites and Fe^{2+} ions on the B site are inverted between the A and B sites. This degree of inversion can vary between 0 and 1 in more complex spinels. Magnetically, Mn_3O_4 is a ferrimagnet with an ordering temperature $T_C = 43$ K [7], Co_3O_4 is an antiferromagnet with Néel temperature $T_N = 30$ K [8], and Fe_3O_4 is a room temperature ferrimagnet with ordering temperature $T_C = 800$ K [9]. Such large differences in their magnetism illustrate how changes in magnetic ions lead to enormous differences in the magnetic properties of the spinels. The physics behind these differences is explained in the original papers listed above and references quoted in these publications.

There are certain topics of fundamental interest related to magnetic spinels which unfortunately did not get included in this book as separate chapters. Since these topics are considered to be quite important, a brief description of these topics is given here for the benefit of the readers of this book. The first such topic is the nature of magnetism in the group of normal spinels such as $ZnFe_2O_4$ and $CdFe_2O_4$ in which the A sites are occupied by nonmagnetic ions only, such as Zn^{2+} and Cd^{2+} in the above cases. A 1956 paper by Anderson [10] showed that in such a case where only the B sites are occupied by cations with magnetic moments, in the above case by Fe^{3+}, magnetic ordering is not possible at any nonzero temperature because of magnetic frustration. This indeed is observed in the above cases [6, 11]. Another recent example of such magnetic frustration is the defect spinel $MgMnO_3$ = (¾){Mg^{2+}}{$Mg^{2+}{}_{1/3}Mn^{4+}{}_{4/3}$ $V_{1/3}$}O_4 with V as vacancy on the octahedral sites; this material showed the presence of some short-range magnetic ordering but without a magnetic transition associated with long-range ordering [12].

Another topic of considerable importance but not covered in any of the chapters included here is the special area of defect spinels, the prime example of which is the mineral maghemite with the chemical formula γ-Fe_2O_3. For γ-Fe_2O_3, not all the usual octahedral B sites are occupied, and the unit cell contains (32/3) formula units arranged as (32/3) (γ-Fe_2O_3) = [8 Fe^{3+}][(40/3)Fe^{3+}.(8/3) V]O_{32}. Just like Fe_3O_4, γ-Fe_2O_3 is also a ferrimagnet at room temperature [6, 13]. However, in γ-Fe_2O_3, ferrimagnetism results from the difference in the number of antiferromagnetically aligned Fe^{3+} ions on the A and B sites, 8 and (40/3) in this case. In contrast, in magnetite with ionic distribution as $8Fe_3O_4$ = 8[Fe^{3+}][$Fe^{2+}Fe^{3+}$]O_4, ferrimagnetism results from the magnetic moments of 8 Fe^{2+} ions on the B sites since the magnetic moment on the antiferromagnetically aligned 8 Fe^{3+} ions each on A and B sites cancels out. For many applications, γ-Fe_2O_3 is sometimes preferred over Fe_3O_4 since Fe^{3+} ions are in the S-state in which spin-orbit coupling does not make any meaningful contribution to its magnetic moment. Among other known examples of defect spinels is $MgMnO_3$ which has already been mentioned in the previous paragraph, nanoparticles of $ZnMnO_3$ with the ionic arrangement as $4ZnMnO_{3-\delta}$ = 3 [$Zn_{1/3}{}^{2+}$ $Mn_{2/3}{}^{3+}Mn_{2/3}{}^{4+}$ $V_{1/3}$]$O_{11/3}$ with δ =0.25 [14] and nanoparticles of $Li_{0.5}Mg_{0.5}MnO_3$ [15]. The nature of magnetic ordering in the latter two cases is not yet certain since the reported measurements were done only on nanoparticles of about 20 nm size, and so the observations were likely dominated by finite size and surface effects. The magnetic properties of bulk system may indeed be different.

Another area which deserves mention and is the focus of major current interest involves the biomedical applications of magnetic nanoparticles, particularly nanoparticles of maghemite (γ-Fe_2O_3) and magnetite (Fe_3O_4) as drug delivery agents, because of their apparent biocompatibility and biodegradability [16–19]. For such applications, surfaces of the nanoparticles are functionalized and attached with appropriate therapeutic molecules. The room temperature super-paramagnetism of magnetic nanoparticles such as γ-Fe_2O_3 [13] and Fe_3O_4 [20] also allows their use for enhancing contrast in magnetic resonance imaging (MRI) [16–19].

This book on magnetic spinels contains 13 chapters written by different groups from around the world. These chapters deal with both the fundamental properties of the spinels as well as some of their applications. These chapters, each dealing with specific properties and applications, are intended to serve not only as a review of the specific research area but also present recent developments in that specific area. As editor, I have reviewed each chapter and made recommendations for changes which were generally accepted by the authors, and I thank these authors for their diligence and contributions to this project. Nevertheless, a uniform style of writing was not possible under this scenario, and therefore the individual style of each author will be evident in these chapters. Also, there is obvious lack of linking between different chapters unlike in a text book written by one or two authors. Nevertheless, the 13 chapters are loosely grouped into four parts based on the overall similarities of the topics in each part. Part 1 has three chapters dealing with the synthesis and properties of spinel ferrites and their nanoparticles, Part 2 contains three chapters on the electronic and magnetic properties of the spinels, Part 3 has two chapters on spinel-based ferrofluids, and Part 4 has five chapters dealing with the synthesis of various spinels and their applications. In each chapter, the authors have endeavored to bring out the current status of research in that particular area so that these chapters are good starting points for those readers who wish to pursue further research in that area. As noted above, although not all aspects of the magnetic spinels could be covered in these chapters, the topics that are covered have sufficient diversity to allow the readers to have good appreciation of the very interesting properties of the magnetic spinels and their many current and potential future applications. Different experimental techniques for the synthesis of specific spinels along with their structural and fundamental magnetic properties are covered in these chapters. Also, some of the basic properties of magnetic nanoparticles and how and why these properties differ from those of their bulk counterparts are covered. Therefore, it is my hope that the material presented in these chapters would be of interest to both new graduate students just starting research in spinels as well as to more seasoned researchers.

References

[1] Snoek, J. L. "New Developments in Ferromagnetic Materials" (Elsevier, Amsterdam, 1949).
[2] Gorter, E.W. Philips Res. Rept. 9, 295 (1954).
[3] Smit, J. and Wijn, H. P. J. "Ferrites" (John Wiley and Sons, Inc. New York, 1959).
[4] Harris, V. G. et al. J. Magn. Magn. Mater. 321, 2035 (2009).
[5] Néel, L. Ann. Phys. (Paris) 3, 137 (1948).
[6] Morrish, A. H. "The Physical Principles of Magnetism", (IEEE Press, New York, 2001).
[7] Srinivasan, G. and Seehra, M. S. Phys. Rev. B 28, 1 (1983)
[8] Dutta, P., Seehra, M. S., Thota, S. and Kumar, J. J. Phys.: Condens. Matter 20, 015218 (2008).
[9] Krycka, K. L., Borchers J. A., Booth R. A., Ijiri, Y., Hasz, K., Rhyne, J. J. and Majetich S. A., Phys. Rev. Lett. 113, 147203 (2014).
[10] Anderson, P. W. Phys. Rev. 102, 1008 (1956).
[11] Schiessl, W. et al. Phys. Rev. B 53, 9143 (1996).
[12] Seehra, M. S., Singh V. and Thota, S. J. Appl. Phys. 110, 113907 (2011).
[13] Pisane, K. L., Despeaux, E. C. and Seehra, M. S. J. Magn. Magn. Mater. 384, 148 (2015).

[14] Rall, J. D., Thota, S. Kumar, J. and Seehra, M. S. Appl. Phys. Lett. 100, 252407 (2012).
[15] Singh, V., Seehra, M. S., Manivannan, A. and Kumta, P. N., J. Appl. Phys. 111, 07E302. (2012).
[16] Thanh, N. T. K.(editor)," Magnetic Nanoparticles: From Fabrication to Clinical Applications" (CRC Press, Boca Raton, FL, USA, 2012).
[17] Revia, R. A. and Zhang, M. Materials Today, 19, 157 (2016).
[18] Pankhurst, Q. A. , Thanh, N. K. T., Jones, S. K. and Dobson, J., J. Phys. D.: Appl. Phys. 42, 224001 (2009).
[19] Wang C. et al. J. Control. Release. 136, 82 (2012).
[20] Dutta, C., Pal. S., Seehra, M. S., Shah N. and Huffman, G. P., J. Appl. Phys.105, 07B501 (2009).

Mohindar Singh Seehra
Department of Physics and Astronomy
West Virginia University
Morgantown, USA

Section 1

Synthesis, Properties and Applications

Chapter 1

Spinel Ferrite Nanoparticles: Correlation of Structure and Magnetism

Barbara Pacakova, Simona Kubickova,
Alice Reznickova, Daniel Niznansky and
Jana Vejpravova

Additional information is available at the end of the chapter

http://dx.doi.org/10.5772/66074

Abstract

This chapter focuses on the relationship between structural and magnetic properties of cubic spinel ferrite MFe_2O_4 (M = Mg, Mn, Fe, Co, Ni, Cu and Zn) nanoparticles (NPs). First, a brief overview of the preparation methods yielding well-developed NPs is given. Then, key parameters of magnetic NPs representing their structural and magnetic properties are summarized with link to the relevant methods of characterization. Peculiar features of magnetism in real systems of the NPs at atomic, single-particle, and mesoscopic level, respectively, are also discussed. Finally, the significant part of the chapter is devoted to the discussion of the structural and magnetic properties of the NPs in the context of the relevant preparation routes. Future outlooks in the field profiting from tailoring of the NP properties by doping or design of core-shell spinel-only particles are given.

Keywords: cubic spinel ferrite nanoparticles, magnetic properties, core-shell structure, particle size, spin canting, Mössbauer spectroscopy, magnetic susceptibility, size effect, superparamagnetism, magnetic anisotropy

1. Introduction

Spinel ferrite nanoparticles (NPs) are in the spotlight of current nanoscience due to immense application potential. Very interesting aspects of the spinel ferrite NPs are their excellent magnetic properties often accompanied with other functional properties, such as catalytic activity. Moreover, the magnetic response of the NPs can be tuned by particle size and shape up to some extent. Consequently, various spinel ferrite NPs are suggested as universal and multifunctional materials for exploitation in biomedicine [1–4], magnetic recording, catalysis [5–8] including magnetically separable catalysts [9–12], sensing [13–16] and beyond ($MgFe_2O_4$ in Li ion batteries [17, 18], or investigation of dopamine [19]). Thus it is of ultimate interest to get control over their functional properties, which requires in-depth understanding of the correlation between their structural and magnetic order. For example, the particle size and shape are extremely important both in biomedical imaging using Magnetic Resonance Imaging (MRI) [20] and therapies by means of magnetic field-assisted hyperthermia [21].

The chapter aims to summarize the most important aspects of magnetism of cubic spinel ferrite nanoparticles (MFe_2O_4, M = Mg, Mn, Fe, Co, Ni, Cu, and Zn) in context of their crystal and magnetic structure. The factors that drive magnetic performance of the spinel ferrite NPs can be recognized on three levels: on the atomic level (degree of inversion and the presence of defects), at single-particle level as balance between the crystallographically and magnetically ordered fractions of the NP (single-domain and multi-domain NPs, core-shell structure, and beyond), and at mesoscopic level by means of mutual interparticle interactions and size distribution phenomena. All these effects are strongly linked to the preparation routes of the NPs. In general, each preparation method provides rather similar NPs by means of morphology and crystalline order. Thus the "three-level" concept, which is the focal motif of the chapter, can be applied to all cubic spinel ferrite NPs.

2. Brief overview of preparation methods of spinel ferrite nanoparticles

In this section, selected methods of the NP preparation are summarized. Explicitly, the wet methods yielding well-defined NPs, either isolated or embedded in a matrix, are accented. The reason is that only such samples can be sufficiently characterized and the factors defined within the "three level" concept can be disentangled. Outstanding reviews on the specific method(s) with further details and references are also included [22–25].

Coprecipitation method is the archetype route, which can be used for preparation of all cubic spinel ferrite NPs: $Fe_3O_4/\gamma\text{-}Fe_2O_3$ [24, 26], $MgFe_2O_4$ [27], etc. In general, two water-soluble metallic salts are coprecipitated by a base. The reaction can be partly controlled in order to improve characteristics of the NPs [26, 28, 29]; however, it is generally reported as a facile method yielding polydispersed NPs with lower crystallinity and consequently less significant magnetic properties.

The family of decomposition routes includes wet approaches based on decomposition of metal organic precursors in high-boiling solvents, typically in the presence of coating agents (all [30], $Fe_3O_4/\gamma\text{-}Fe_2O_3$ [24, 31], $CoFe_2O_4$ [32, 33], $NiFe_2O_4$ [34], $ZnFe_2O_4$ [35]). The most common organic complexes used for decomposition are metal oleates and acetylacetonates. The decomposition methods yield highly crystalline particles close to monodisperse limit with very good magnetic properties. However, the reaction conditions must be controlled in correspondence of the growth model suggested by Cheng et al. [22]. They can be also tailored to produce NPs of different shapes ($CoFe_2O_4$ [36, 37], $Fe_3O_4/\gamma\text{-}Fe_2O_3$ [38, 39]). Higher-order assemblies of the NPs can be also achieved by varying the ratio of the precursor and reaction temperature ($CoFe_2O_4$ [40]). Alternatively, the decomposition takes place in high-pressure vessel (autoclave) [41, 42].

A large group of preparation protocols is based on solvothermal treatments, in aqueous conditions termed as hydrothermal. The preparation can be carried out either in a simple single-solvent system ($MgFe_2O_4$ [43], $NiFe_2O_4$ [44, 45]), mixture of solvents ($MnFe_2O_4$ [46], $ZnFe_2O_4$ [47]), surfactant-assisted routes ($ZnFe_2O_4$ [48]), or in multicomponent systems, such as water-alcohol-fatty acid (Fe_3O_4/γ-Fe_2O_3 [49], $CoFe_2O_4$ [50]). The solvothermal routes are often carried out at elevated pressure and can be also maintained in supercritical conditions ($MgFe_2O_4$ [51]). The NPs prepared by this class of methods are in general of very good crystallinity, in some cases competitive to the NPs obtained by the decomposition routes.

Spinel ferrite NPs can be also obtained with the help of normal or reverse micelle methods, often referred as microemulsion routes (all [52, 53], Fe_3O_4/γ-Fe_2O_3 [54], $MgFe_2O_4$ [55–57], $CoFe_2O_4$ [58], $NiFe_2O_4$ [59], $CoFe_2O_4$ and $MnFe_2O_4$ [60], $ZnFe_2O_4$ [61]). This approach takes advantage of the defined size of the micelle given by the ratio of the microemulsion components (water-organic phase-surfactant) according to the equilibrium phase diagram [62]. The micelles serve as nano-reactors, which exchange the constituents dissolved in the water phase during the reaction and self-limit the maximum size of the NPs. The as-prepared NPs are often subjected to thermal posttreatment, which improves the NP crystallinity and enhances magnetic properties. However, such NPs are no more dispersible in liquid phase and their applications are thus limited. On the other hand, the microemulsion technique can be used for preparation of the NPs of a defined shape [63] and mixed ferrites [64].

A modified polyol method is also used for preparation of spinel ferrite NPs [22, 24, 65–67]. While in the standard route the polyol acts as a solvent and sometimes reducing or complexing agent for metal ions, for preparation of the ferrite NPs, the reaction of 1,2-alkanediols and metal acetylacetonates in high-boiling solvents is the most common variant.

Sol-gel chemistry is a handy approach to produce spinel ferrite NPs. The common tactic is the growth of the NPs in porous silica matrix yielding well-developed NPs embedded in the transparent matrix. The route requires annealing of the gel; however, the particle size can be sufficiently varied by the annealing temperature. Different spinel ferrites can be prepared (Fe_3O_4/γ-Fe_2O_3 [68], $CoFe_2O_4$ [69–71], $MnFe_2O_4$ [72], $NiFe_2O_4$ [73–75]).

NPs of spinel ferrites are also prepared with assistance of microwaves [76, 77], ultrasound ($CoFe_2O_4$ [78], $MgFe_2O_4$ [79]), combustion routes ($MgFe_2O_4$ [80]), or mechanical treatments ($MgFe_2O_4$ [81], $MnFe_2O_4$ [82], $NiFe_2O_4$ [83]); however, the as-prepared NPs often require heat treatment, and the resulting samples with sufficient crystallinity are better classified as fine powders. Less common methods such as the use of electrochemical synthesis for the γ-Fe_2O_3 NPs [84] or $NiFe_2O_4$ [85] and synthesis employing ionic liquids for cubic magnetite NPs [86] were recently reported. NPs with size in the multicore limit were obtained by disaccharide-assisted seed growth [87]. Recently, combination of stop-flow lithography and coprecipitation was reported [88]. Typical TEM images of spinel ferrite NPs prepared by the most common routes are shown in **Figure 1**.

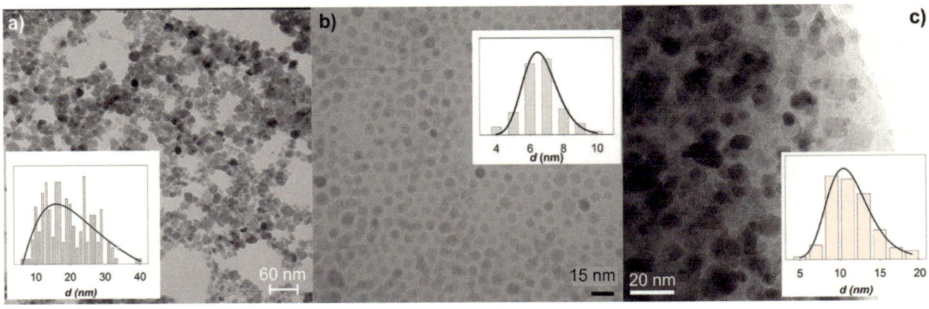

Figure 1. TEM images of the spinel NPs prepared by different methods, with distributions of particle diameters in the inset. (a) γ-Fe_2O_3 NPs prepared by coprecipitation technique. (b) $CoFe_2O_4$ NPs prepared by decomposition of oleic precursor. (c) $ZnFe_2O_4$ NPs prepared by sol-gel method.

The preparation methods described above can be successfully applied to preparation of core-shell NPs: $CoFe_2O_4$@MFe_2O_4; M = Ni, Cu, Zn or γ-Fe_2O_3 [89], and $MnFe_2O_4$@γ-Fe_2O_3 [90]; $CoFe_2O_4$@$ZnFe_2O_4$ [91]; $CuFe_2O_4$@$MgFe_2O_4$ [92]; or other mixed ferrites NPs [93, 94]. A natural core-shell structure is obtained for magnetite NPs due to topotactic oxidation to maghemite, which is mirrored, for example, in varying heating efficiency [95]. As a final remark, the selection of a particular preparation route yielding either a single core or multicore NPs is crucial and must be considered in the context of a specific application [96].

3. Characterization of magnetic nanoparticles: parameters and methods

In this section, the most important parameters characterizing structural and magnetic properties of NPs are introduced. Overview of the key experimental methods used for their evaluation is also included. For straightforwardness, details on the theoretical models and related formalism are not given, but relevant references are included. More details on the topic can be found in a comprehensive work by Koksharov [97].

3.1. Basic structural and magnetic characterization

The most important parameter is the particle size itself, usually attributed to the diameter of a single NP. The first-choice technique for determination of the particle size is the transmission electron microscopy (TEM), which gives the real (or physical) particle size, d_{TEM}. As the NPs of spinel ferrites are usually spherical, cubic, octahedral, or symmetric star-like objects, the value is a reasonable measure of the NP dimension as it gives information on the principal dimensions of those objects. Analysis of the TEM images also provides particle-size distribution, sometimes expressed as polydispersity index (PDI = $\sigma(d_{TEM})/\langle d_{TEM}\rangle$). The direct TEM observation gives information on aggregation, chaining of particles, and other morphological specifics. Using high-resolution TEM (HR TEM), internal structure of the NPs can be inspected, for example, the thickness of disordered surface layer and defects can be identified.

http://dx.doi.org/10.5772/66074

Particle size can be also determined using powder X-ray diffraction (XRD). The profile of the diffraction peak contains information about the so-called crystallite size, D_{hkl}, and the microstrain (arise from the presence of vacancies, dislocation, stacking faults, or poor crystallinity of the material). Generally, the experimental profile is the convolution of the instrumental profile caused by the experimental setup and the physical profile caused by the intrinsic properties of the measured material [98].

The physical profile is the convolution of the two dominant contributions caused by the small D_{hkl} and by the microstrain. The D_{hkl} is defined as a coherently diffracting length in a crystallographic direction [hkl] that is parallel to the diffraction vector (surface normal) [98]. Assuming the spherical NPs with random orientation of individual [hkl] directions, the D_{hkl} determines the diameter of the coherently diffracting domain; in other words it is the diameter of the crystalline part of the NP, the d_{XRD}. For highly symmetric shapes expected for spinel ferrite NPs, the coherently diffracting domain can be sufficiently described by a sphere or an ellipsoid in the case of flat crystallites.

Other important parameters characterizing magnetic NPs are related to formation of a single-domain state. In order to decrease the magnetostatic energy that is associated with the dipolar fields, the ferromagnetic (or ferrimagnetic)-ordered crystal is divided into the magnetic domains. Within each of the domain, the magnetization, M reaches the saturation. The domain creation depends on the competition between the reduction of the magnetostatic energy and the energy required to form the domain walls separating the adjacent domains. The size of the domain wall is a balance between the exchange energy that tries to unwind the domain wall and the magnetocrystalline anisotropy with the opposite effect.

In the magnetic NPs, the typical dimensions are comparable with the thickness of the domain; thus, at some critical size, it is energetically favorable for the NP to become single domain. The critical dimension ranging from 10^{-7} to 10^{-8} m is strictly specific to each magnetic spinel ferrite [99].

In small magnetic NPs reaching the single-domain regime, the paramagnetic-like behavior can be observed even below the Curie temperature, T_c. The state is therefore called the superparamagnetism (SPM) as the whole particle behaves as one giant spin (superspin) consisting of the atomic magnetic moments; thus, the magnetic moment of the whole NP is 10^2 to 10^5 times larger than the atomic moment. The magnetization follows the behavior of the Langevin function. The theory of SPM and superspin relaxation of the NPs was treated by C. P. Bean, J.D. Livingston and M. Knobel. et al. [100, 101]. The key parameters representing the magnetic properties of single-domain NPs are blocking temperature, T_B, and superspin or NP magnetic moment, μ_m. The T_B is related to the particle size through its volume, V as:

$$T_B = K_{eff} V / (a k_B) \tag{1}$$

where K_{eff} is the effective anisotropy constant. Parameter a is given by the measurement time, τ_m as $a = \ln(\tau_m / \tau_0)$, $a = 25$ for the SPM systems with relaxation time $\tau_0 = 10^{-12}$ s (see the following paragraphs) and $\tau_m = 100$ s [101, 102].

The μ_m is related to the saturation magnetization, M_s, which is defined as the maximum allowed magnetization at given temperature (all spins are aligned along the field direction) and often deviates from a theoretical bulk value. For ideal NPs (physical volume is identical with the volume where the magnetic structure is like in the bulk spinel), the dependence of μ_m on M_s can be written as $\mu_m = M_s V$.

Another important parameter is the relaxation time, τ of the NP superspin. For a particle with uniaxial anisotropy, the superspin relaxation corresponds to the flip between two equilibrium states separated by an energy barrier $K_{eff}V$, which can be overcome by the thermal fluctuations at the T_B. The superspin relaxation in the SPM systems is described by the Néel-Arrhenius law as [103, 104]:

$$\tau = \tau_0 \exp(E_A / k_B T) \tag{2}$$

where E_A is the anisotropy energy and other variables and constants have usual meaning.

Below the T_B, the NPs are in the so-called blocked state analogous to the ordered state (such as ferromagnetic or ferrimagnetic), and the magnetic moments are fixed into the direction of the easy axis and can only fluctuate around these directions. The T_B also depends on the time window of the measurements, τ_m. If the $\tau_m > \tau$, the NPs have enough time to fluctuate and the SPM state can be observed. On the other hand, if $\tau_m < \tau$, the blocked regime is observed. Thus, determination of the T_B is dependent on the used experimental technique (10^{-8} s for Mössbauer spectroscopy (MS), 1 s for magnetic measurements, 10–10^{-3} s for a.c. susceptibility measurements).

A very important parameter characterizing the blocked state is the coercivity, H_c (or coercive field) as it gives information on opening of the hysteresis loop. Depending on the dominant anisotropy term, the H_c value reaches values in fractions of $2K_{eff}/M_s$ [101, 105]. In general, the coercivity (and also remanence) of NPs with nonspherical shapes shows complex angular dependence due to the shape anisotropy [106]. In very small particles, the coercivity is an interplay between the surface disorder and surface anisotropy [107].

The typical magnetic measurements of the NP magnetic parameters yielding the above-described parameters can be summarized as follows: temperature dependence of the magnetization in low external applied field, the so-called zero field cooled curve (ZFC) and field cooled curve (FC); field dependence of the magnetization at fixed temperatures, the so-called magnetization isotherm (or hysteresis loop in the blocked state); and the a.c. susceptibility measurement. The ZFC-FC protocol reveals the value of the T_B, while the analysis of the magnetization isotherms in the SPM state serves for determination of the μ_m. From this value, the so-called magnetic size of a NP, d_{mag} (size of the magnetically ordered part), can be determined.

A unique tool used in characterization of spinel ferrite NPs is the Mössbauer spectroscopy. It is a dual probe both for structure and magnetism at local level based on recoilless resonant absorption of γ radiation. In general, information on coordination surroundings of the iron cations, their valence, degree of inversion of the spinel structure, and orientation of spins on the cubic spinel sub-lattices can be obtained [108, 109].

The small spinel ferrite NPs exhibit relaxation time in order of 10^{-9} s that is close to the time window of the MS (10^{-8} s) allowing the study of relaxation of the NPs by means of MS [109, 110]. Furthermore, the big advantage of the MS is that it is not restricted to the well crystalline samples; thus, a non-well crystalline NP can be also investigated using MS. Finally, the so-called spin canting angle, usually attributed to the presence of the surface spins, can be estimated [111].

3.2. Real effects in magnetic nanoparticles

3.2.1. Size distribution

All real systems of the NPs exhibit an intrinsic size distribution, which must be considered in evaluation and interpretation of structural and magnetic data. The most common is the log-normal distribution (see **Figure 1**); however, Gaussian distribution has been also reported [112–114]. In the case of the TEM observation for the d_{TEM}, the NPs can be termed depending on the value of the PDI as monodisperse (PDI < 0.05–0.1), highly uniform (PDI < 0.2), and polydisperse (PDI > 0.2). Similar classification might be applied to the distribution of d_{XRD}; however, such in-depth analysis is usually not included in common Rietveld treatment of the XRD data. On the other hand, the role of size distribution by means of magnetic size, d_{mag}, and superspin values is extremely important for evaluation of magnetic properties. The mean magnetic moment per single NP, μ_m, and distribution width, σ, can be derived from the experimental data, $\mu_m = \mu_0 \exp(\sigma^2 / 2)$, as the magnetization as a function of the applied field, H, and temperature, T, in SPM state can be described as a weighted sum of Langevin functions [69, 115, 116]:

$$M(H,T) = \int_0^\infty \mu L\left(\frac{\mu H}{k_B T}\right) f_L(\mu)\,d\mu \qquad (3)$$

where $L(x)$ represents the Langevin function and $f_L(\mu)$ is the log-normal distribution of magnetic moments μ. The NP size distribution also affects the character of ZFC and FC curves as it is mirrored in distribution of the T_B and K_{eff}, and suited models must be applied to obtain median values, T_{Bm} and distribution width σ as relevant parameters [117–120].

One of the possible approaches evaluating the T_B distribution is based on refinement of the ZFC temperature dependence of magnetization, $M_{ZFC}(T)$ which is given by equation [101, 121, 122]:

$$M_{ZFC}(T) \propto \frac{M_S^2 H}{3K_{eff}}\left[\frac{25}{t}\int t_B f(t_B)\,dt_B + \int f(t_B)\,dt_B\right] \qquad (4)$$

where $t_B = T_B/T_{Bm}$ is the reduced blocking temperature of individual NPs and $f(t_B)$ is the log-normal distribution function of reduced blocking temperatures. The first term in Eq. (4) represents contribution of the NPs in the SPM state, whereas the second term belongs to the NPs in blocked state.

Typical examples of magnetization isotherms and ZFC-FC curves influenced by the particle-size distribution are shown in **Figure 2**, presenting unhysteretic magnetization isotherms (Langevin curves) for different values of μ and σ and ZFC-FC curves for different values of T_B and σ.

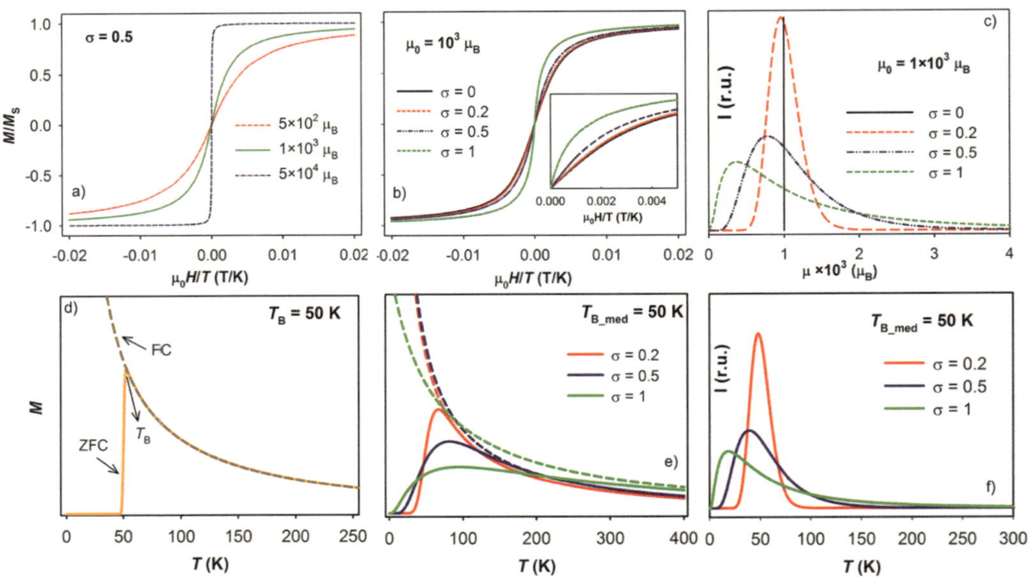

Figure 2. Model Langevin and ZFC-FC curves for selected NP magnetic moments and blocking temperature. (a) Langevin curves for magnetic moments with different orders of magnitude and $\sigma = 0.5$. (b) Evolution of Langevin curve for different magnetic moment distributions visualized in (c). (d) Ideal ZFC-FC curves for NP without size distribution. (e) Evolution of the ZFC-FC curve for fixed T_B and different distribution widths (f).

3.2.2. *Spin canting phenomenon and surface effects*

Decreasing the NP size, the number of atoms located at the surface dramatically increases. Thus the surface spins become dominant in the magnetic properties of the whole NP. The atoms at the surface exhibit lower coordination numbers originating from breaking of symmetry of the lattice at the surface.

Moreover, the exchange bonds are broken resulting in the spin disorder and frustration at the surface leading to the undesirable effects such as low saturation magnetization of the NP and the unsaturation of the magnetization in the high magnetic applied field [123]. To explain these effects, J. M. D. Coey proposed the so-called core-shell model in which the NP consists of a core with the normal spin arrangement and the disordered shell, where the spins are inclined at random angles to the surface, the so-called spin canting angle [123] (see **Figure 3**). The spin canting angle in general depends on the number of the magnetic nearest neighbors connecting with the reduced symmetry and dangling bonds. Other effects such as the interparticle interactions play role [124]. The spin canting angle can be determined with the help of in-field Mössbauer spectroscopy (IFMS); an example is given in **Figure 4**.

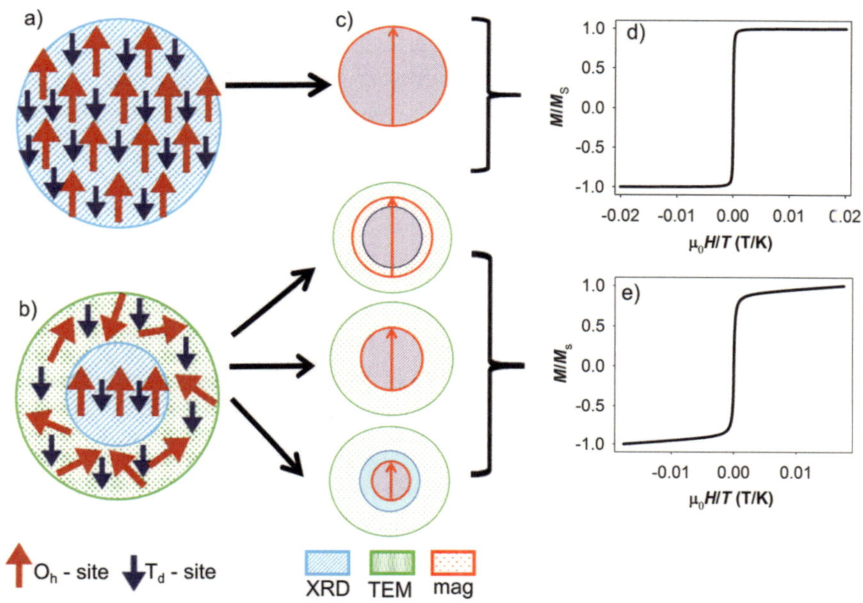

Figure 3. Scheme of the internal structure of ideal (a) and magnetic core-shell (b) structure of NP. (c) Scheme of the ideal and core-shell NP with the model of NP diameters determined by TEM, X-ray diffraction, and magnetic measurements. (d and e) Model Langevin curves for the ideal and core-shell NP with paramagnetic contribution due to the disordered spins in the NP shell.

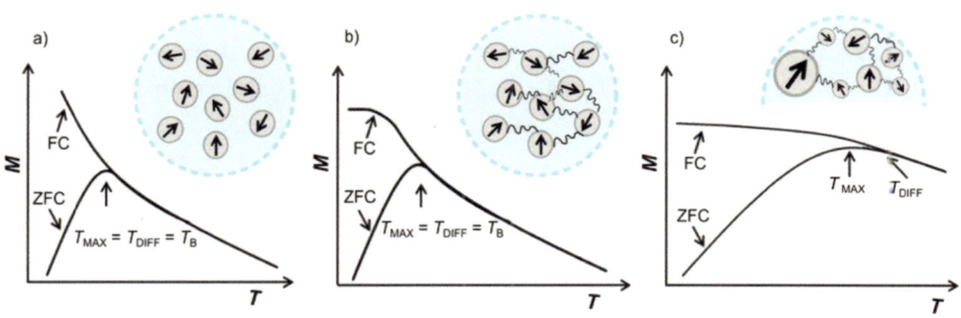

Figure 4. Schematic representation of the typical zero field cooled curve (ZFC) and field cooled curves (FC) for different types of NP ensembles. a) uniform NPs with negligible interparticle interactions, b) uniform NPs with "intermediate" interparticle interactions, c) NPs with non-negligible particle size distribution and strong interparticle interactions. The blocking temperature, T_B and the irreversibility temperature, T_{DIFF} typical for strongly interacting regime are shown.

However, the spin canting is not a unique property of the surface spins, and several works point to the volume nature of the effect [125–127]. Thus the surface effects in the NPs together with the origin of the spin canting angle are still discussed within the scientific community [109, 128–130].

Another consequence of the increased number of the surface atoms is the dominance of surface term to the anisotropy energy, usually expressed as a sum: $K_v + (6/d)K_s$, where K_v is the bulk value of the K_{eff} and K_s describes the contribution from the surface spins originated by structural deviations and spin frustration on the surface. Depending on the NP shape, the surface anisotropy may contain non-negligible admixture of higher-order Néel terms [130]. In real systems, the K_{eff} is additionally modified by the presence of other effects, mainly interparticle interactions described below.

3.2.3. *Interparticle interactions*

The interparticle interactions play a very important role in the magnetic response of the NPs, because they are usually not enough spatially separated to follow the behavior of an ideal SPM system. In general, two types of interaction can be observed: 1) the exchange interaction that affects mainly the surface spins of the NPs in close proximity thus can be neglected in most cases and 2) the long-range order dipolar interaction that is the dominant due to the high magnetic moment of the NPs [131].

The NP systems can be tentatively divided into the weakly interacting systems (the representatives are much diluted ferrofluids or NPs embedded in matrix in small concentration) and strongly interacting system with the powder samples as representatives. The strength of the interparticle interactions is given by the magnitude of the superspins and interparticle distance, in reality by the concentration of the NPs in ferrofluids, thickness of the NP coating, or matrix-to-NP ratio. The interparticle interactions affect all parameters characterizing the single-domain state. Furthermore, the strong interparticle interactions can result in the collective magnetic state at low temperature that resembles the typical physical properties of spin glasses [104], termed as superspin glasses in the case of strongly interacting SPM species [132–134].

In weakly interacting system, the dipolar interaction is treated as a perturbation to the SPM model within the Vogel-Fulcher law [104], the NP relaxation time is then written as:

$$\tau = \tau_0 \exp\left(E_A / k_B \left(T - T_0\right)\right). \tag{5}$$

The effect on the T_B is described by two models giving contradictory results on the relaxation times—the Hansen-Morup model (HM) [135] and the Dormann-Bessais-Fiorani model (DBF) [129]. The decrease of the T_B is predicted by the HM model, while its increase was obtained by the DBF model. So far, there have been no clear experimental evidences for a preference of one of these models. Some authors suggested that a phenomenological correction to the T_B in the weakly interacting systems could be used in the same way as it is done fore the relaxation time by adding the phenomenological constant T_0 to the T_B of the SPM system [102, 136, 137]. A different approach treats weak interparticle interactions as additional magnetic field acting on a single NP, when correction to the external magnetic field, $(1-H/H_K)^a$ is added to Eq. (1) [101].

In the case of strong interactions, the collective state of the NP condensates below a characteristic temperature—the so-called glass-transition temperature, T_g—and the equation for the relaxation time is usually given by scaling law for critical spin dynamics [131, 132]:

$$\tau = \tau_0 \left(T_m / T_g - 1 \right)^{z\nu} \tag{6}$$

where T_m is the temperature of the maximum at the a.c. susceptibility curve and $z\nu$ is the dynamical critical component. However; strongly interacting systems do not necessarily fulfill criteria for the so-called superspin-glass systems obeying Eq. (6). Then, one of the approaches dealing with the effect of strong interactions on shift of the T_B is treated within the random anisotropy model (RAM) [101, 138–140]. RAM predicts the increase of interparticle interactions with decreasing correlation length, L which is a measure of average distance at which the magnetization fluctuations within the NP system are correlated. Then the K_{eff} and particle volume V are averaged to the number N of the NP involved in the interactions, introducing new K_L and V_L variables, and consequently, the formula for the T_B is modified to:

$$T_B = K_L V_L N^{1/2} \tag{7}$$

The heart of the problem of calculating the T_B for interacting systems within the RAM model is the correct evaluation of the K_L and V_L of NP system.

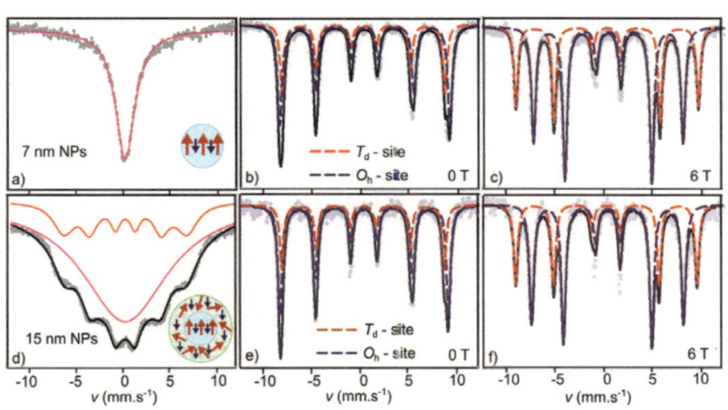

Figure 5. Typical MS spectra of almost ideal and core-shell NPs. The first column shows comparison of room-temperature MS spectra for the perfectly crystalline and ordered 7nm -Fe2O3 NPs (a) and core-shell 15 nm with 7 nm crystalline cores (d). The pink line is the fit of the spectra attributed to the fraction of NPs in SPM state. The middle and right column shows evolution of MS at low temperatures (4 K) in 0 T (b and e) and 6 T(c and f), respectively. Splitting of the lines attributed to the octahedral and tetrahedral positions can be disentangled after application of external magnetic field (c and f). Peak widening due to the disordered magnetic spins in the NP shell (d) is observable on the 4 K spectra (e and f), especially for the 1st and 6th lines.

The presence of interparticle interactions (as well as the particle-size distribution) is usually evidenced on the ZFC-FC curves; typical examples of medium and strongly interacting ensembles of NPs in comparison to the ideal noninteracting case are given in **Figure 5**. In real samples, all effects are present with variable contribution, and in some cases, both the size distribution and interparticle interactions must be addressed to describe the magnetic response of the samples properly [141, 142] (**Figure 5**).

4. Synergy of structural and magnetic probes

In order to provide complete insight into properties of magnetic NPs, synergy of structural and magnetic probes is essential. At the atomic and single-particle level, the complementarity of the (HR) TEM and XRD provides information on phase composition, the presence and type of defects, and particle sizes: d_{XRD} and d_{TEM}. The analysis of MS gives important knowledge on the degree of inversion and spin canting, which is then considered for interpretation of the magnetization data. Moreover, the particle-size distribution obtained from the TEM should be confronted with the superspin distribution obtained by the analysis of the Langevin curves; this analysis also yields the magnetic size, d_{mag}. Using the three different particle-size parameters (d_{mag}, d_{XRD}, d_{TEM}), the concept of the core-shell model of NP can be extended as the core-shell structure of the spins is often not identical with the crystallographically ordered-disordered part of the NP. The reason is that the spin frustration and disorder usually occur at volume larger than the size of the crystalline part. Comparing the d_{mag}, d_{XRD}, and d_{TEM} values, a very good estimate of the particle crystallinity and degree of spin order is obtained. A schematic representation of the crystallographic (structural) and magnetic core-shell model structures together with typical magnetization isotherms in the SPM state are shown in **Figure 3**.

At the mesoscopic level, the influence of interparticle interactions should not be neglected. For that purpose, morphology of the NP ensembles observed by the TEM gives estimate of mutual interparticle distance. The relevance of the interaction regime can be corroborated by a.c. susceptibility experiments, which yields characteristic relaxation times of the superspins, τ. As discussed above, those are strongly reformed because of the interactions. Finally, the effect of the μ, T_B, and K_{eff} distribution must be then carefully disentangled in order to estimate the pure contribution of the interaction.

5. Impact of preparation and strategies of tuning magnetic properties

The intrinsic NP parameters at all levels are imprinted during the preparation process. In this section, a brief discussion of this issue is given in the view of the "three-level" concept considering the structural and spin order in the unit cell and coordination polyhedra, single-particle, and NP ensemble level. Strategies profiting from the control over the imprint of the real effects by substitution or formation of artificial core-shell structures are also mentioned.

The degree of inversion, δ of the spinel structure is found to be significantly influenced by the preparation of the spinel ferrite NPs. In bulk, the normal or inverse spinel structure usually

http://dx.doi.org/10.5772/66074

dominates. However, the degree of inversion in the NPs is often close to 0.5 and the mixed spinel structure is the most common. For example, the $NiFe_2O_4$ is a typical inverse spinel, while in NPs obtained by the sol-gel method, the δ value of 0.6 was reported [143]. A very similar values were observed for sol-gel-prepared NPs of $CoFe_2O_4$ [69] (inverse spinel in bulk) and of $ZnFe_2O_4$ with normal spinel bulk structure [144]. The cation distribution in NPs prepared by coprecipitation method also often corresponds to mixed spinel structure as was demonstrated for $ZnFe_2O_4$ [145] and $MnFe_2O_4$ [146]. Moreover, the δ value can be controlled in the NPs prepared by the polyol method [65] and using tailored solvothermal protocols [147]. In addition, the stoichiometry of the NPs is not always matching the expected $M^{2+}/M^{3+}/O^{2-}$ ratio (1:2:4), e.g., as reported for NPs prepared by hydrothermal method [50].

The presence of defects, mainly by means of oxygen vacancies, is believed to be another important factor driving magnetic properties of the NPs. It was shown that they dominate the properties of the NPs obtained by mechanochemical processes [148], and it was also demonstrated that the level of defects can be influenced by vacuum annealing [149–151]. A specific issue is related to the presence of the Verwey transition in the Fe_3O_4 NPs [152] as the topotactic oxidation from magnetite to maghemite is a rapid process in common environments. Consequently, experimental investigations of the iron oxide NPs with size below 20 nm do not evidence the transition [26, 153]. Recently, the Verwey transition was observed in the NPs with a size of 6 nm, which were kept under inert atmosphere, and thus their oxidation was prevented [154].

The most significant and discussed issue is the spin order at single-particle level and its surface to volume nature. Most works report the dominance of surface spin frustration and suggest the presence of the magnetically dead layer. The increased contribution of the frustrated spins is attributed mainly to size effect, low crystallinity, and surface roughness, dominating in the NPs obtained by coprecipitation method [26, 155–165]. The spin canting in the surface layer was also observed in diluted ferrofluids, which confirms the nature of the effect on single-particle level [166]. However, the surface spin structure can be reformed when the NPs are in close proximity [131]. Significant increase of the amount of disordered spins was reported for hollow NPs of $NiFe_2O_4$ thanks to the additional inner surface [167]. On the other hand, the spin canting was also considered as volume effect, which occurs due to ion order-disorder in the spinel structure [127, 168] or pinning of the spins on internal defects in single NPs [125].

Focusing on the mesoscopic effects, the NP size distribution and interparticle interactions will be addressed. The particle-size distribution is found to be very sensitive to the preparation method used. The NPs with almost monodisperse character are obtained by the decomposition route; however, the parameters of the reaction must be carefully controlled. For example, the prolongation of the reaction time leads both to larger NPs but also increased size distribution [33, 169]. Similar effect was observed for increasing concentration of the oleic acid or oleylamine [33]. Other techniques provide NPs with PDI over 0.2, and the size distribution must be then considered in analysis of the magnetic measurements [69, 116, 142]. However, it is worth mentioning that the narrow-size distribution of the d_{TEM} does not automatically imply the same value of the d_{XRD} or d_{mag}, as shown, e.g., for maghemite NPs [125]. In majority of real samples, interparticle interactions contribute to the magnetic properties. In most cases, the samples are studied in form of powders, which contain NPs in very close contact. Consequently, the response of such systems is always in the limit of the medium to strong interactions and is almost invariant to the preparation route used, and the interaction strength for a given NP size is given by minimum distance between the NPs, in other words by the thickness of the surface coating [42, 170–172]. Upon specific conditions, well-defined aggregates are formed [173], as reported, e.g., for preparations in microemulsion [174], by decomposition method [175] and by controlled encapsulation into phospholipides [176]. Such assemblies attracted interest due to considerably enhanced heating properties in hyperthermia [177], which is associated with the enhancement of the single-object (aggregate) anisotropy. In dense ensembles of the NPs, the onset of collective relaxation is also corroborated by significant increase of the relaxation time [178–184]; analogous consequence was observed in the aggregates [185]. However, the influence of the intra- and inter-aggregate interactions is not explicitly decoupled. Recent studies also suggest a strong influence of the reformed particle energy barrier on the details of the aging dynamics, memory behavior, and apparent superspin dimensionality of the particles [132].

In spite of the fact that the surface effects, defects, and interparticle interactions are believed to be contra-productive factors as they in general decrease the value of saturation magnetization [26], they were recognized as potential enhancers of effective magnetocrystalline anisotropy, reflected, for example, in increase of the hysteresis losses [186]. Consequently, attempts to prepare smart NPs based on artificial core-shell structure, e.g., $NiFe_2O_4$@γ-Fe_2O_3 [187], $ZnFe_2O_4$@γ-Fe_2O_3 [188], and Co,Fe_2/Ni,Fe_2O_4 [189], appeared recently. Tri-magnetic multi-shell structures prepared by high-temperature decomposition of the metal oleates were also reported [190].

Alternative strategy is the tuning of magnetic properties of the spinel ferrite NPs via site-specific occupation of the spinel lattice. This is a straightforward approach as the relevant metal ions can substitute each other in the spinel structure easily. In this case, however, the site occupation must be carefully evaluated and controlled. Successful preparation and basic investigation of structure and magnetic properties of the NPs of Mn-doped $CuFe_2O_4$ ferrite [191], Zn-doped $MnFe_2O_4$ [192] and $NiFe_2O_4$ [193, 194], Co-doped $NiFe_2O_4$ [195] and $ZnFe_2O_4$ [196], and Cr-doped $CoFe_2O_4$ [197] were reported. Recently, doping of spinel ferrites by large cations was suggested as a promising way to increase the effective magnetic anisotropy. La-doped $CoFe_2O_4$ [198], Sr-doped $MgFe_2O_4$ [199], and Ce-doped $NiFe_2O_4$ [200] or $ZnFe_2O_4$ [201] were prepared. For the doped samples, the most promising are the polyol, sol-gel, or microemulsion methods as they do not require identical decomposition temperatures of metal precursors like the organic-based routes, allow rather good control over homogeneity of doping, and yield samples with sufficiently low particle-size distribution.

6. Conclusions and outlooks

The core message of the chapter is to emphasize the importance of structural and spin order mirrored in magnetic properties of well-defined spinel ferrite nanoparticles (NPs). The

http://dx.doi.org/10.5772/66074

correlation between the specific preparation route to the typical structural and magnetic parameters of the particles is given, and the suitability of the resulting NPs in the context of possible applications is evaluated. Explicitly the meaning of different particle sizes obtained by different characterization methods, related to the degree of structural and spin order, is emphasized in the context of the magnetic properties. In order to wrap up the given subject, let's outline future outlooks in the field. The research of fine magnetic particles is progressively developing thanks to high demand on their practical exploitation, mainly in biomedicine. The forthcoming trend in customization of the magnetic NPs is obviously converging to control of the required magnetic properties at single-particle level by adjustment of the synthetic protocols, which lead to fine tuning of the particle size, shape, and degree of order [169, 202]. For example, enhancement of the specific absorption rate in NPs can be achieved in natural or arbitrary core-shell structures [203], via coupling of magnetically soft and hard ferrites for maximization of hysteresis losses [204] or by doping-driven enhancement of heat generation [205]. Finally, smart self-assembling strategies leading to superstructures [206], which can be even induced by magnetic field [207], seem to be a powerful tool for managing the magnetic response of the NPs at mesoscopic level.

Acknowledgements

The authors gratefully acknowledge Dr. Puerto Morales from the Instituto de Ciencia de Materiales de Madrid for her generous support and for sharing his expertise in synthesis of uniform iron oxide nanoparticles and Prof. Carla Cannas from the Università degli studi di Cagliari for sharing her knowledge in synthesis of core-shell spinel ferrite nanoparticles. The research was carried out thanks to the support of the Czech Science Foundation, project no. 15-01953S, and 7FP program project MULTIFUN (no. 262,943), cofinanced by the Ministry of Education, Youth, and Sports (project no. 7E12057). Magnetic measurements were performed in MLTL (http://mltl.eu/), which is supported within the program of Czech Research Infrastructures (project no. LM2011025).

Author details

Barbara Pacakova[1], Simona Kubickova[1], Alice Reznickova[1], Daniel Niznansky[2] and Jana Vejpravova[1,2*]

*Address all correspondence to: vejpravo@fzu.cz

1 Department of Magnetic Nanosystems, Institute of Physics of the Czech Academy of Sciences, Prague, Czech Republic

2 Department of Inorganic Chemistry, Faculty of Science, Charles University in Prague, Prague, Czech Republic

References

[1] Bauer LM, Situ SF, Griswold MA, Samia ACS. Nanoscale 2016;8:12162–9. doi:10.1039/c6nr01877g.

[2] Kanagesan S, Bin Ab Aziz S, Hashim M, Ismail I, Tamilselvan S, Alitheen NBBM, et al. Molecules 2016;21:312. doi:10.3390/molecules21030312.

[3] Peng Y, Wang Z, Liu W, Zhang H, Zuo W, Tang H, et al. Dalt Trans 2015;44:12871–7. doi:10.1039/c5dt01585e.

[4] Bae S, Lee SW, Takemura Y. Appl Phys Lett 2006;89:252503. doi:10.1063/1.2420769.

[5] Ghahremanzadeh R, Rashid Z, Zarnani AH, Naeimi H. Appl Catal A Gen 2013;467:270–8. doi:10.1016/j.apcata.2013.07.029.

[6] Zhang L, Wu Y. J Nanomater 2013;2013:640940–6. doi:10.1155/2013/640940.

[7] Kumar AS, Thulasiram B, Laxmi SB, Rawat VS, Sreedhar B. Tetrahedron 2014;70:6059–67. doi:10.1016/j.tet.2014.01.051.

[8] Pan L, Xu B. JOM 2013;65:695–701. doi:10.1007/s11837-013-0607-2.

[9] El-Remaily MAEAAA, Abu-Dief AM. Tetrahedron 2015;71:2579–84. doi:10.1016/j.tet.2015.02.057.

[10] Tyrpekl V, Vejpravova JP, Roca AG, Murafa N, Szatmary L, Niznansky D. Appl Surf Sci 2011;257:4844–8. doi:10.1016/j.apsusc.2010.12.110.

[11] Tasca JE, Ponzinibbio A, Diaz G, Bravo RD, Lavat A, Gonzalez MG. Top Catal 2010;53:1087–90. doi:10.1007/s11244-010-9538-0.

[12] Wang ZL. RSC Adv 2015;5:5563–6. doi:10.1039/c4ra14486d.

[13] Ensafi AA, Allafchian AR, Mohammadzadeh R. Anal Sci 2012;28:705–10. doi:10.2116/analsci.28.705.

[14] Liu YL, Liu ZM, Yang Y, Yang HF, Shen GL, Yu RQ. Sens Actuat B Chem 2005;107:600–4. doi:10.1016/j.snb.2004.11.026.

[15] Vignesh RH, Sankar KV, Amaresh S, Lee YS, Selvan RK. Sens Actuat B Chem 2015;220:50–8. doi:10.1016/j.snb.2015.04.115.

[16] Sun Z, Liu L, Ha DZ, Pan W. Sens Actuat B Chem 2007;125:144–8. doi:10.1016/j.snb.2007.01.050.

[17] Pan Y, Zhang Y, Wei X, Yuan C, Yin J, Cao D, et al. Electrochim Acta 2013;109:89–94. doi:10.1016/j.electacta.2013.07.026.

[18] Zhang W, Hou X, Lin Z, Yao L, Wang X, Gao Y, et al. J Mater Sci Electron 2015;26:9535–45. doi:10.1007/s10854-015-3616-9.

[19] Reddy S, Swamy BEK, Chandra U, Mahathesha KR, Sathisha TV, Jayadevappa H. Anal Methods 2011;3:2792–6. doi:10.1039/c1ay05483j.

[20] Smolensky ED, Park HYE, Zhou Y, Rolla GA, Marjanska M, Botta M, et al. J Mater Chem B 2013;1:2818–28. doi:10.1039/c3tb00369h.

[21] Martinez-Boubeta C, Simeonidis K, Makridis A, Angelakeris M, Iglesias O, Guardia P, et al. Sci Rep 2013;3:1652. doi:10.1038/srep01652.

[22] Cheng K, Frey NA, Sun S, Peng S, Cheng K, Sun S. Chem Soc Rev 2009;38:2532–42. doi:10.1039/b815548h.

[23] Salabas EL. Structural and Magnetic Investigations of Magnetic Nanoparticles and Core-Shell Colloids. Universität Duisburg-Essen, 2004. doi:http://purl.oclc.org/NET/duett-03182005-143720.

[24] Tartaj P, Morales M del PM, Veintemillas-Verdaguer S, González-Carreño T, Serna CJ. J Phys D Appl Phys 2009;42:224002. doi:10.1088/0022-3727/42/22/224002.

[25] Wu Z, Yang S, Wu W. Nanoscale 2016;8:1237–59. doi:10.1039/C5NR07681A.

[26] Roca AG, Niznansky D, Poltierova-Vejpravova J, Bittova B, Gonzalez-Fernandez MA, Serna CJ, et al. J Appl Phys 2009;105:114309. doi:10.1063/1.3133228.

[27] Chen Q, Rondinone AJ, Chakoumakos BC, Zhang ZJ. J Magn Magn Mater 1999;194:1–7. doi:10.1016/S0304-8853(98)00585-X.

[28] Roca AG, Morales MP, Serna CJ. IEEE Trans Magn 2006;42:3025–9. doi:10.1109/TMAG.2006.880111.

[29] Suppiah DD, Abd Hamid SB. J Magn Magn Mater 2016;414:204–8. doi:10.1016/j.jmmm.2016.04.072.

[30] Sun S, Zeng H, Robinson DB, Raoux S, Rice PM, Wang SX, Li G. J Am Chem Soc 2004;126:273–9. doi:10.1002/asia.201300068.

[31] Tartaj P, Morales MP, Veintemillas-Verdaguer S, Gonzales-Carre OT, Serna CJ. J Phys D Appl Phys 2003;36:R182–97. doi:10.1088/0022-3727/36/13/202.

[32] López-Ortega A, Lottini E, Fernández CDJ, Sangregorio C. Chem Mater 2015;27:4048–56. doi:10.1021/acs.chemmater.5b01034.

[33] Lu LT, Dung NT, Tung LD, Thanh CT, Quy OK, Chuc NV, et al. Nanoscale 2015;7:19596–610. doi:10.1039/c5nr04266f.

[34] Baruwati B, Manorama SV. Mater Chem Phys 2008;112:631–6. doi:10.1016/j.matchemphys.2008.06.018.

[35] Maiti D, Saha A, Devi PS. Phys Chem Chem Phys 2016;18:1439–50. doi:10.1039/c5cp05840f.

[36] Bao N, Shen L, An W, Padhan P, Turner CH, Gupta A. Chem Mater 2009;21:3458–68. doi:10.1021/cm901033m.

[37] Zhang K, Zuo W, Wang Z, Liu J, Li T, Wang B, et al. RSC Adv 2015;5:10632–40. doi: 10.1039/C4RA15675G.

[38] Guardia P, Labarta A, Batlle X. J Phys Chem C 2011;115:390–6. doi:10.1021/jp1084982.

[39] Liang WI, Zhang X, Bustillo K, Chiu CH, Wu WW, Xu J, et al. Chem Mater 2015;27:8146–52. doi:10.1021/acs.chemmater.5b03930.

[40] Cannas C, Ardu A, Peddis D, Sangregorio C, Piccaluga G, Musinu A. J Colloid Interface Sci 2010;343:415–22. doi:10.1016/j.jcis.2009.12.007.

[41] Douglas FJ, MacLaren DA, Murrie M. RSC Adv 2012;2:8027–35. doi:10.1039/c2ra20494k.

[42] Repko A, Vejpravova J, Vackova T, Zakutna D, Niznansky D. J Magn Magn Mater 2015;390:142–51. doi:10.1016/j.jmmm.2015.04.090.

[43] Nonkumwong J, Ananta S, Jantaratana P, Phumying S, Maensiri S, Srisombat L. J Magn Magn Mater 2015;381:226–34. doi:10.1016/j.jmmm.2015.01.001.

[44] Almeida TP, Fay M, Zhu Y, Brown PD. J Nanosci Nanotechnol 2012;12:8797–800. doi: 10.1166/jnn.2012.6467.

[45] Liu Q, Huang H, Lai L, Sun J, Shang T, Zhou Q, et al. J Mater Sci 2009;44:1187–91. doi: 10.1007/s10853-009-3268-3.

[46] Aslibeiki B, Kameli P, Ehsani MH, Salamati H, Muscas G, Agostinelli E, et al. J Magn Magn Mater 2016;399:236–44. doi:10.1016/j.jmmm.2015.09.081.

[47] Guo P, Cui L, Wang Y, Lv M, Wang B, Zhao XS. Langmuir 2013;29:8997–9003. doi: 10.1021/la401627x.

[48] Rameshbabu R, Ramesh R, Kanagesan S, Karthigeyan A, Ponnusamy S. J Supercond Nov Magn 2014;27:1499–502. doi:10.1007/s10948-013-2466-z.

[49] Repko A, Niznansky D, Matulkova I, Kalbac M, Vejpravova J. J Nanopart Res 2013;15. doi:10.1007/s11051-013-1767-2.

[50] Repko A, Niznansky D, Poltierova-Vejpravova J. J Nanopart Res 2011;13:5021–31. doi: 10.1007/s11051-011-0483-z.

[51] Sasaki T, Ohara S, Naka T, Vejpravova J, Sechovsky V, Umetsu M, et al. J Supercrit Fluids 2010;53:92–4. doi:10.1016/j.supflu.2009.11.005.

[52] Mathew DS, Juang RS. Chem Eng J 2007;129:51–65. doi:10.1016/j.cej.2006.11.001.

[53] Vestal CR, Zhang ZJ. Int J Nanotechnol 2004;1:240–63. doi:10.1504/IJNT.2004.003727.

[54] Makovec D, Kosak A, Znidarsic A, Drofenik M. J Magn Magn Mater 2005;289:32–5. doi: 10.1016/j.jmmm.2004.11.010.

[55] Holec P, Plocek J, Niznansky D, Vejpravova JP. J Sol-Gel Sci Technol 2009;51:301–5. doi: 10.1007/s10971-009-1962-x.

[56] Chandradass J, Jadhav AH, Kim H. Appl Surf Sci 2012;258:3315–20. doi:10.1016/j.apsusc.2011.11.092.

[57] Chandradass J, Kim KH. J Alloys Compd 2011;509:L59–62. doi:10.1016/j.jallcom.2010.10.090.

[58] Laokul P, Arthan S, Maensiri S, Swatsitang E. J Supercond Nov Magn 2015;28:2483–9. doi:10.1007/s10948-015-3068-8.

[59] Coskun M, Korkmaz M, Firat T, Jaffari GH, Shah SI. J Appl Phys 2010;107. doi: 10.1063/1.3359426.

[60] Vestal CR, Zhang ZJ. Nano Lett 2003;3:1739–43. doi:10.1021/nl034816k.

[61] Ahn Y, Choi EJ, Kim S, An DH, Kang KU, Lee BG, et al. J Korean Phys Soc 2002;41:123–8.

[62] Aubery C, Solans C, Prevost S, Gradzielski M, Sanchez-Dominguez M. Langmuir 2013;29:1779–89. doi:10.1021/la303817w.

[63] Nguyen TD. Nanoscale 2013;5:9455–82. doi:10.1039/c3nr01810e.

[64] Pemartin K, Solans C, Alvarez-Quintana J, Sanchez-Dominguez M. Colloids Surf A Physicochem Eng Asp 2014;451:161–71. doi:10.1016/j.colsurfa.2014.03.036.

[65] Ammar S, Jouini N, Fievet F, Stephan O, Marhic C, Richard M, et al. J Non Cryst Sol 2004;345:658–62. doi:10.1016/j.jnoncrysol.2004.08.162.

[66] Sui J, Zhang C, Li J, Cai W. J Nanosci Nanotechnol 2012;12:3867–72. doi:10.1166/jnn.2012.5872.

[67] Zhao H, Liu R, Zhang Q, Wang Q. Mater Res Bull 2016;75:172–7. doi:10.1016/j.materresbull.2015.11.052.

[68] Dormann JL, Viart N, Rehspringer JL, Ezzir A, Niznansky D. Hyperfine Interact 1998;112:89–92. doi:10.1023/A:1011088611227.

[69] Vejpravova J, Sechovsky V, Plocek J, Niznansky D, Hutlova A, Rehspringer JL. J Appl Phys 2005;97. doi:10.1063/1.1929849.

[70] Vejpravova J, Plocek J, Niznansky D, Hutlova A, Rehspringer JL, Sechovsky V. IEEE Trans Magn 2005;41:3469–71. doi:10.1109/TMAG.2005.854879.

[71] Hutlova A, Niznansky D, Rehspringer JL, Estournes C, Kurmoo M. Adv Mater 2003;15:1622. doi:10.1002/adma.200305305.

[72] Rivas P, Sagredo V, Rossi F, Pernechele C, Solzi M, Pena O. IEEE Trans Magn 2013;49:4568–71. doi:10.1109/TMAG.2013.2262039.

[73] Carta D, Loche D, Mountjoy G, Navarra G, Corrias A. J Phys Chem C 2008;112:15623–30. doi:10.1021/jp803982k.

[74] Chakraverty S, Mandal K, Mitra S, Chattopadhyay S, Kumar S. Jpn J Appl Phys Part I 2004;43:7782–7. doi:10.1143/JJAP.43.7782.

[75] Mitra S, Mandal K, Kumar PA. J Magn Magn Mater 2006;306:254–9. doi:10.1016/j.jmmm.2006.03.024.

[76] Phuruangrat A, Kuntalue B, Thongtem S, Thongtem T. Mater Lett 2016;167:65–8. doi:10.1016/j.matlet.2016.01.005.

[77] Galvao WS, Freire RM, Ribeiro TS, Vasconcelos IF, Costa LS, Freire VN, et al. J Nanopart Res 2014;16:2803–6. doi:10.1007/s11051-014-2803-6.

[78] Abbas M, Rao BP, Islam MN, Kim KW, Naga SM, Takahashi M, et al. Ceram Int 2014;40:3269–76. doi:10.1016/j.ceramint.2013.09.109.

[79] Barati MR, Selomulya C, Suzuki K. J Appl Phys 2014;115:17B522. doi:10.1063/1.4867751.

[80] Khot VM, Salunkhe AB, Thorat ND, Phadatare MR, Pawar SH. J Magn Magn Mater 2013;332:48–51. doi:10.1016/j.jmmm.2012.12.010.

[81] Pavlovic MB, Jovalekic C, Nikolic AS, Manojlovic D, Sojic N. Mater Sci Technol 2010;26:968–74. doi:10.1179/174328409X415020.

[82] Osmokrovic P, Jovalekic C, Manojlovic D, Pavlovic MB. J Optoelectron Adv Mater 2006;8:312–4.

[83] Zhou ZH, Xue JM, Wang J, Chan HSO, Yu T, Shen ZX. J Appl Phys 2002;91:6015–20. doi:10.1063/1.1462853.

[84] Starowicz M, Starowicz P, Zukrowski J, Przewoznik J, Lemanski A, Kapusta C, et al. J Nanopart Res 2011;13:7167–76. doi:10.1007/s11051-011-0631-5.

[85] Galindo R, Mazario E, Gutierrez S, Morales MP, Herrasti P. J Alloys Compd 2012;536:S241–4. doi:10.1016/j.jallcom.2011.12.061.

[86] Sundrarajan M, Ramalakshmi M. E J Chem 2012;9:1070–6. doi:10.1155/2012/541254.

[87] Tanaka T, Shimazu R, Nagai H, Tada M, Nakagawa T, Sandhu A, et al. J Magn Magn Mater 2009;321:1417–20. doi:10.1016/j.jmmm.2009.02.054.

[88] Suh SK, Yuet K, Hwang DK, Bong KW, Doyle PS, Hatton TA. J Am Chem Soc 2012;134:7337–43. doi:10.1021/ja209245v.

[89] Gomes JA, Azevedo GM, Depeyrot J, Mestnik-Filho J, Paula FLO, Tourinho FA, et al. J Phys Chem C 2012;116:24281–91. doi:10.1021/jp3055069.

[90] Goncalves ES, Cornejo DR, Oliveira CLP, Figueiredo Neto AM, Depeyrot J, Tourinho FA, et al. Phys Rev E 2015;91:42317. doi:10.1103/PhysRevE.91.042317.

[91] Masala O, Hoffman D, Sundaram N, Page K, Proffen T, Lawes G, et al. Sol State Sci 2006;8:1015–22. doi:10.1016/j.solidstatesciences.2006.04.014.

[92] Khedr MH, Bahgat M, El Rouby WMA. Mater Technol 2008;23:27–32. doi: 10.1179/175355508X266872.

[93] Morrison SA, Cahill CL, Carpenter EE, Calvin S, Harris VG. J Nanosci Nanotechnol 2005;5:1323–44. doi:10.1166/jnn.2005.303.

[94] Naik PP, Tangsali RB, Sonaye B, Sugur S. J Nano Res 2013;24:194–202. doi:10.4028/www.scientific.net/JNanoR.24.194.

[95] Lee SC, Fu CM, Chang FH. Appl Phys Lett 2013;103:163104. doi:10.1063/1.4825270.

[96] Gutiérrez L, Costo R, Grüttner C, Westphal F, Gehrke N, Heinke D, et al. Dalt Trans 2015:2943–52. doi:10.1039/c4dt03013c.

[97] Koksharov YA. Magnetic Nanoparticles, Wiley-VCH Verlag; 2009. p. 197–254.

[98] Rodriguez-Carvajal J, Roisnel T. In: Andersson, Y and Mittemeijer, EJ and Welzel U, editor. Eur Powder Diffr EPDIC 8, vol. 443–4, 2004, p. 123–6.

[99] Coey JMD. Magnetism and Magnetic Materials. 1st ed. New York: Cambridge University Press; 2009.

[100] Bean CP, Livingston JD. J Appl Phys 1959;30:S120–9. doi:http://dx.doi.org/10.1063/1.2185850.

[101] Knobel M, Nunes WC, Socolovsky LM, De Biasi E, Vargas JM, Denardin JC. J Nanosci Nanotechnol 2008;8:2836–57. doi:10.1166/jnn.2008.017.

[102] Singh V, Seehra MS, Bonevich J. J Appl Phys 2009;105:19–22. doi:10.1063/1.3073949.

[103] Dormann JL, Fiorani D, Cherkaoui R, Tronc E, Lucari F, Orazio FD, et al. J Magn Magn Mater 1999;203:23–7. doi:10.1016/S0304-8853(99)00180-8.

[104] Shtrikman S, Wohlfarth EP. Phys Lett 1981;85:467–70. doi:10.1016/0375-9601(81)90441-2.

[105] Cullity BD, Graham CD. Introduction to Magnetic Materials. New Jersey: John Wiley & Sons, Inc.; 2009.

[106] Corona RM, Altbir D, Escrig J. J Magn Magn Mater 2012;324:3824–8. doi:10.1016/j.jmmm.2012.06.022.

[107] Zhou CG, Schulthess TC, Landau DP. J Appl Phys 2006;99. doi:10.1063/1.2172557.

[108] Long GJ. Mössbauer Spectroscopy Applied to Inorganic Chemistry. 1st ed. New York: Plenum Publ. Corp.; 1984. doi:10.1007/978-1-4899-0462-1.

[109] Mørup S. Hyperfine Interact 1990;60:959–74. doi:10.1007/BF02399910.

[110] Gutlich P, Bill E, Trautwein AX. Mossbauer Spectroscopy and Transition Metal Chemistry: Fundamentals and Applications. 1st ed. Springer-Verlag Berlin Heidelberg; 2011. doi:10.1007/978-3-540-88428-6.

[111] Gonser U, Aubertin F, Stenger S, Fischer H, Smirnov G, Klingehoffer G. Hyperfine Interact 1991;67:701–9. doi:10.1007/BF02398222.

[112] Buhrman RA, Granqvist CG. Bull Am Phys Soc 1976;21:229.

[113] Granqvist CG, Buhrman RA. J Appl Phys 1976;47:2200–19. doi:10.1063/1.322870.

[114] Buhrman RA, Granqvist CG. J Appl Phys 1976;47:2220–2. doi:10.1063/1.322871.

[115] Fonseca FC, Goya GF, Jardim RF, Muccillo R, Carreño NLV, Longo E, et al. Phys Rev B 2002;66:104406. doi:10.1103/PhysRevB.66.104406.

[116] Ferrari E, da Silva F, Knobel M. Phys Rev B 1997;56:6086–93. doi:10.1103/PhysRevB.56.6086.

[117] Rondinone AJ, Samia ACS, Zhang ZJ. J Phys Chem B 1999;103:6876–80. doi:10.1021/jp9912307.

[118] Mamiya H, Ohnuma A, Nakatani I, Furubayashim T. IEEE Trans Magn 2005;41:3394–6. doi:10.1109/TMAG.2005.855205.

[119] Lee TH, Choi KY, Kim GH, Suh BJ, Jang ZH. Phys Rev B 2014;90:184411. doi:10.1103/PhysRevB.90.184411.

[120] Chowdhury D, Mookerj A. Phys B C 1984;124:255–8. doi:10.1016/0378-4363(84)90083-4.

[121] Hansen MF, Morup S. J Magn Magn Mater 1998;184:267–74. doi:10.1016/S0304-8853(97)01165-7.

[122] El-Hilo M, O'Grady K, Chantrell RW. J Magn Magn Mater 1992;114:295–306. doi:10.1016/0304-8853(92)90272-P.

[123] Coey JMD. Phys Rev Lett 1971;27:1970–2. doi:10.1103/PhysRevLett.27.1140.

[124] Tronc E, Fiorani D, Noguès M, Testa AM, Lucari F, D'Orazio F, et al. J Magn Magn Mater 2003;262:6–14. doi:10.1016/S0304-8853(03)00011-8.

[125] Kubickova S, Niznansky D, Morales MP, Salas G, Vejpravova J. Appl Phys Lett 2014;104:223105. doi:10.1063/1.4881331.

[126] Morrish AH, Haneda K. J Magn Magn Mater 1983;35:105–13. doi:10.1016/0304-8853(83)90468-7.

[127] Morales MP, Serna CJ, Bodker F, Morup S. J Phys Condens Matter 1997;9:5461–7. doi:10.1088/0953-8984/9/25/013.

[128] Mørup S, Brok E, Frandsen C. J Nanomater 2013;2013:720629. doi:10.1155/2013/720629.

[129] Dormann JL, Tronc E. In: Prigogine I, Rice SA, editors. Adv Chem Phys. Vol. 98. 1st ed., John Wiley & Sons, Inc., Hoboken, NJ, USA; 1997.

[130] Garanin DA, Kachkachi H. Phys Rev Lett 2003;90:65504. doi:10.1103/PhysRevLett.90.065504.

[131] Mørup S, Hansen MF, Frandsen C. Beilstein J Nanotech 2010;1:182–90. doi:10.3762/bjnano.1.22.

[132] Andersson MS, Angel De TJ, Lee SS, Normile PS, Nordblad P, Mathieu R Phys Rev B 2016;93:54407. doi:10.1103/PhysRevB.93.054407.

[133] Andersson MS, Mathieu R, Normile PS, Lee SS, Singh G, Nordblad P, et al. Mater Res Express 2016;3:45015. doi:10.1088/2053-1591/3/4/045015.

[134] Andersson MS, Mathieu R, Lee SS, Normile PS, Singh G, Nordblad P, et al. Nanotechnology 2015;26:475703. doi:10.1088/0957-4484/26/47/475703.

[135] Hansen MF, Mørup S. J Magn Magn Mater 1998;184:262–74. doi:10.1016/S0304-8853(97)01165-7.

[136] Seehra MS, Pisane KL. J Phys Chem Sol 2016;93:79–81. doi:10.1016/j.jpcs.2016.02.009.

[137] Pisane KL, Despeaux EC, Seehra MS. J Magn Magn Mater 2015;384:148–54. doi:10.1016/j.jmmm.2015.02.038.

[138] Herzer G. NATO Sci Ser II Math Phys Chem 2003;184:15. doi:10.1007/1-4020-2965-9_2.

[139] Loffler J, Braun H, Wagner W. Phys Rev Lett 2000;85:1990–3. doi:10.1103/PhysRevLett.85.1990.

[140] Luo W, Nagel SR, Rosenbaum TF, Rosensweig RE. Phys Rev Lett 1991;67:2721–4. doi.org/10.1103/PhysRevLett.67.2721

[141] Masunaga SH, Jardim RF, Correia MJ, Figueiredo W. J Phys Chem C 2016;120:765–70. doi:10.1021/acs.jpcc.5b10933.

[142] Pacakova B, Mantlikova A, Niznansky D, Kubickova S, Vejpravova J. J Phys Condens Matter 2016;28:206004–15. doi:10.1088/0953-8984/28/20/206004.

[143] Atif M, Nadeem M, Siddique M. Appl Phys A 2015;120:571–8. doi:10.1007/s00339-015-9216-y.

[144] Li FS, Wang L, Wang JB, Zhou QG, Zhou XZ, Kunkel HP, et al. J Magn Magn Mater 2004;268:332–9. doi:10.1016/S0304-8853(03)00544-4.

[145] Mozaffari M, Arani ME, Amighian J. J Magn Magn Mater 2010;322:3240–4. doi:10.1016/j.jmmm.2010.05.053.

[146] Chinnasamy CN, Yang A, Yoon SD, Hsu K, Shultz MD, Carpenter EE, et al. J Appl Phys 2007;101:09M509. doi:10.1063/1.2710218.

[147] Vamvakidis K, Katsikini M, Sakellari D, Paloura EC, Kalogirou O, Dendrinou-Samara C. Dalt Trans 2014;43:12754–65. doi:10.1039/c4dt00162a.

[148] Goya GF, Rechenberg HR. J Magn Magn Mater 1999;196:191–2. doi:10.1016/S0304-8853(98)00723-9.

[149] Ayyappan S, Raja SP, Venkateswaran C, Philip J, Raj B. Appl Phys Lett 2010;96:143106. doi:10.1063/1.3374332.

[150] Ayyappan S, Paneerselvam G, Antony MP, Philip J. Mater Chem Phys 2011;128:400–4. doi:10.1016/j.matchemphys.2011.03.012.

[151] Philip J, Gnanaprakash G, Panneerselvam G, Antony MP, Jayakumar T, Raj B. J Appl Phys 2007;102:54305. doi:10.1063/1.2777168.

[152] Walz F. J Phys Condens Matter 2002;14:R285. doi:10.1088/0953-8984/14/12/203.

[153] Mitra A, Mohapatra J, Meena SS, Tomy C V, Aslam M. J Phys Chem C 2014;118:19356–62. doi:10.1021/jp501652e.

[154] Lee J, Kwon SG, Park JG, Hyeon T. Nano Lett 2015;15:4337–42. doi:10.1021/acs.nanolett.5b00331.

[155] Kodama RH, Berkowitz AE, McNiff EJ, Foner S. Phys Rev Lett 1996;77:394–7. doi:10.1103/PhysRevLett.77.394.

[156] Mitra S, Mandal K, Choi ES. IEEE Trans Magn 2008;44:2974–7. doi:10.1109/TMAG.2008.2002477.

[157] Nathani H, Gubbala S, Misra RDK. Mater Sci Eng B Sol State 2005;121:126–36. doi:10.1016/j.mseb.2005.03.016.

[158] Ahlawat A, Sathe VG, Reddy VR, Gupta A. J Magn Magn Mater 2011;323:2049–54. doi:10.1016/j.jmmm.2011.03.017.

[159] Oshtrakh MI, Ushakov MV, Senthilkumar B, Selvan RK, Sanjeeviraja C, Felner I, et al. Hyperfine Interact 2013;219:7–12. doi:10.1007/s10751-012-0660-1.

[160] Lestari KR, Yoo P, Kim DH, Liu C, Lee BW. J Korean Phys Soc 2015;66:651–5. doi:10.3938/jkps.66.651.

[161] Mendonca EC, Jesus CBR, Folly WSD, Meneses CT, Duque JGS. J Supercond Nov Magn 2013;26:2329–31. doi:10.1007/s10948-012-1426-3.

[162] Mendonca EC, Jesus CBR, Folly WSD, Meneses CT, Duque JGS, Coelho AA. J Appl Phys 2012;111:53917. doi:10.1063/1.3691792.

[163] Wang L, Zhou QG, Li FS. Phys Status Sol B 2004;241:377–82. doi:10.1002/pssb.200301923.

[164] Bakuzis AF, Morais PC, Pelegrini F. J Appl Phys 1999;85:7480–2. doi:10.1063/1.369383.

[165] Peddis D, Yaacoub N, Ferretti M, Martinelli A, Piccaluga G, Musinu A, et al. J Phys Condens Matter 2011;23:426004. doi:10.1088/0953-8984/23/42/426004.

[166] Sousa MH, Hasmonay E, Depeyrot J, Tourinho FA, Bacri JC, Dubois E, et al. J Magn Magn Mater 2002;242:572–4. doi:10.1016/S0304-8853(01)01122-2.

[167] Jaffari GH, Ceylan A, Ni C, Shah SI. J Appl Phys 2010;107:13910. doi:10.1063/1.3277041.

[168] Morales MP, Veintemillas-Verdaguer S, Montero MI, Serna CJ, Roig A, Casas L, et al. Chem Mater 1999;11:3058–64. doi:10.1021/cm991018f.

[169] Salas G, Casado C, Teran FJ, Miranda R, Serna CJ, Morales MP. J Mater Chem 2012;22:21065–75. doi:10.1039/C2JM34402E.

[170] Bullita S, Casu A, Casula MF, Concas G, Congiu F, Corrias A, et al. Phys Chem Chem Phys 2014;16:4843–52. doi:10.1039/c3cp54291b.

[171] Coskun M, Korkmaz M. J Nanopart Res 2014;16:2316. doi:10.1007/s11051-014-2316-3.

[172] Bittova B, Poltierova-Vejpravova J, Roca AG, Morales MP, Tyrpekl V. J Phys Conf Ser 2010;200:72012. doi:10.1088/1742-6596/200/7/072012.

[173] Vindedahl AM, Strehlau JH, Arnold WA, Penn RL. ES Nano 2016;3:494–505. doi: 10.1039/C5EN00215J.

[174] Vejpravova JP, Tyrpekl V, Danis S, Niznansky D, Sechovsky V. J Magn Magn Mater 2010;322:1872–5. doi:10.1016/j.jmmm.2009.12.044.

[175] Peddis D, Cannas C, Musinu A, Piccaluga G. Chemistry 2009;15:7822–9. doi:10.1002/chem.200802513.

[176] Larsen BA, Haag MA, Serkova NJ, Shroyer KR, Stoldt CR. Nanotechnology 2008;19:265102. doi:10.1088/0957-4484/19/26/265102.

[177] Demortiere A, Panissod P, Pichon BP, Pourroy G, Guillon D, Donnio B, et al. Nanoscale 2011;3:225–32. doi:10.1039/c0nr00521e.

[178] Tackett RJ, Bhuiya AW, Botez CE. Nanotechnology 2009;20:445705. doi:10.1088/0957-4484/20/44/445705.

[179] Aslibeiki B, Kameli P, Salamati H. J Appl Phys 2016;119:63901. doi:10.1063/1.4941388.

[180] Aslibeiki B, Kameli P, Salamati H, Eshraghi M, Tahmasebi T. J Magn Magn Mater 2010;322:2929–34. doi:10.1016/j.jmmm.2010.05.007.

[181] Balaji G, Wilde G, Weissmuller J, Gajbhiye NS, Sankaranarayanan VK. Phys Status Sol B Basic Sol State Phys 2004;241:1589–92. doi:10.1002/pssb.200304671.

[182] Balaji G, Gajbhiye NS, Wilde G, Weissmuller J. J Magn Magn Mater 2002;242:617–20. doi:10.1016/S0304-8853(01)01043-5.

[183] Fannin PC, Vincent D, Massart G, Perov P, Neveu S. Eur Phys J Appl Phys 1999;8:247–51. doi:10.1051/epjap:1999252.

[184] Bittova B, Poltierova Vejpravova J, Morales MP, Roca AG, Mantlikova A. J Magn Magn Mater 2012;324:1182–8. doi:10.1016/j.jmmm.2011.11.005.

[185] Bittova B, Vejpravova JP, Morales MP, Roca AG, Niznansky D, Mantlikova A. Nano 2012;7:1250004. doi:10.1142/S179329201250004X.

[186] Vasilakaki M, Binns C, Trohidou KN. Nanoscale 2015;7:7753–62. doi:10.1039/c4nr07576e.

[187] Sousa EC, Rechenberg HR, Depeyrot J, Gomes JA, Aquino R, Tourinho FA, et al. J Appl Phys 2009;106:93901. doi:10.1063/1.3245326.

[188] Li J, Wang A, Lin Y, Liu X, Fu J, Lin L. J Magn Magn Mater 2013;330:96–100. doi:10.1016/j.jmmm.2012.10.048.

[189] Almeida TP, Moro F, Fay MW, Zhu Y, Brown PD. J Nanopart Res 2014;16:2395. doi:10.1007/s11051-014-2395-1.

[190] Gavrilov-Isaac V, Neveu S, Dupuis V, Taverna D, Gloter A, Cabuil V. Small 2015;11:2614–8. doi:10.1002/smll.201402845.

[191] Rai A, Banerjee M. J Nanosci Nanotechnol 2008;8:4172–5. doi:10.1166/jnn.2008.AN28.

[192] Beji Z, Hanini A, Smiri LS, Gavard J, Kacem K, Villain F, et al. Chem Mater 2010;22:5420–9. doi:10.1021/cm1001708.

[193] Jadhav SS, Shirsath SE, Toksha BG, Patange SM, Shengule DR, Jadhav KM. Phys B Condensed Matter 2010;405:2610–4. doi:10.1016/j.physb.2010.03.008.

[194] Kavas H, Baykal A, Toprak MS, Koeseoglu Y, Sertkol M, Aktas B. J Alloys Compd 2009;479:49–55. doi:10.1016/j.jallcom.2009.01.014.

[195] Kubickova S, Vejpravova J, Holec P, Niznansky D. J Magn Magn Mater 2013;334:102–6. doi:10.1016/j.jmmm.2013.01.005.

[196] Mehran E, Shayesteh SF, Nasehnia F. J Supercond Nov Magn 2016;29:1241–7. doi:10.1007/s10948-016-3415-4.

[197] Vestal CR, Zhang ZJ. Chem Mater 2002;14:3817–22. doi:10.1021/cm020112k.

[198] Burianova S, Vejpravova JP, Holec P, Plocek J, Niznansky D. J Appl Phys 2011;110:73902. doi:10.1063/1.3642992.

[199] Loganathan A, Kumar K. Appl Nanosci 2016;6:629–39. doi:10.1007/s13204-015-0480-0.

[200] Dixit G, Singh JP, Srivastava RC, Agrawal HM. J Magn Magn Mater 2013;345:65–71. doi:10.1016/j.jmmm.2013.05.060.

[201] Kuai S, Nan Z. J Alloys Compd 2014;602:228–34. doi:10.1016/j.jallcom.2014.03.049.

[202] Moya C, Morales MDP, Batlle X, Labarta A. Phys Chem Chem Phys 2015;17:13143–9. doi:10.1039/C5CP01052G.

[203] Phadatare MR, Meshram JV, Gurav KV, Kim JH, Pawar SH. J Phys D Appl Phys 2016;49:95004. doi:10.1088/0022-3727/49/9/095004.

[204] Zhang Q, Castellanos-Rubio I, Munshi R, Orue I, Pelaz B, Gries KI, et al. Chem Mater 2015;27:7380–7. doi:10.1021/acs.chemmater.5b03261.

[205] Mameli V, Musinu A, Ardu A, Ennas G, Peddis D, Niznansky D, et al. Nanoscale 2016;8:10124–37. doi:10.1039/C6NR01303A.

[206] Singh G, Chan H, Baskin A, Gelman E, Repnin N, Kral P, et al. Science (80-) 2014;345:1149–53. doi:10.1126/science.1254132.

[207] Singh G, Chan H, Udayabhaskararao T, Gelman E, Peddis D, Baskin A, et al. Faraday Discuss 2015;181:403–21. doi:10.1039/c4fd00265b.

Chapter 2

Lithium Ferrite: Synthesis, Structural Characterization and Electromagnetic Properties

Sílvia Soreto, Manuel Graça, Manuel Valente and
Luís Costa

Additional information is available at the end of the chapter

http://dx.doi.org/10.5772/110790

Abstract

Lithium ferrite ($LiFe_5O_8$) is a cubic ferrite, belongs to the group of soft ferrite materials with a square hysteresis loop, with high Curie temperature and magnetization. The spinel structure of $LiFe_5O_8$ has two crystalline forms: ordered, β-$LiFe_5O_8$ (*Fd3m* space group) and disordered, α-$LiFe_5O_8$ ($P4_132/P4_332$ space group). It has numerous technological applications in microwave devices, computer memory chip, magnetic recording, radio frequency coil fabrication, transformer cores, rod antennas, magnetic liquids among others. It is also a promising candidate for cathode in rechargeable lithium batteries. In this work, the dc electrical conductivity, the impedance spectroscopy and the magnetization of Li_2O-Fe_2O_3 powders, with [Li]/[Fe]=1/5 (mol), heat-treated at several temperatures, are studied and related to their structure and morphology. The structural data were obtained by X-ray diffraction and Raman spectroscopy, and the morphology by scanning electron microscopy. The impedance spectroscopy was analysed in function of temperature and frequency, and it was observed that the dielectric properties are highly dependent on the microstructure of the samples. The dc magnetic susceptibility was recorded with a vibrating sample magnetometer, under zero field cooled and field cooled sequences, between 5-300 K. Typical hysteresis curves were obtained and the saturation magnetization increases with increase in heat-treatment temperature.

Keywords: lithium ferrite, electric and dielectric properties, magnetic properties

1. Introduction

Technologically, ferrites are very important due to their interesting magnetic and electrical properties which can be exploited for applications in high-capacity batteries, electrochromic displays, wastewater cleaning, low magnetization ferrofluids, intercalation electrodes in rechargeable batteries and as strong oxidizing agents [1–5]. Ferrites crystallize in three structure: (i) cubic spinel structure with the general formula $MO.Fe_2O_3$ (M = Mn^{2+}, Fe^{2+}, Co^{2+}, Ni^{2+}, Cu^{2+}, Zn^{2+}, Mg^{2+}); (ii) hexagonal ferrites with the formula $MO.6Fe_2O_3$ (M = Ba^{2+}, Ca^{2+}, Sr^{2+}); and (iii) and garnets with the formula $2M_2O_3.5Fe_2O_3$ where M is a cation with (3+) charge such as Y or another rare earth [6].

The cubic lithium ferrite (**Figure 1**), spinel $LiFe_5O_8$ [7], is one of the most important ferrites, and it belongs to the soft ferrite materials group, with high Curie temperature (620°C) [8], square hysteresis loop and high magnetization. The spinel lithium ferrite has been widely studied and confirmed to have two crystalline forms: β-$LiFe_5O_8$ (*Fd3m* space group), known as the disordered $LiFe_5O_8$, and the α-$LiFe_5O_8$ (space group $P4_132/P4_332$), called ordered spinel phase. The first one is obtained by the rapid quenching of the samples from temperatures above 800°C to room temperature. Upon slow cooling and below 750°C, an ordered phase is obtained [9]. In the ordered form, α-$LiFe_5O_8$, the octahedral *12d* and tetrahedral *8c* sites are occupied by iron ions, Fe^{3+}, and the octahedral *4b* positions are occupied by lithium ions, Li^+, in the cubic primitive cell. The disordered structure, β-$LiFe_5O_8$, has an inverse spinel structure, where the tetrahedral *8a* positions are occupied by Fe^{3+} ions and the ions Li^+ and Fe^{3+} are randomly distributed over the *16d* octahedral positions [10, 11].

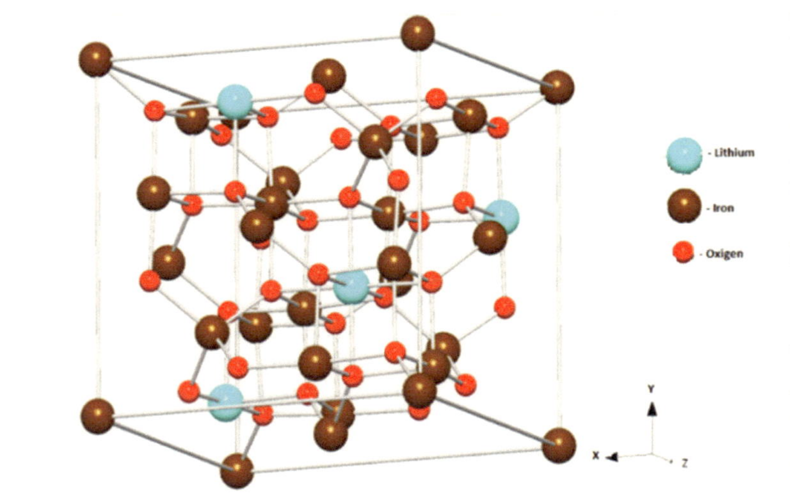

Figure 1. Cubic crystal of $LiFe_5O_8$ with space group P4132 (213).

To prepare $LiFe_5O_8$ by solid-state reaction method, high temperatures (>1200°C) are needed, which is a major issue due to the volatility of lithium above 1000°C. Therefore, the prepared material has a low quality because the sintering process normally leads to low specific surface areas [12], affecting its electrical and magnetic properties. Several chemical methods have been used for synthesis, such as co-precipitation, glass crystallization, hydrothermal, mechanical alloying and sol-gel methods. Therefore, the preparation of $LiFe_5O_8$ at low temperatures is a subject of interest.

In order to improve the material properties, in this work lithium ferrite was prepared at a low temperature, by the solid-state reaction method using iron and lithium nitrates as precursors. First, the use of nitrates as base materials and, second, the use of the high-energy planetary ball milling to prepare the base material for further heat treatments are not conventional route and denote improvements which have not been reported in the literature. The electric and magnetic properties of the ferrite prepared by this method have been already published [5, 13, 14]. With the main aim to enhance the electric and magnetic properties of $LiFe_5O_8$ and determine the best temperature for the thermal treatment, the powders after the ball milling are heat-treated at different temperatures.

The dependence of the particle size with the sintering temperature was also studied, and these studies were correlated with the measured electrical, dielectric and magnetic properties. The obtained results were analysed and compared with those presented in the literature.

2. Methods and procedures

As described in detail in our recent papers [5, 12, 13], the solid-state method was used to prepare lithium ferrite ($LiFe_5O_8$) powders starting with iron (III) nitrate ($Fe(NO_3)_3.9H_2O$) and lithium nitrate ($LiNO_3$ (Merck KGaA, Darmstadt, Germany). Considering the 1:5 required stoichiometry of Li:Fe, appropriate amounts of the two starting materials were mixed and homogenized in a planetary ball mill system (Fritsch Pulverisette 7.0) at 250 rpm for 1 h using equal volumes of balls to powders. Following this, 10 mL of ethanol was added to the mixture followed by additional ball milling at 500 rpm for 3 h, stopping the system for 5 min every hour to reduce over-heating. Next, in order to evaporate the ethanol, the vessel with the mixture was placed in a furnace at 80°C for 24 h.

The next step in the synthesis procedure was heat treatment of the powder at nine temperatures between 200 and 1400°C (see **Figure 2**) at 10°C/min in two steps: first at 100°C for 1 h and the second at desired temperature for 4 h. After this heat treatment, the samples were structurally characterized using the techniques of X-ray diffraction (XRD), micro-Raman spectroscopy and scanning electron microscopy (SEM). The equipment used for the structural characterization was X'Pert MPD Philips diffractometer (CuKα radiation, λ = 1.54060 Å) for X-ray diffraction, an HR-800-UV Jobin Yvon Horiba spectrometer (532 nm laser line) for micro-Raman spectroscopy using a microscope objective (50×) and a Hitachi S4100-1 SEM system for SEM images with the samples covered with carbon before microscopic observation. Further details on these procedures are given in Refs. [5, 12, 13]. Electrical and magnetic measurements on the samples were performed following the structural characterization.

For the measurements of electrical conductivity and impedance, the samples were pressed into 2-mm-thick discs with the opposite sides of the discs painted with silver paste for electrical contacts. The measurements were performed in helium atmosphere in order to improve the heat transfer and eliminate the moisture. The dc conductivity (σ_{dc}) was measured from 100 to 360 K employing a Keithley Model 617 electrometer using applied voltage of 100 V. Employing an Agilent 4294A precision impedance analyser in the Cp-Rp configuration, the measurements of impedance in the frequency range of 100 Hz–2 MHz were performed from 200 to 360 K.

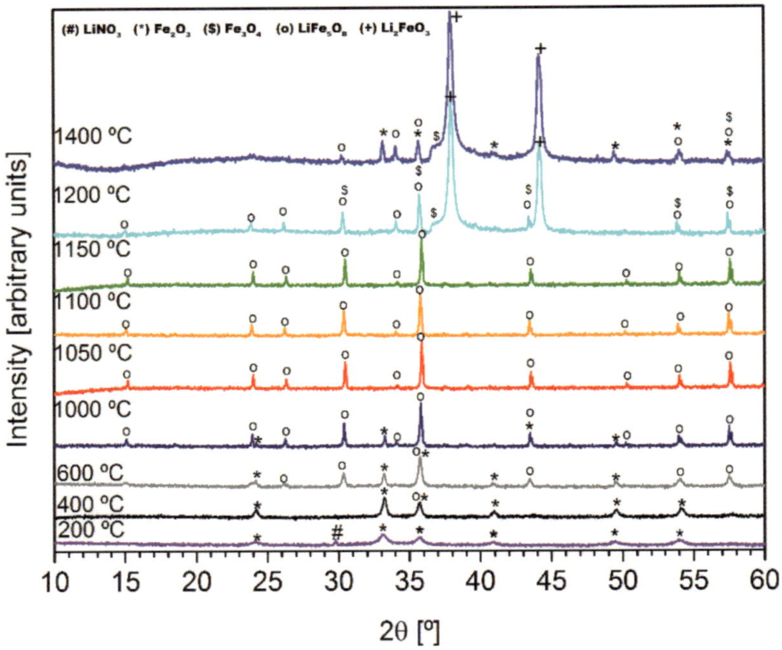

Figure 2. XRD patterns of the powders heat-treated between 400 and 1400°C: (*) Fe_2O_3; (o) $LiFe_5O_8$; (+) Li_2FeO_3; (#) $LiNO_3$. Adapted from Refs. [13, 14].

The magnetic properties of the samples were measured using a vibrating sample magnetometer (VSM). The dc magnetic susceptibility was measured from 5 to 300 K under two protocols: (i) cooling the sample in zero field (ZFC) and (ii) cooling the sample with the magnetic field applied (FC). Typical hysteresis curves were obtained at several temperatures.

3. Results and discussion

3.1. Structural and morphological measurements

Figure 2 shows the XRD patterns of the samples after each heat treatment. The XRD spectra of the sample treated at 200°C shows the presence of lithium nitrate crystal phase. This phase disappears with the increase in the heat-treatment temperature. The samples heat-treated at 200 and 400°C present the α-Fe_2O_3 as the major crystalline phase. The non-detection by the XRD of the lithium nitrate phase in the sample treated at 400°C suggests the existence of an amorphous phase containing mainly lithium ions. This result also revealed that the planetary ball milling process did not induce the formation of new crystalline phases before the heat treatments.

A low amount of lithium ferrite is found in the sample with heat treatment at 600°C, but the main diffraction peaks are attributed to the α-Fe_2O_3 phase. The formation of $LiFe_5O_8$ crystal phase can be attributed to the reaction between α-Fe_2O_3 and free Li^+ ions. In samples with heat treatments at 1050, 1100 and 1150°C, all diffraction peaks can be assigned to the lithium ferrite crystal phase as described in Eq. (1) with loses of NO_x.

$$5\text{Fe}\left(\text{NO}_3\right)_3 + \text{LiNO}_3 \xrightarrow{\Delta} \text{LiFe}_5\text{O}_8 + y\,\text{NO}_x \quad (1)$$

In the sample treated at 1200°C besides lithium ferrite peaks, the lithium ferrate (Li_2FeO_3) phase is also detected. The presence of this phase can be explained through the chemical reaction, Eq. (2), where the peak characteristics of Fe_3O_4 are also present.

$$15\,\text{Fe}\left(\text{NO}_3\right)_3 + 3\,\text{LiNO}_3 \xrightarrow{\Delta} \text{LiFe}_5\text{O}_8 + \text{Li}_2\text{FeO}_3 + 3\text{Fe}_3\text{O}_4 + y\,\text{NO}_x \quad (2)$$

In the sample heat-treated at 1400°C, besides the peak characteristics of magnetite, lithium ferrite and lithium ferrate crystal phases also present peak characteristics of hematite, Eq. (3). In this sample, the major phase is attributed to lithium ferrate crystal phase showing broad diffraction peaks.

$$15\,\text{Fe}\left(\text{NO}_3\right)_3 + 3\,\text{LiNO}_3 \xrightarrow{\Delta} \text{LiFe}_5\text{O}_8 + \text{Li}_2\text{FeO}_3 + \text{Fe}_3\text{O}_4 + 3\text{Fe}_2\text{O}_3 + y\,\text{NO}_x \quad (3)$$

According to Wolska et al. [7] results, for heat treatment above 1000°C the disordered $LiFe_5O_8$ phase can be formed. However, in our results, the $LiFe_5O_8$ phase detected was only the ordered one (α-$LiFe_5O_8$) with the space group $P4_132/P4_332$.

To determine the crystallite size, L_c, of lithium ferrite crystal phase, the Debye-Scherrer equation was used:

$$L_c = \frac{N\,\lambda}{\beta\cos\theta} \quad (4)$$

Here, β is full width half maximum of the diffracted peaks, λ is the wavelength of X-ray radiation, θ is the angle of diffraction and N is a numerical factor frequently referred to as

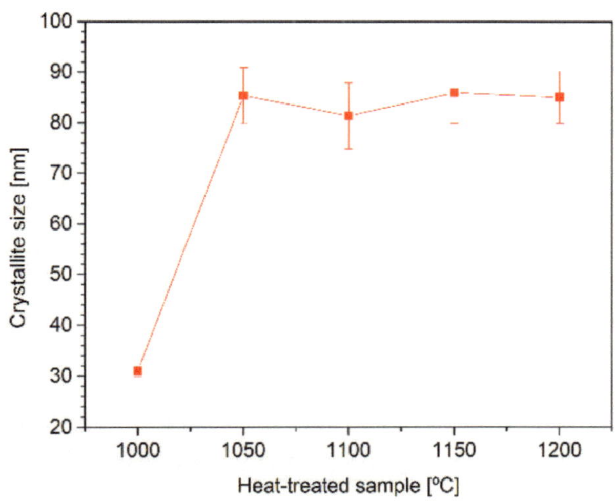

Figure 3. Crystallite size of $LiFe_5O_8$ phase in samples heat-treated at 1000, 1050, 1100, 1150 and 1200°C.

the crystallite-shape factor [15, 16] and being $N = 0.9$ a good approximation in the absence of detailed information [17].

Substituting the relevant data from XRD profile measurement, the average crystallite sizes and its errors bars are shown in **Figure 3**.

A crystallite size of about 85 nm was observed in the samples heat-treated at 1050, 1100, 1150°C, which only presents $LiFe_5O_8$ crystal phase, and in the sample heat-treated at 1200°C. The sample treated at 1000°C has the lowest crystallite size, 31 ± 1 nm.

The samples presenting $LiFe_5O_8$ as major crystal phase, that is, heat-treated at temperatures between 400 and 1400°C, were also analysed using Raman spectroscopy (**Figure 4**).

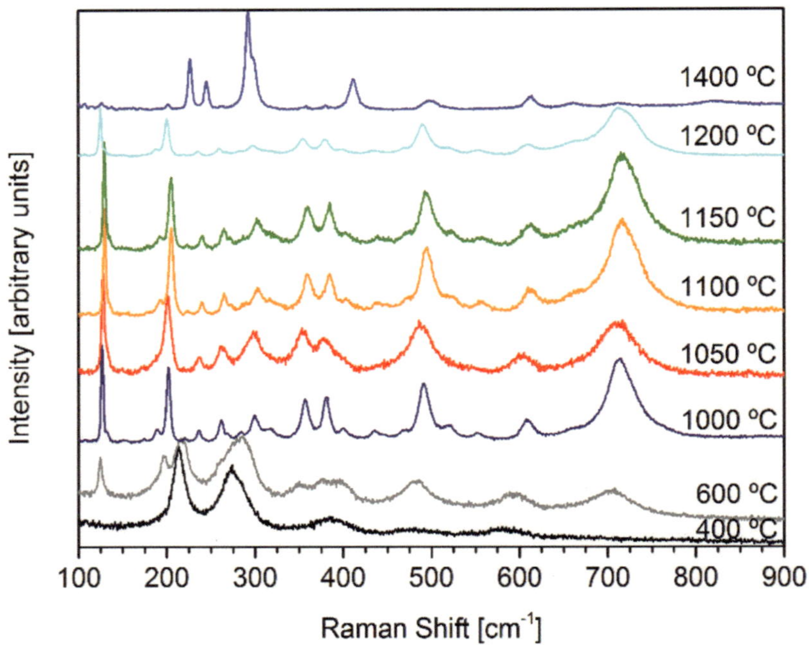

Figure 4. Raman spectra for the samples with heat treatments between 400 and 1400°C. Adapted from Refs. [13, 14].

According to the Raman spectroscopy spectra, all the samples show the vibration mode characteristic of both ordered and disordered lithium ferrite phases.

For the samples treated between 1000 and 1400°C, the vibrational peaks at 199–206 and 237–241 cm^{-1} indicate the presence of the ordered α-$LiFe_5O_8$ phase [18]. All the vibrational peaks of the $LiFe_5O_8$ phase are given in **Table 1**. In the samples heat-treated at 600, 1000 and 1400°C, besides the vibrational modes that mark the presence of lithium ferrite, the vibrational mode characteristic of α-Fe_2O_3 is also present [19].

According to the Raman spectra, for the sample heat-treated at 1200°C and crossing with the XRD results, we can infer about the vibration modes of the lithium ferrate (Li_2FeO_3). The

http://dx.doi.org/10.5772/110790

Sample								Assigned to
400°C	**600°C**	**1000°C**	**1050°C**	**1100°C**	**1150°C**	**1200°C**	**1400°C**	
						125	125	Li_2FeO_3
	126	126	128	130	130			$LiFe_5O_8$
							137	α-Fe_2O_3
		170	170	170	172	171		$LiFe_5O_8$
						188		Fe_3O_4 19]
		195		196	195			$LiFe_5O_8$
	199	202	201	206	206	201	202	α-$LiFe_5O_8$ [7, 18]
214	217		B					α-Fe_2O_3 [19]
		223	220	226	221		226	$LiFe_5O_8$
		237	239	241	241	239	245	α-$LiFe_5O_8$ [7, 18]
		263	267	266	267	262	262	α-$LiFe_5O_8$ [7, 18] β-$LiFe_5O_8$ [7, 18]
274	287	286					292	α-Fe_2O_3 [19]
						298	298	Fe_3O_4 [19]
		300	301	306	305	301		α-$LiFe_5O_8$ [6, 17] β-$LiFe_5O_8$ [7]
		320						α-$LiFe_5O_8$ [7, 18] β-$LiFe_5O_8$ [7, 18]
	353	358	358	361	362	357	357	α-$LiFe_5O_8$ [7, 18] β-$LiFe_5O_8$ [7, 18]
388	379	383	382	386	385	382	380	α-$LiFe_5O_8$ [7, 18] β-$LiFe_5O_8$ [7, 18]
	400	405		407	410	409	410	α-$LiFe_5O_8$ [7, 18] β-$LiFe_5O_8$ [7, 18]
						437		Li_2FeO_3
		439		438	441			α-$LiFe_5O_8$ [7, 18] β-$LiFe_5O_8$ [7, 18]
		475		475	475	473		α-$LiFe_5O_8$ [7, 18] β-$LiFe_5O_8$ [7, 18]
486	489	492	492	496	496	494	498	α-$LiFe_5O_8$ [7, 18] β-$LiFe_5O_8$ 7, 18]
		523		526	526	523		$LiFe_5O_3$
		555		560	558	558		α-$LiFe_5O_8$ 7, 18] β-$LiFe_5O_8$ [7, 18]
590								not identified
	603							not identified
		611	610	612	614	612	613	α-$LiFe_5O_8$ [7, 18] β-$LiFe_5O_8$ [7, 18]
		663		660	660		661	α-$LiFe_5O_8$ [7, 18]
						668		Fe_3O_4 [19]
	707	715	719	719	719	718		α-$LiFe_5O_8$ [7, 18] β-$LiFe_5O_8$ [7, 18]

Table 1. Raman peaks identification for the different samples.

assignment of vibrational modes related with lithium ferrite and magnetite (Fe_3O_4) [20] was also made. In the particular case of lithium ferrate, it is interesting to focus that the vibrational band with higher intensity is centred at 125 cm^{-1} and the bands at 437 and 523 cm^{-1} show lower intensity. **Table 1** shows all vibration modes present for each sample.

In the morphological analysis of the samples (**Figure 5**), the increasing of the particle size is clearly visible from 100 nm for the sample heat-treated at 400°C to 4 μm approximately for the sample heat-treated at 1400°C.

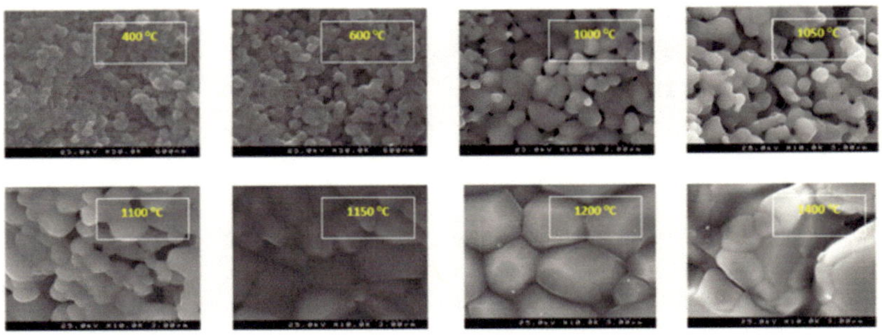

Figure 5. SEM micrographs for samples with heat treatments at 200, 400, 600, 1000, 1050, 1100, 1150, 1200 and 1400°C. Adapted from Refs. [5, 13, 14].

The aggregation of the spherical grains is quite evident, related to the Fe_2O_3, with the increase in the heat-treatment temperature. The formation of the $LiFe_5O_8$ phase shows a different microstructure, with prismatic grains clearly visible in the sample with heat treatment at 1150°C. In the sample treated at 1400°C, the formation of small grains attached to the prismatic grains is evident. As the shape of these grains is spherical, it could be attributed to the formation of Fe_2O_3, once this shape is the same as the one present in the samples treated at low temperatures. This aggregation to the prismatic particles also seems to appear in the sample treated at 1200°C. These results support the ones obtained by XRD and Raman measurements.

3.2. Electrical and dielectric measurements

The electrical measurements were performed on the samples heat-treated at temperatures between 400 and 1200°C. To interpret the temperature dependence of the dc conductivity, σdc data, the Arrhenius expression [21] has been used:

$$\sigma_{dc} = \sigma_0 exp\left(-\frac{E_{a(dc)}}{KT}\right) \tag{5}$$

Here σ_0 is a pre-exponential factor, $E_{a(dc)}$ the dc activation energy, K the Boltzmann constant and T the temperature. **Figure 6** shows the experimental data and the fit for all the treated samples.

The maximum of the conductivity is obtained for the sample heat-treated at 1100°C, which according to the XRD pattern only contains the $LiFe_5O_8$ crystal phase. For the other samples,

http://dx.doi.org/10.5772/110790

which also contain hematite, the conductivity increases with the temperature of heat treatment, and the activation energy decreases. This behaviour can be explained by the decrease in the hematite phase until the treatment temperature reaches 1100°C. The lower conductivity of the sample treated at 1200°C compared to the one treated at 1100°C can be related to the presence of lithium ferrate crystal.

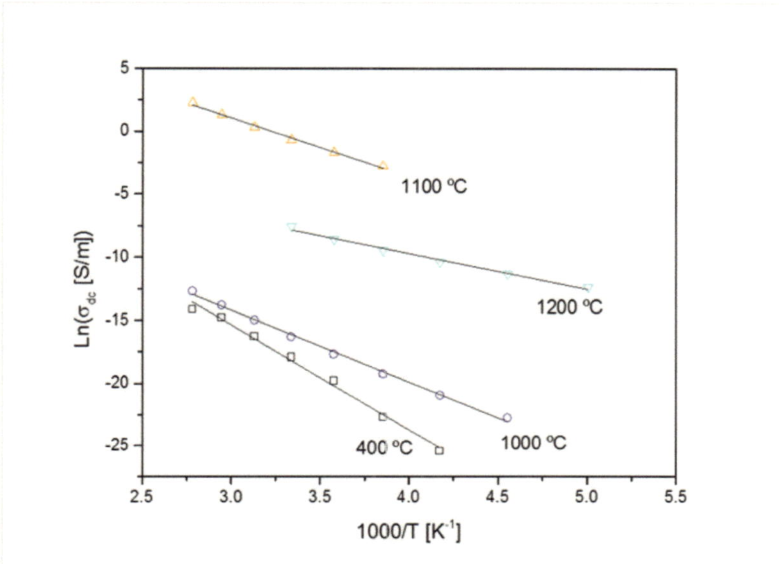

Figure 6. Log (σ_{dc}) versus 1000/T.

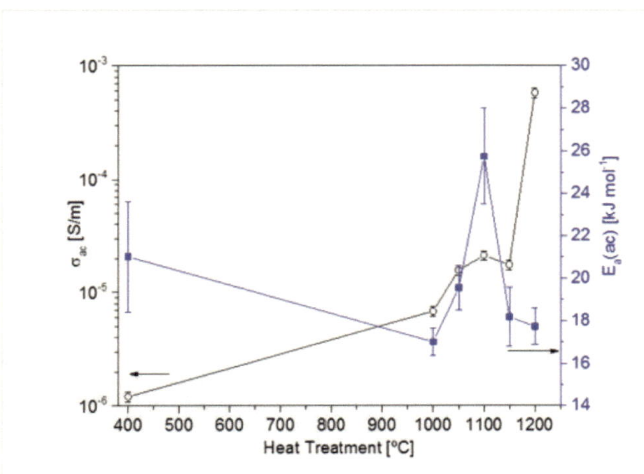

Figure 7. Dependence of the ac activation energy ($E_{a(ac)}$) and ac conductivity (σ_{ac}), at 300 K and 100 kHz, with heat-treatment temperatures.

Figure 7 shows the behaviour of the ac conductivity with the temperature of heat treatment. For samples treated at 1000, 1050, 1100 and 1150°C, there is a similar tendency between *ac* activation energy and conductivity. Similar to the dc conductivity variation of Eq. (3), the Arrhenius behaviour also fits correctly the ac data:

$$\sigma_{ac} = \sigma_0 exp\left(-\frac{E_{a(ac)}}{KT}\right) \tag{6}$$

Table 2 shows a comparison of the ac and dc activation energy for each sample.

Heating temperature (°C)	$E_{a(ac)} \pm \Delta E_{a(ac)}$ **(kJ/mol)**	$E_{a(dc)} \pm \Delta E_{a(dc)}$ **(kJ/mol)**
400	21.0 ± 2.6	69.4 ± 0.7
1000	17.0 ± 0.6	47.4 ± 0.5
1050	20 ± 1	
1100	26 ± 2	39.1 ± 1.6
1150	18 ± 1	
1200	17.7 ± 0.9	23.6 ± 0.2

Table 2. The *dc* and *ac* activation energies for the samples.

The dielectric measurements of C_p and R_p allowed us to calculate the real, ε′, and the imaginary, ε″, part of the complex permittivity using the equations [22]:

$$\varepsilon' = C_p \frac{d}{A\,\varepsilon_0} \tag{7}$$

$$\varepsilon'' = \frac{d}{\omega\, R_p\, \varepsilon_0} \tag{8}$$

Here d represents the sample thickness, A the electrode area, ε_0 the empty space permittivity and ω the angular frequency, respectively. These relations are only valid if $d \ll A$. The frequency dependence of the real part of the complex permittivity, ε′, and of the imaginary part of the complex permittivity, ε″, at constant temperature, $T = 300$ K, is shown in **Figure 8**.

According to **Figure 8**, the sample heat-treated at 1200°C presents the higher values of ε′, but the ε″ results are not suitable for their intended use due to their high losses, tanδ = 1.12 at 300 K and 1 kHz. On the other hand, the samples with heat treatments at 1050, 1100 and 1150°C have high dielectric constant and low losses at the same conditions of temperature and frequency: 0.86, 0.40 and 0.84, respectively.

Again, the behaviour is similar to that one observed with the other measurements. The increase in the complex permittivity with the treatment temperature is due to the structural and morphologic changes observed. This behaviour occurs for other ferrites [23, 24], where an increase in ε′ with the temperature of the heat treatment is observed. The crystallite size of the particles (**Figure 3**) influences the dielectric response, that is, the increase in the crystallite size, which is maximum for the samples treated at 1050, 1100, 1150 and 1200°C, leads to an increase in the dielectric constant.

http://dx.doi.org/10.5772/110790

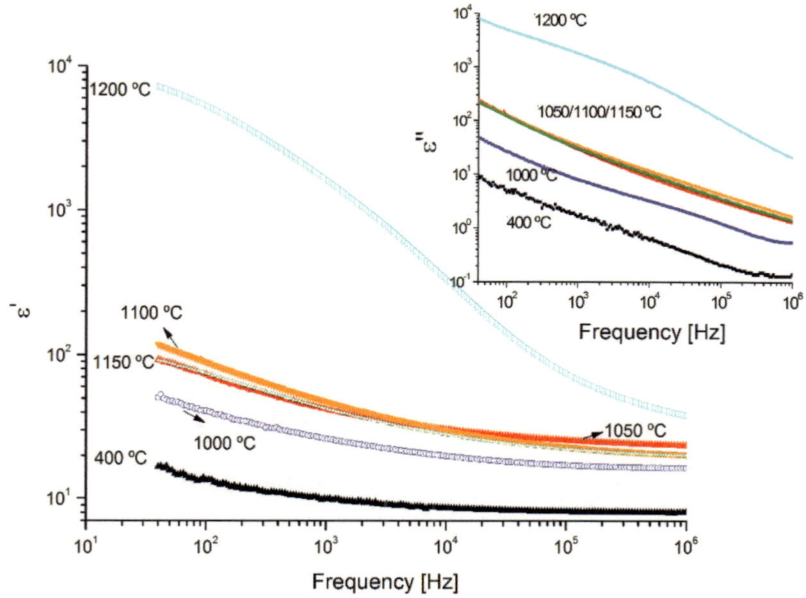

Figure 8. Frequency dependence of the complex permittivity, at T = 300 K [14].

The modulus formalism, where M* = 1/ε*, proposed by Macedo et al. [25] was used to study the dielectric response. **Figure 9** shows the frequency dependence of the imaginary part of the modulus (M″) for the sample that was heat-treated at 1050°C.

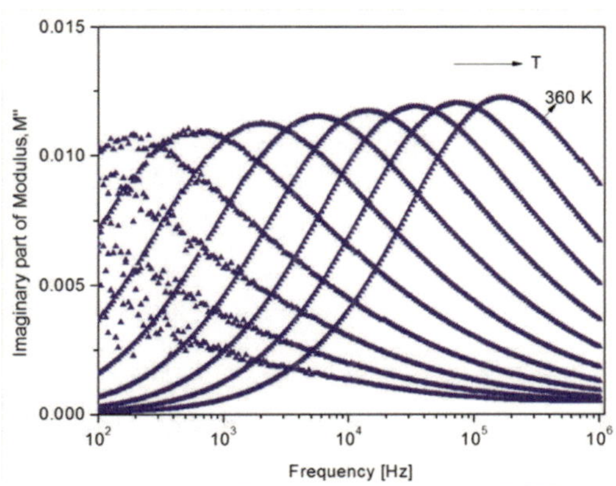

Figure 9. Imaginary part of the complex modulus, M″, for the sample treated at 1050°C, in the temperature range between 260 and 360 K, in steps of 10 K [14].

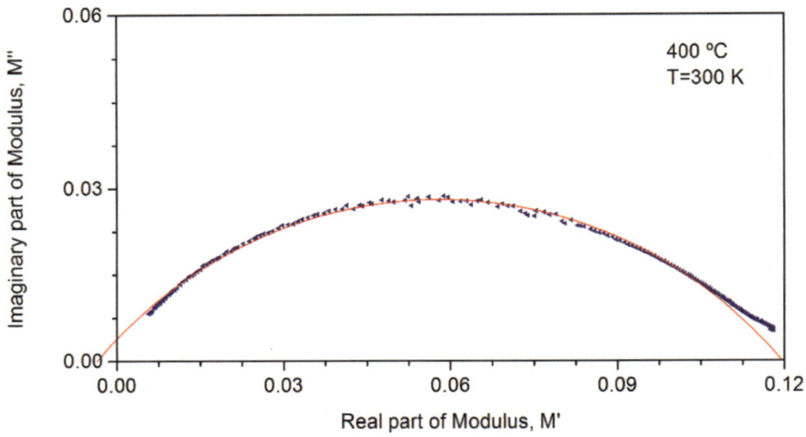

Figure 10. Nyquist plot fit for the sample heat-treated at 400°C for a temperature of measurement of 300 K [5].

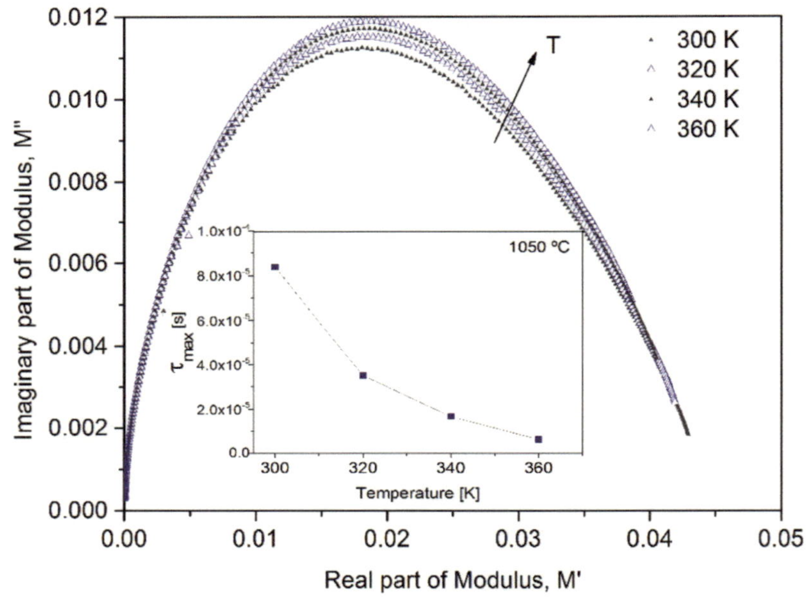

Figure 11. Nyquist plot for the sample treated at 1050°C for a range of temperatures between 300 and 360 K. Adapted from Ref. [14].

For all samples, a relaxation process is visible, as the one shown in **Figure 10**, which has a shape of a decentred semicircle, with its centre situated below the abscissa axis. **Figure 11** shows sample heat-treated at 1050°C, the evolution of the Nyquist plot with the temperature. From these results, the maximum of the relaxation time, τ_{max}, (**Figure 11** inset) shows a typical behaviour, that is, a decrease in the τ_{max} with increase in the temperature. This Cole-Cole analysis can also be made with the magnetic ac susceptibility data as the work presented by Wang and Seehra [26].

This profile indicates that the simple exponential decay, corresponding to a Debye relaxation, is inappropriate to describe the relaxation phenomena and should be replaced by an empirical model, such as Cole-Cole analysis [27]:

$$M^* = M_\infty + \frac{\Delta M}{1+(iw\tau_M)^{1-\alpha_M}} \tag{9}$$

Eq. (9) represents a modification of the Debye equation since for $\alpha_M = 0$, the Debye model corresponding to a single relaxation time is recovered. In Eq. (9), α_M is a parameter between 0 and 1, which reflects the dipole interaction in the system; also M_∞ is the relaxed modulus; ΔM is the modulus relaxation strength; and τ_M is the relaxation time.

According to the values of the exponent parameter α_M, the system is a non-Debye, and for all samples, α_M decreases with the temperature. It decreases from about 0.56 to 0.50 in the sample treated at 400°C and from about 0.43 to 0.30 at 1200°C treated one. For modulus relaxation strength ΔM, the behaviour is similar to exponent parameter α_M, decreasing with temperature for each heat treatment (**Figure 12**).

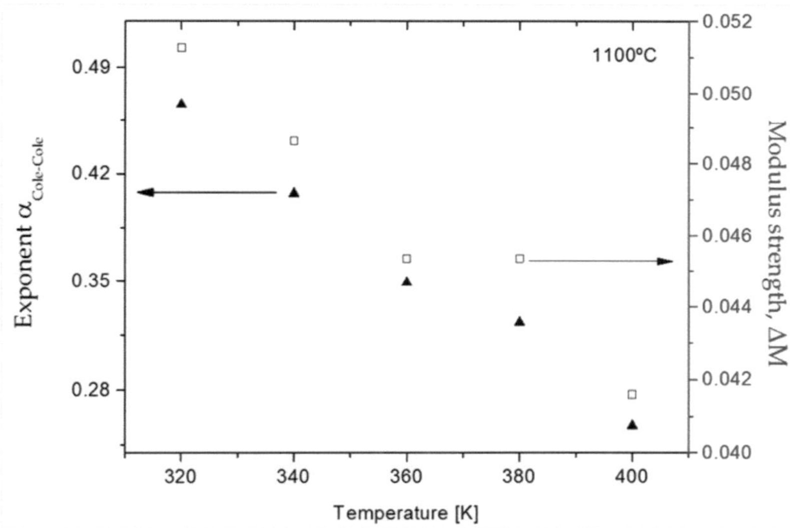

Figure 12. Temperature dependence of the $\alpha_{Cole\text{-}Cole}$ parameter and ΔM for sample heat-treated at 1100°C [14].

The relaxation frequency, $f_{max} = \frac{1}{2\pi\tau}$, can be expressed by the Arrhenius law, where

$$f_{max} \propto exp\left(-\frac{E_a}{KT}\right) \tag{10}$$

In Eq. (10), E_a is the activation energy for the relaxation process. The logarithmic representation of the relaxation frequency versus the inverse temperature, the relaxation map, allows to obtain the activation energy (**Figure 13**). Also in this process, the activation energy decreases with the temperature of the treatment unless for the samples treated at 1050, 1100 and 1150°C which increases. This may be related to structural changes, since at this heat treatment temperature only the crystalline phase of lithium ferrite is present. The same tendency as the one observed in the ac conductivity regime, in the samples with only one crystal phase, $LiFe_5O_8$. **Table 3** resumes all the activation energies of the relaxation process.

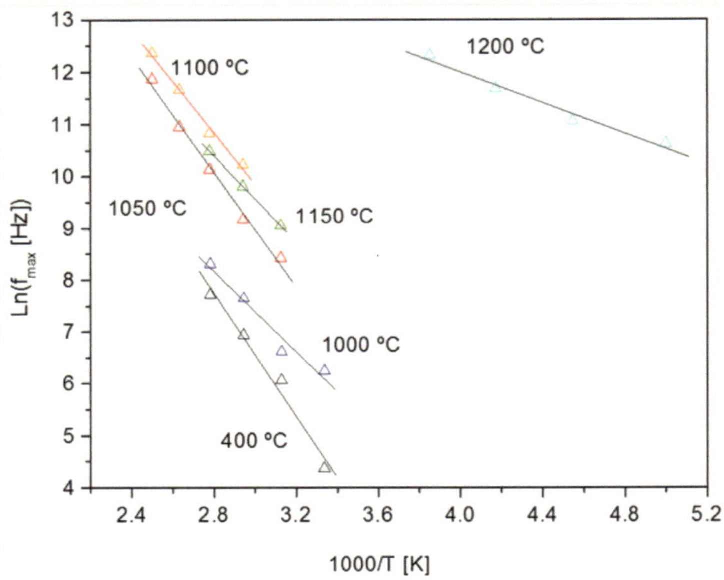

Figure 13. Ln (f_{max}) versus 1000/T [14].

Heat-treated sample (°C)	$E_a \pm \Delta E_a$ (kJ/mol)
400	21 ± 3
1000	32 ± 2
1050	46.0 ± 0.8
1100	37.3 ± 0.8
1150	34.45 ± 0.01
1200	12.2 ± 0.7
Adapted from Ref. [14].	

Table 3. The activation energy for the relaxation process for all the samples investigated here.

3.3. Magnetic measurements

The magnetic measurements were performed on the samples heat-treated at 200, 400, 600, 1000, 1050, 1100, 1150, 1200 and 1400°C. Initially, the measurements were performed in ZFC condi-

http://dx.doi.org/10.5772/110790

tions with an applied magnetic field (*B*) of 0.1 T, from 5 K up to 300 K. At 300 K, magnetization (*M*) versus *B* measurements were performed. Then, FC measurements were performed from 300 K down to 5 K under the magnetic field of 0.1 T. Again, at 5 K, *M* versus *B* measurements were also performed. This experimental sequence is important to explain why after the M versus B at 300 K, an increase of magnetic susceptibility in the FC curves is observed, which can be ascribed to the remnant magnetization of the sample (**Figure 14a** and **b**).

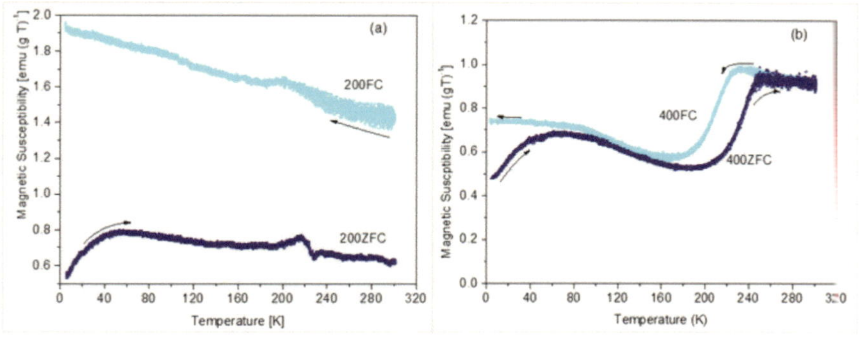

Figure 14. Magnetic susceptibility versus temperature, recorded under zero field cooled (ZFC) and field cooled (FC) sequences with B = 0.1 T, between 5 and 300 K, for the samples with a heat treatment of 200°C (a) and 400°C (b). Adapted from Ref. [13].

In **Figure 14a**, in the sample treated at 200°C, the results of the magnetic susceptibility recorded under ZFC and FC show the presence of a blocking temperature (T_B) between 50 and 70 K. This blocking temperature was observed with the translation of the FC curve into the ZFC curve, where T_B is the temperature which separates the FC and ZFC curves. **Figure 14b**, shows that in the sample treated at 400 °C, T_B is slightly higher, around 70 K, than in the sample treated at 200 °C. This difference can be related to the particles size, which is higher for the sample heat-treated at 400°C (**Figure 5**). The dependence of the magnetic susceptibility on temperature is shown in **Figure 15**.

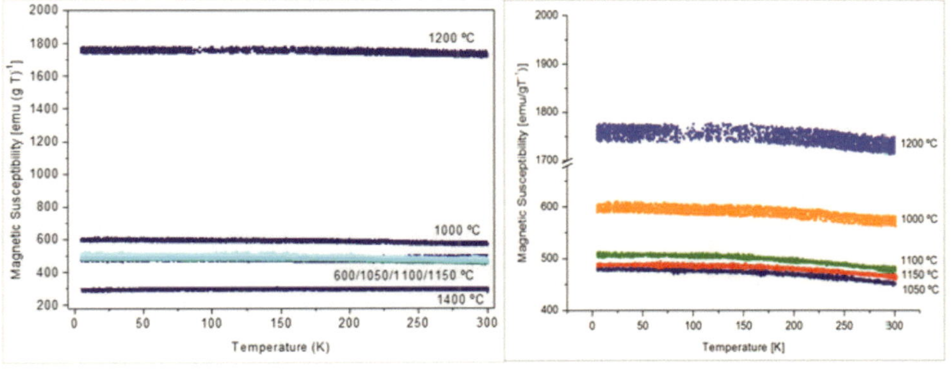

Figure 15. Magnetic susceptibility versus temperature, recorded under field cooled (FC) with an applied magnetic field of 0.1 T, between 5 and 300 K. Adapted from Ref. [13].

With the exception of samples treated at 200 and 400°C, a decrease in the magnetic susceptibility with increase in the temperature of measurement is observed in the remaining samples. This behaviour of the dependence of the susceptibility with the temperature was expected because it is characteristic of ferromagnetic materials. This feature will change to a non-magnetic order (paramagnetic characteristic) above Curie temperature, which is, according to Iliev et al. [8], about 893 K for lithium ferrite This property was not observed because in our experimental procedure the maximum temperature of measurement was 300 K (**Figure 15**). The sample heat-treated at 1200°C shows the highest magnetic susceptibility.

As noted earlier, the XRD diffraction patterns of the samples heat-treated at 1050, 1100 and 1150°C are characteristic of a single lithium ferrite phase (**Figure 2**). The SEM micrographs of these samples (**Figure 5**) show an increase in the grain size, changing from about 100 nm for sample treated at 400°C to 4 μm for sample treated at 1400°C. This promotes the increase in the probability of the random distribution of magnetic moments and therefore an increase in the soft magnetic type response.

Figure 16a and **b** shows the magnetization versus applied field, measured at 5 K, for all the samples. It should be noticed that **Figure 16b** is presented only to show a better graphical visualization of the magnetization of the samples treated between 1000 and 1200°C. The samples heat-treated at 200 and 400°C present a low magnetization, due to the fact that the major phase is antiferromagnetic α-Fe_2O_3.

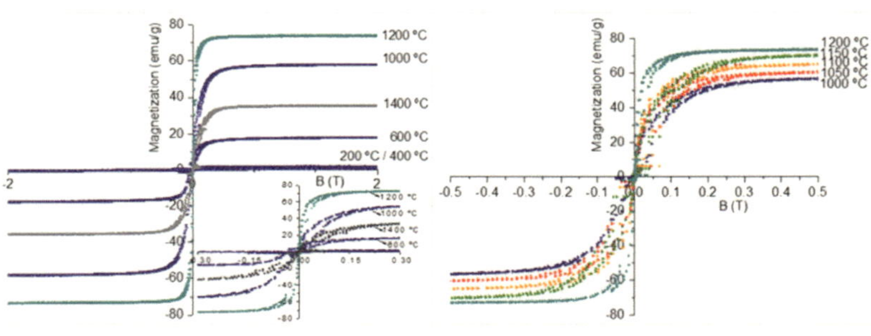

Figure 16. Magnetic hysteresis curves (T = 5 K). Adapted from Ref. [13].

In the literature [28, 29], the magnetization of the lithium ferrite is around 60 emu/g, which is rather lower than the one obtained in this work (**Figure 16**). In these samples, the generation of the lithium ferrite phase takes to the decrease in the contribution of the α-Fe_2O_3 particles with low magnetization. XRD and Raman results confirm that the samples heat-treated at temperatures between 1000 and 1200°C have the ordered phase α-$LiFe_5O_8$ as the major phase. Very likely, this ordered ferrite phase is developed from the thermal-activated reaction between α-Fe_2O_3 and "free" lithium ions present in the lattice, suggesting the existence of some amorphous phase, whose amount decreases with the increase in the heat-treatment temperature. The presence of this lithium ferrite phase also promotes the increase in the magnetization (**Figure 16**), showing a maximum of 73 emu/g, for the sample treated at 1200°C, which is a

http://dx.doi.org/10.5772/110790

similar value to the one obtained by Singhal [30] on lithium ferrite prepared by aerosol route. This high magnetization is attributed to the formation of Fe_3O_4 (M_s = 92 emu/g [31]) as proved by XRD results. For this sample, the minimum magnetic field that saturates the sample is around 0.1 T, meaning that this sample magnetizes easily [13]. The further increase in the heat-treatment temperature of 1400°C also promotes the development of hematite, α-Fe_2O_3 (**Figure 2**), whose magnetization of saturation is around ~10 emu/g [32, 33]. This should be the reason for the observed decrease in magnetization from sample treated at 1200 (73 emu/g) to sample treated at 1400°C (30 emu/g) (**Figure 16**). Moreover, the contribution of Li_2FeO_3 to the absolute magnetization should be lower than the one of $LiFe_5O_8$ in accordance with Hessien et al. [34] who states that the increase in the molar ratio between Fe^{3+} and Li^+ leads to lower magnetization values.

Shirsath et al. [35] have studied this particular type of ferrite, prepared using also nitrates as initial materials but following a sol-gel method. Comparing their results with ours and taking into account the fact that the followed method was the solid-state route, the magnitudes of the magnetization for our samples are higher, nearly 73 and 55 emu/g for the sample heat-treated at 1200 and 1000°C respectively. The values for the same treatment temperatures, obtained by Shirsath et al. through the sol-gel method, are around 55 and 45 emu/g, respectively, and related to the particle size. In the referred work, as the particle size increases, the magnetization decreases. For the sample treated at 600°C, which has also a large amount of hematite (**Figure 2**), the results are similar for both methods.

4. Conclusions

The $LiFe_5O_8$ crystal phase was obtained using nitrates as raw materials by an easy and relatively inexpensive route. From the structural and morphological results, the lithium ferrite crystal phase obtained is the ordered one, α-$LiFe_5O_8$, and is mostly present in the samples with thermal treatment from 1000 to 1200°C. Heat treatments above 1150°C promote the formation of the Li_2FeO_3 and Fe_3O_4 crystal phases. Spherical grains related to α-Fe_2O_3 were detected in the sample heat-treated at 1400°C. The formation of lithium ferrite phases leads to a different microstructure (prismatic grains), whose size increases with the increase in heat-treatment temperature, obtaining a maximum for 1200°C. In the sample treated at 1400°C, SEM micrograph shows modification in the surface that is related to the lithium ferrate crystal phase and the appearance of α-Fe_2O_3.

Electrical measurements suggest that the samples with heat treatments at 1000, 1050 and 1150°C have a good dielectric response; however, the sample treated at 1100°C is more suitable for application in electronic devices.

Magnetic analysis confirms that the presence of α-Fe_2O_3 that leads to a decrease in the magnetization as a function of the applied field. The presence of α-$LiFe_5O_8$ and Fe_3O_4 contributes to the maximum magnetization magnitude observed (73 emu/g at the sample treated at 1200°C). The magnetic susceptibility as a function of temperature has a maximum value of nearly 1750 emu $(gT)^{-1}$ for the sample treated at 1200°C confirming that this sample has a good magnetic response.

Acknowledgements

This work was financed by FEDER funds through the COMPETE 2020 Programme and National Funds through FCT—Portuguese Foundation for Science and Technology under the project UID/CTM/50025/2013. S. Soreto Teixeira also acknowledges to the FCT for a Ph D. Grant SFRH/BD/105211/2014.

Author details

Sílvia Soreto*, Manuel Graça, Manuel Valente and Luís Costa

*Address all correspondence to: silvia.soreto@ua.pt

Department of Physics, I3N - University of Aveiro, Aveiro, Portugal

References

[1] Louh R-F, Reynolds TG, Buchanan RC. Ferrite Ceramics. In: Buchanan RC, editor. Ceramic Materials for Electronics. first ed. New York: Dekker, Marcel; 1986. p. 323–76.

[2] Pullar RC. Hexagonal ferrites: a review of the synthesis, properties and applications of hexaferrite ceramics. Prog Mater Sci [Internet]. Elsevier Ltd; 2012 Sep [cited 2014 Jun 3];57(7):1191–334. Available from: http://linkinghub.elsevier.com/retrieve/pii/S0079642512000369

[3] Rubinger CPL, Gouveia DX, Nunes JF, Salgueiro CCM, Paiva JAC, Graça MPF, et al. Microwave dielectric properties of $NiFeO_4$ nanoparticles ferrites. Microw Opt Technol Lett. 2007;49(6):1341–3.

[4] Snelling EC. Soft ferrites Properties and Applications. second ed. London: Butterwoeth & Co. (Publishers) Ltd; 1988.

[5] Teixeira SS, Graça MPF, Costa LC. Dielectric, morphological and structural properties of lithium ferrite powders prepared by solid state method. J Non Cryst Solids [Internet]. Elsevier B.V.; 2012 Aug [cited 2014 Jun 24];358:1924–9. Available from: http://linkinghub.elsevier.com/retrieve/pii/S0022309312003183

[6] Gingasu D, Mindru I, Patron L, Stoleriu S. Synthesis of lithium ferrites from polymetallic carboxylates. J Serbian Chem Soc. 2008;73(10):979–88.

[7] Wolska E, Piszora P, Nowicki W, Darul J. Vibrational spectra of lithium ferrites: infrared spectroscopic studies of Mn-substituted $LiFe_5O_8$. Int J Inorg Mater. 2001;3:503–7.

[8] Iliev MN, Ivanov VG, Todorov ND, Marinova V, Abrashev MV, Petrova R, et al. Lattice dynamics of the α and β phases of $LiFe_5O_8$. Phys Rev B [Internet]. 2011 May [cited 2014 Jun 24];83:174111-1–7. Available from: http://link.aps.org/doi/10.1103/PhysRevB.83.174111

[9] Ahniyaz A, Fujiwara T, Song S, Yoshimura S. Low temperature preparation of β-$LiFe_5O_8$ fine particles by hydrothermal ball milling. Solid State Ionics. 2002;151:419–23.

[10] Dey S, Roy A, Das D, Ghose J. Preparation and characterization of nanocrystalline disordered lithium ferrite by citrate precursor method. J Magn Magn Mater [Internet]. 2004 Mar [cited 2014 Jun 24];270(1–2):224–9. Available from: http://linkinghub.elsevier.com/retrieve/pii/S030488530300711X

[11] Jović NG, Masadeh AS, Kremenovic AS, Antic BV, Blanusa JL, Cvjeticanin ND, et al. Effects of thermal annealing on structural and magnetic properties of lithium ferrite nanoparticles. J Phys Chem C [Internet]. 2009 [cited 2014 Jun 24];113:20559–67. Available from: http://pubs.acs.org/doi/abs/10.1021/jp907559y

[12] Widatallah HM, Berry FJ. The influence of mechanical milling and subsequent calcination on the formation of lithium ferrites. J Solid State Chem. 2002;164:230–6.

[13] Teixeira SS, Graça MPF, Costa LC, Valente MA. Study of the influence of thermal treatment on the magnetic properties of lithium ferrite prepared by wet ball-milling using nitrates as raw material. Mater Sci Eng B [Internet]. 2014 Aug [cited 2014 Jun 24];186:83–8. Available from: http://linkinghub.elsevier.com/retrieve/pii/S0921510714000762

[14] Teixeira SS, Graça MPF, Costa LC. Dielectric and structural properties of lithium ferrites. Spectrosc Lett [Internet]. 2014 May 28 [cited 2014 Jun 24];47(5):356–62. Available from: http://www.tandfonline.com/doi/abs/10.1080/00387010.2013.840316

[15] Jones FW. The Measurement of Particle Size by the X-Ray Method. R Soc London A. 1938;166(924):16–43.

[16] Patterson AL. The Scherrer formula for I-Ray particle size determination. Phys Rev. 1939;56:978–82.

[17] Devesa S, Graça MP, Henry F, Costa LC. Microwave dielectric properties of (Bi1-xFex) NbO_4 ceramics prepared by the sol–gel method. Ceram Int. 2015;41(6):8186–90.

[18] Cook W, Manley M. Raman characterization of a- and b-$LiFe_5O_8$ prepared through a solid-state reaction pathway. J Solid State Chem [Internet]. Elsevier; 2010;183(2):322–6. Available from: http://dx.doi.org/10.1016/j.jssc.2009.11.011

[19] Nasibulin AG, Simas R, Jiang H, Tian Y, Mudimela PR, Shandakov SD, et al. Simple and rapid synthesis of ambient conditions α-Fe_2O_3 nanowires under ambient conditions. Nano Res. 2009;2:373–9.

[20] Shebanova ON, Lazor P. Raman spectroscopic study of magnetite ($FeFe_2O_4$): a new assignment for the vibrational spectrum. J Solid State Chem. 2003;174:424–30.

[21] Graça MPF, Valente MA, Da Silva MF. The electric behavior of a lithium-niobate-phosphate glass and glass-ceramics. J Mater Sci. 2006;41:1137–44.

[22] Jonscher AK. Dielectric Relaxation in Solids. London: Chelsea Dielectrics Press; 1983.

[23] Shitre AR, Kawade VB, Bichile GK, Jadhav KM. X-ray diffraction and dielectric study of $CoCo_{1-x}Cd_xFe_{2-x}Cr_xO_4$ ferrite system. Mater Lett. 2002;56:188–93.

[24] Singh AK, Goel TC, Mendiratta RG, Thakur OP, Prakash C. Dielectric properties of Mn-substituted Ni–Zn ferrites. J Appl Phys. 2002;91:6626–30.

[25] Macedo BP, Moynihan CT, Bose R. Role of ionic diffusion in polarization vitreous conductors. Phys Chem Glasses. 1972;13:171–179.

[26] Wang Z, Seehra MS. Ising-like chain magnetism, Arrhenius magnetic relaxation, and case against 3D magnetic ordering in β-manganese phthalocyanine ($C_{32}H_{16}$ MnN_8). J Phys Condens Matter [Internet]. IOP Publishing; 2016;28(136002):1–9. Available from: http://stacks.iop.org/0953-8984/28/i=13/a=136002?key=crossref.5ce814db016d8c3680972e0aa65d1762

[27] Cole KS, Cole RH. Dispersion and absorption in dielectrics I. alternating current characteristics. J Chem Phys. 1941;9:341–51.

[28] Jovic N, Antic B, Goya GF, Spasojevic V. Magnetic properties of lithium ferrite nanoparticles with a core/shell structure. Curr Nanosci. 2012;8(5):1–8.

[29] Verma S, Joy PA, Verma S, Joy PA. Magnetic properties of superparamagnetic lithium ferrite nanoparticles. J Appl Phys. 2005;98(124312):1–9.

[30] Singhal S. Cation distribution in lithium ferrite ($LiFe_5O_8$) prepared via aerosol route. J Electromagn Anal Appl. 2010;2(1):51–5.

[31] Mikhaylova M, Kim DK, Bobrysheva N, Osmolowsky M, Semenov V, Tsakalakos T, et al. Superparamagnetism of magnetite nanoparticles: dependence on surface modification. Langmuir. 2004;20:2472–7.

[32] Wang WW, Zhu YJ, Ruan ML. Microwave-assisted synthesis and magnetic property of magnetite and hematite nanoparticles. J Nanoparticle Res. 2007;9:419–26.

[33] Skomski R, Coey JMD. Permanent Magnetism. Taylor & Francis, editor. Bristol: Institute of Physics Publishing Ltd.; 1999.

[34] Hessien MM. Synthesis and characterization of lithium ferrite by oxalate precursor route. J Magn Magn Mater. 2008;320:2800–7.

[35] Shirsath SE, Kadam RH, Gaikwad AS, Ghasemi A, Morisako A. Effect of sintering temperature and the particle size on the structural and magnetic properties of nanocrystalline $Li_{0.5}Fe_{2.5}O_4$. J Magn Magn Mater [Internet]. Elsevier; 2011;323:3104–8. Available from: http://dx.doi.org/10.1016/j.jmmm.2011.06.065

Chapter 3

Optimizing Processing Conditions to Produce Cobalt Ferrite Nanoparticles of Desired Size and Magnetic Properties

Oscar Perales-Pérez and Yarilyn Cedeño-Mattei

Additional information is available at the end of the chapter

http://dx.doi.org/10.5772/66842

Abstract

The excellent chemical stability, good mechanical hardness, and a large positive first order magnetocrystalline anisotropy constant of cobalt ferrite ($CoFe_2O_4$) make it a promising candidate for magneto-optical recording media. For practical applications, the capability to control particle size at the nanoscale is required in addition to precise control of the composition and structure of $CoFe_2O_4$. It has been well established that a fine tuning in cobalt ferrite nanocrystal size within the magnetic single domain region would lead to the achievement of extremely high coercivity values at room temperature. The development of a size-sensitive phase separation method for cobalt ferrite that is based on a selective dissolution of the superparamagnetic fraction and subsequent size-sensitive magnetic separation of single-domain nanoparticles is presented. The attained room temperature coercivity value (11.9 kOe) was mainly attributed to the enlargement of the average crystal size within the single domain region coupled with the removal of the superparamagnetic fraction. The strong influence of crystal size, ferrite composition, and cation distribution in the ferrite lattice on the corresponding magnetic properties at the nanoscale was also confirmed. The superparamagnetic and magnetic single domain limits were experimentally determined.

Keywords: cobalt ferrite, nanocrystals, size-controlled synthesis, high coercivity, magnetic properties

1. Introduction

Technological applications of cobalt ferrite ($CoFe_2O_4$) nanocrystals include their potential use in ferrofluids, hyperthermia, and biological treatment agents due to their unusual properties

[1, 2]. In addition, $CoFe_2O_4$ possesses excellent chemical stability, good mechanical hardness, and a large positive first order crystalline anisotropy constant, making it a promising candidate for magneto-optical recording media also [3]. Control on particle size and shape, ion distribution, and/or structure could allow a fine-tuning of the magnetic properties of ferrites [4–7]. In this regard, magnetic nanocrystals exhibit strong size-dependent properties that may provide valuable information on estimating the scaling limits of magnetic storage while contributing to the development of high-density data storage devices.

To the best of our knowledge, there is still a lack of a systematic effort to restrict the growth of single nanocrystals within the magnetic single domain region where enhancement of coercivity could be achieved. Theoretically, the single domain region ranges between 5 and 40 nm [8]. Maximum coercivity value of around 5.3 kOe [9] has also been reported for 40 nm nanocrystals. Since coercivity is strongly dependent on particle size, any attempt to achieve higher coercivity values in cobalt ferrite must consider the development of alternative approaches in order to obtain more homogeneous crystal sizes with less or null presence of superparamagnetic particles, which have near-zero coercivity.

It is desired to select a synthesis procedure capable of producing nanocrystals with a narrow size distribution due to the above-mentioned direct dependence of coercivity on crystal size. Synthesis approaches such as reverse micelles and thermal decomposition meet this criterion but the excessive consumption of resources, namely, synthesis reagents and experimental time, must be minimized. The use of toxic solvents and surfactants make these processes less attractive. On the other hand, aqueous-based synthesis routes (e.g., coprecipitation method) are environmentally friendly; the experimental time and reagents consumption are minimum and, most important, can allow a fine-tuning of crystal size.

Although coercivity is mainly governed by the magnetocrystalline anisotropy energy, contribution from surface anisotropy becomes important, particularly on the nanoscale. Additionally, the strain anisotropy and shape anisotropy as well as the ferrite composition will also contribute to the magnetic anisotropy and superexchange interaction energies.

Based on the above considerations, the present chapter addresses the experimental optimization of the synthesis parameters leading to the precise control of the coercivity of cobalt ferrite nanocrystals. In particular, the systematic study on composition, size, and cation distribution allowed the determination of the experimental limits of the single and multidomain regions as a function of the ferrite crystal size.

2. Theoretical background

2.1. Ferrites

The spinel structure derives from the mineral $MgAl_2O_4$ whose structure was elucidated in 1915. Analogous to it, the spinel structure in ferrites has the general formula MFe_2O_4, where M corresponds to a divalent metal (i.e. Co^{2+}, Mn^{2+}, Ni^{2+}, and Zn^{2+}). The spinel lattice [10] is composed of a close-packed oxygen arrangement with 32 oxygen atoms forming the unit

http://dx.doi.org/10.5772/66842

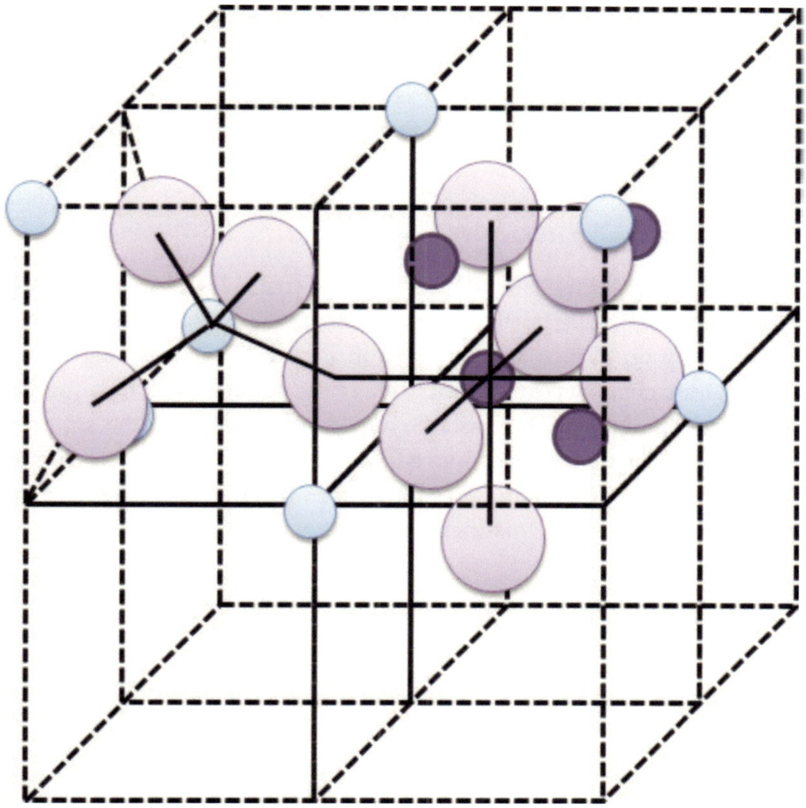

Figure 1. The spinel structure [10].

cell and includes two types of atomic arrangements, called A and B sites, where the cations are accommodated. The site A is tetrahedrally and the B site is octahedrally coordinated by oxygen atoms; the spinel unit cell contains 64 tetrahedral sites, only 8 being occupied, and 32 octahedral sites, with half of them occupied. More specifically, a spinel unit cell can be considered to consist of two types of subcells (**Figure 1**) that alternate in a three-dimensional array. The magnetic properties exhibited by this kind of structures are also influenced by the interaction of cations on the A and B sites; for instance, large angles and short distances between cations would be conducive to the enhancement of the exchange coupling and hence to the corresponding magnetic properties. The angles in the spinel structure for the A-A, B-B, and A-B interactions are 79.63°, 90°/125.3°, and 125.15°/154.57°, respectively, thus A-B exchange interactions are the strongest.

2.2. The cobalt ferrite

Cubic ferrites could be classified as normal, inverse, or mixed spinels. In the case of a normal spinel, the divalent ions occupy all the tetrahedral sites, whereas the trivalent ions do the same with the octahedral sites. An example of a material with a normal spinel structure is $ZnFe_2O_4$. In the case of inverse spinels such as cobalt ferrite, all tetrahedral sites are occupied by the trivalent ion (Fe^{3+}) whereas octahedral sites are equally occupied by divalent (Co^{2+}) and trivalent ions (Fe^{3+}) [11]. The magnetic moment = 5 μ_B from octahedral Fe^{3+} ions is antiparallel to the magnetic moment = 5 μ_B from the tetrahedral Fe^{3+} ions, thus compensating each other. Accordingly, the magnetization in cobalt ferrite is attributed to the magnetic moment = 3 μ_B provided by Co^{2+} ions on the B-sites.

2.3. Magnetic properties and particle size

There is a relationship between coercivity and particle size [12] that is based on the presence of magnetic domains. Magnetic particles would behave as single domain (SD) or multidomain (MD) depending on particle size (as seen in **Figure 2**). This SD region is subdivided into two main regions characterized by their magnetic stability and they are known as the unstable and the stable regions. The unstable region corresponds to the superparamagnetic particles (SPM). It is a size-dependent issue where the thermal energy overcomes the anisotropy energy causing the spins rotate and randomize acting as paramagnetic (do not retain magnetization and coercivity after magnetic field removal).

As seen in Eq. (1), the critical radius below which particles behave as a single domain can be calculated by [13, 14]:

$$r_c = \frac{9\left(A K_u\right)^{\frac{1}{2}}}{\mu_0 M_s^{\ 2}} \tag{1}$$

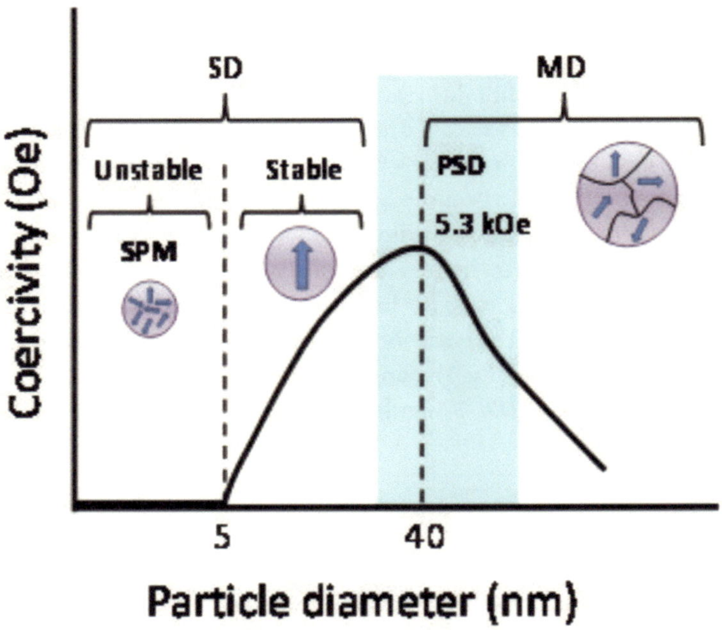

Figure 2. Dependence of coercivity with particle diameter for cobalt ferrite.

http://dx.doi.org/10.5772/66842

A is the exchange stiffness (constant characteristic of the material related to the critical temperature for magnetic ordering), K_u is the uniaxial anisotropy constant; μ_0 is the vacuum permeability, and M_s the saturation magnetization. For the stable single domain region, a simple model developed by Stoner and Wohlfarth [15] assumes coherent spin rotation whereby all spins within the single-domain particle are collinear (the magnetization is uniform) and rotate in unison. The model predicts the field strength necessary to reverse the spin orientation direction or coercivity. In this stable region, the magnetocrystalline energy gradually overcomes the thermal energy and becomes responsible for the increase in coercivity. The only way to magnetize a material is to rotate the spins, a process that involves high energies and, hence, high coercivity. Thus, SD grains are magnetically hard and have high coercivity and remanent magnetization. In the cobalt ferrite case, the limit between the single and multidomain has been established to be around 40 nm with an expected coercivity of 5.3 kOe at room temperature. For the multidomain region, domain walls are present and coercivity tends to decrease with an increase in particle size [15]. In turn, the magnetization process in a multidomain, particle is controlled by domain wall movement, which is an energetically easier process.

Based on the above considerations, a precise control on the particle size will allow a fine tuning of the magnetic properties at the nanoscale. In this regard, our modified coprecipitation method described here allows the required control in crystal size based on enhancement of the heterogeneous nucleation mechanism. Consequently, a control in nucleation and growth steps could be achieved, which are key points for crystal size control.

3. Methodology

3.1. Conventional coprecipitation method

All reagents used in the synthesis and treatment of cobalt ferrite nanocrystals were of analytical grade and used without further purification. Ferrite nanocrystals are synthesized by the cconventional coprecipitation method. In the conventional approach, an aqueous solution containing Co (II) and Fe(III) ions was rapidly contacted with an excess of hydroxide (OH^-) ions. The hydrolysis reaction in the presence of an excess of OH^- ions leads to the formation of a paramagnetic Fe-Co hydroxide, which undergoes dehydration and atomic rearrangement conducive to a ferrite structure with no need of further annealing, according to:

$$2Fe^{(3+)} + Co^{(2+)} + 8OH^- \underset{\Delta}{\rightarrow} Fe_2\,Co(OH)_8 \underset{\Delta}{\rightarrow} CoFe_2O_4 + 4H_2O \tag{2}$$

The nucleation rate is quite high at the beginning of the precipitation process whereas the excess of OH^- ions provides a net negative surface charge to the nuclei limiting their further growth and aggregation. Under these conditions, polydispersed particles of less than 30 nm in diameter are typically produced.

The reactant solution was mechanically stirred at 500 rpm and intensively heated to accelerate the dissolution/recrystallization stages involved with the ferrite formation. The intensive heating helped to reduce the reaction time dramatically. Cobalt ferrite nanocrystals were washed out with deionized water, dried at 80°C for 24 hours and characterized.

3.2. Modified coprecipitation method

In order to enhance the magnetic properties by controlling the nucleation and growth conditions of the ferrite crystals, the conventional coprecipitation route was modified by controlling the flow rate of addition of the metal ions solution to the alkaline one under intensive heating conditions. For this purpose, a microperistaltic pump with a precise control of flow rate was used.

3.3. Combined acid washing and magnetically assisted size-sensitive separation

3.3.1. Single acid washing process

In order to improve the monodispersity of the as-synthesized cobalt ferrite nanocrystals, they were contacted with HCl solution; it is expected that a higher solubility of the smaller particles when compared to the bigger sized ones, would lead to a narrow size distribution and, therefore, higher coercivity values. Also, the acidic treatment could remove the "magnetically dead" surface in remaining particles. On an experimental basis, the ferrite nanocrystals produced using the above described modified coprecipitation method were then treated with the acidic solution for1 hour.

As-synthesized nanocrystals were first contacted with 10% v/v HCl solution for 30 minutes at a $\frac{0.1\text{ g ferrite}}{10\text{ mL acidicsolution}}$ ratio. At the end of this acid treatment stage, the recovered solids were dispersed in water and submitted for the magnetically assisted separation stage using a commercial N38 (field strength ≥ 1.2 T) neodymium permanent magnet. The particles suspension was contacted with the magnet for 2 minutes at the end of which, settled particles were separated from the suspended ones that remained in the supernatant. The solids in this supernatant represented the first fraction of the size-sensitive separation process. In turn, the magnetically settled particles were processed through the same water dispersion-magnetic separation cycle for two more times. The fractions corresponding to the solids recovered from their corresponding supernatants and the last sediment were submitted for structural and magnetic characterization.

3.3.2. Double acid washing process

Five grams of ferrite synthesized by the above indicated modified coprecipitation method was used in the size-sensitive phase separation process and characterization. The ferrite to acidic solution mass/volume ratio was the same as described in the previous section. After a first acid washing and size-sensitive separation, the settled fraction with the largest amount of particles was used as a starting material for a second size-sensitive separation cycle (second settled fraction in this case). This experimental procedure will be known as combined

http://dx.doi.org/10.5772/66842

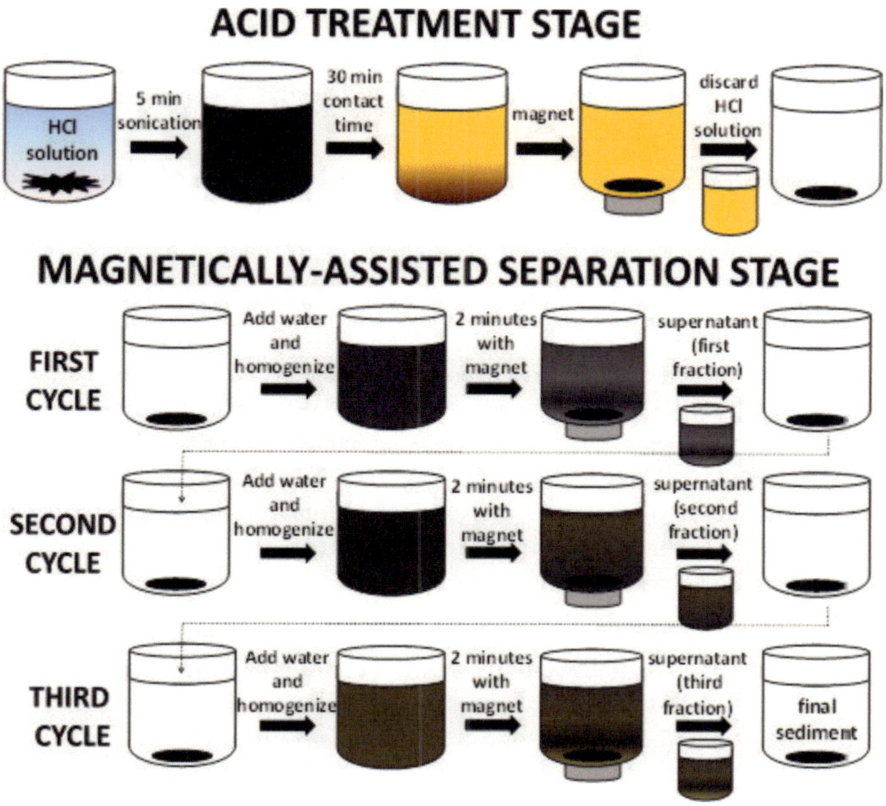

Figure 3. The acid washing and magnetically assisted size-sensitive separation.

acid washing and magnetically assisted size-sensitive separation (**Figure 3**), double acid washing process.

3.4. Effect of the ferrite composition: Variation of the Fe:Co mole ratio

Cobalt ferrite nanocrystals with initial Fe:Co mole ratios of 3:1, 2:1, 1.7:1, and 1.4:1 were produced by the conventional and the modified coprecipitation method (flow rate: 0.67 mL/min). The reaction time was kept constant at 1 hour.

3.5. Inhibition of crystal growth by using surfactants or polymers

As an attempt to determine the superparamagnetic limits in cobalt ferrite nanoparticles, oleic acid sodium salt (Na-oleate) was used in the synthesis of cobalt ferrite in order to inhibit crystal growth. This surfactant was introduced into the same vessel with the sodium hydroxide solution and heated as in the conventional coprecipitation method although for very short reaction times (5 minutes). Prolonged reaction times favored crystal growth. In a modified route, the boiling hydroxide and surfactant solution was removed from the heating source followed by its contact with the ionic solution containing iron (III) and cobalt (II).

4. Results and discussion

4.1. Optimization of synthesis parameters: 2^3 Factorial design

The coprecipitation method employed in the synthesis of cobalt ferrite nanocrystals allows the manipulation of synthesis parameters such as reaction time (A), control on flow rate of the addition of reactants (B), and the NaOH concentration (C). A deep understanding of those parameters will allow a control in the nucleation and growth steps, key points in size-control approaches [8, 16, 17]. In order to increase coercivity, the superparamagnetic limit must be surpassed; this limit has been estimated at 5 nm [18].

It is important to take into account each parameter and the interactions between them in order to tune crystal size and hence, the magnetic properties. A 2^3 factorial design is a useful tool in the study of the synthesis parameters and its interactions. The factorial design considers a low and a high level of each parameter under study. Each level was selected based on previous studies and taking into account the experimental limitations such as the capacity of the peristaltic pump used to control the flow rate of addition of reactants. The low and high levels for each parameter under study are summarized in **Table 1**. The levels were also coded using (-) and (+) signs for low and high, respectively.

In addition to the ABC Design, three replicates at the center of it were used to estimate the experimental error. **Table 1** also summarizes the response variable, i.e., coercivity. A wide range of coercivity values (between 870 and 4626 Oe) was obtained. The 2^3 factorial design suggested that reaction time is not a significant parameter. Taking it into account, the reaction time was set to 1 hour in later experiments. Otherwise, the flow rate of addition of reactants, NaOH concentration, and the interaction between these two parameters are the most significant.

4.2. Combined acid washing and magnetically assisted size-sensitive separation

The modified size-controlled coprecipitation method to synthesize ferrite nanocrystals can allow a fine tuning of the average crystallite size within the single magnetic domain region.

Combinations	A B C Design	A: Reaction time (min)	B: Flow rate (mL/min)	C: NaOH (M)	Hc (Oe)
1	- - -	5	0.85	0.34	4518
A	+ - -	180	0.85	0.34	4626
B	- + -	5	20	0.34	871
AB	+ + -	180	20	0.34	870
C	- - +	5	0.85	0.54	2877
AC	+ - +	180	0.85	0.54	3448
BC	- + +	5	20	0.54	922
ABC	+ + +	180	20	0.54	1007

Table 1. Experimental conditions for the 2^3 factorial design and the corresponding room temperature coercivity, H_c [19].

http://dx.doi.org/10.5772/66842

However, the product is still polydispersed in size [19]. Accordingly, any attempt to achieve higher coercivity values in cobalt ferrite must consider the development of an alternative approach in order to obtain more homogeneous crystal sizes with less or null presence of superparamagnetic particles. In this regard, a rapid and simple size-sensitive phase separation treatment based on the size-dependence of nanoparticles solubility in an aqueous phase was developed. In the case of particles with narrow size distribution, the dissolution behavior will depend on the ferrite degree of inversion and composition [20]. However, when moderately polydispersed nanocrystals are synthesized, the selective dissolution of tiny individual crystals can be expected according to the Ostwald-Freundlich law [21], i.e., particles with smaller diameter will be more soluble than bigger ones. This fact was taken into account to get rid of the superparamagnetic or smaller particles that do make the coercivity of the powders to decrease. The selective dissolution of superparamagnetic particles will be complemented by a magnetically assisted size-separation stage. The separation stage consists of contacting the nanoparticles/acid solution with a magnet in order to separate the settled particles from the suspended (smaller) ones.

4.2.1. Single acid washing process

The corresponding average crystallite size varied from 16 ± 2 nm in the as-synthesized sample to 22 ± 2 nm in the solid fraction recovered after 6 minutes of magnetic separation after treatment with 10% HCl solution. This enlargement of average crystallite size can be attributed to the preferential acidic dissolution of small nanoparticles including the superparamagnetic ones along with an efficient magnetically assisted size classification.

The full set of fractions recovered after treatment with 10% HCl solution is shown in **Figure 4**. The increment in crystal size with an increment in separation time is evident.

Transmission electron microscopy-energy dispersive spectroscopy (TEM-EDS) analyses were performed to measure the elemental composition and calculate the experimental Fe:Co mole ratio. TEM-EDS of cobalt ferrite nanocrystals suggested that selective dissolution of Fe could have taken place after contacting them with the 10% v/v HCl solution. The Fe:Co mole ratio decreased from 1.81:1 in the nontreated sample down to 1.58:1 after the acidic treatment. The selective dissolution of Fe in bulk ferrite was suggested by Figueroa et al. [22]; they attributed the drop in the Fe:Co ratio to the removal of the less crystalline surface layer of iron oxide in cobalt ferrite produced by thermal decomposition. As-synthesized particles show agglomeration while treated particles have better dispersion. It may be a result of electrostatic

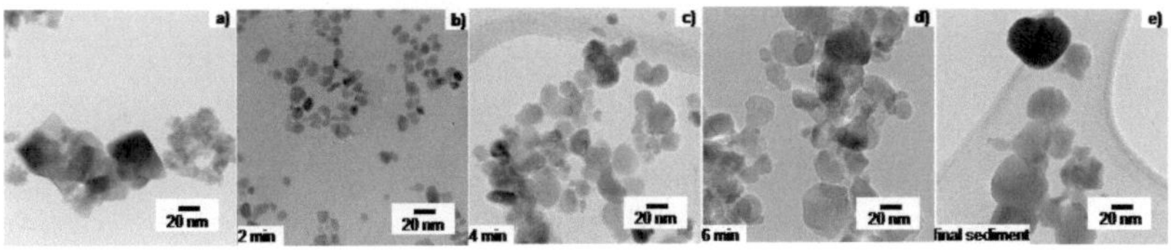

Figure 4. TEM images of $CoFe_2O_4$ powders before (a) and after acidic treatment with 10% HCl followed by magnetically assisted separation at different times (b–e). The images correspond to the solids contained in the fractions collected after 2 minutes (b), 4 minutes (c), and 6 minutes (d) of magnetic separation and the sediment at the end of this stage (e).

repulsion resulting from the chemisorption of charged species, in this case, H^+, on the nanocrystal surfaces.

The coercivity of the as-synthesized ferrite powders was 5.4 kOe. It increased up to 9.4 kOe after treating the powders with 10% w/w HCl. The maximum magnetization of this high coercivity sample is 58 emu/g. The "constricted loops" observed in the central part of the M-H loops suggest a mixture of soft and hard material in the powders. In this case, the soft material corresponds to the superparamagnetic fraction. In turn, the large room temperature coercivity in these powders can be attributed to the enlargement of their average crystal size caused by the removal of the superparamagnetic fraction during the acidic treatment and the subsequent size-sensitive magnetic separation stages. This interpretation is supported by the results provided by TEM and XRD analyses on these samples. Furthermore, the reduction in the Fe:Co atomic ratio in the size-selected cobalt ferrite particles, suggested by TEM-EDS analyses, could also be involved with the drastic change in coercivity. In this case, acidic treatment should have promoted the removal of a poorly crystalline and magnetically disordered surface layer of the ferrite oxide.

Mössbauer spectra for the cobalt ferrite powders without treatment showed a broadened central peak, attributed to the presence of superparamagnetic particles, in addition to the hyperfine splitting typical of magnetically ordered iron species in the ferrite lattice. The relative abundance of the superparamagnetic portion was calculated at 19.3% while the combined three magnetically split sites accounted for 81.7% of the remainder of the cobalt ferrite. No central peak was observed in the Mössbauer spectrum for the sample recovered at the end of the magnetic separation stage, which suggest the complete removal of the superparamagnetic fraction after the acidic treatment and magnetic separation stages. The internal magnetic field for the first site, which was assigned to Fe in the tetrahedral site (Fe^{Tetra1}), was 449.97 kOe with quadrupole splitting of -1.584 mm/s and isomer shift of 1.087 mm/s, while its relative abundance was 29.1%. The other two sites corresponded to Fe in the octahedral ferrite sites (Fe^{Oct1}). The first one was characterized by the internal magnetic field of 481.13 kOe, quadrupole splitting of -0.974 mm/s and isomer shift of 0.024 mm/s with a relative abundance of 41.8%. The second octahedral site, Fe^{Oct2}, was fitted with an internal magnetic field of 391.31 kOe, quadrupole splitting of -0.640 mm/s, isomer shift of 0.085 mm/s, and relative abundance of 29.1%. The fittings are consistent with an inverse ferrite structure whereby Fe^{3+} cations randomly occupied both the tetrahedral and octahedral sites [23, 24]. Based on these considerations and since an equal occupation of the tetrahedral and octahedral sites by Fe^{3+} cations did not take place, an incomplete inversion in synthesized cobalt ferrite nanocrystals can be considered. This partial inversion in ferrite structure can also account for the unusual coercivity value.

4.2.2. The double acid washing process

As an attempt to explore the possibility of further enhancement of the ferrite coercivity by narrowing the size distribution even more, previously acid-washed ferrite powders were retreated by following a similar acid washing/magnetic separation cycles. Starting cobalt ferrite powders were synthesized by the modified coprecipitation method at 0.81 mL/min flow rate and a NaOH concentration of 0.315 M, which corresponds to the optimum concentration determined from the experimental design. A ferrite to acidic solution ratio of 5.0 g ferrite to

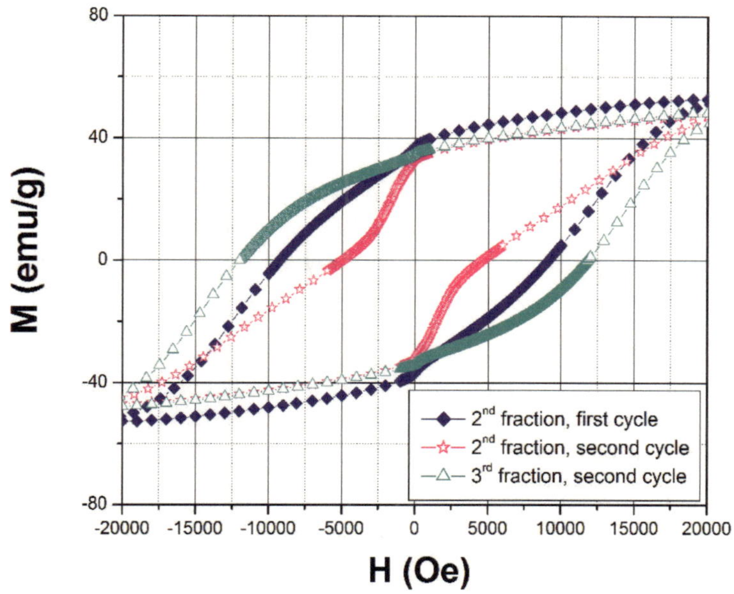

Figure 5. Room temperature M-H measurements of $CoFe_2O_4$ powders submitted to two cycles of acid treatment and magnetically assisted separation stages.

500 mL HCl 10% v/v was used. The coercivity increased starting from 3.3 up to 9.7 kOe for the third fraction after 6 minutes of magnetically assisted settling; this increment in coercivity is around three times its initial value. Again, "constricted" hysteresis loops were observed (**Figure 5**), which are typical of a mixture of soft and hard magnetic materials [25]. In this case, this sort of "necking" in the central part of the loop can be attributed to the coexistence of superparamagnetic single-domain cobalt ferrite nanoparticles. Fractions exhibiting larger coercivity values did not show this "necking" confirming the removal of the superparamagnetic fraction.

The second settled fraction (corresponding to 4 minutes of settling time) was submitted to the second cycle of our size-sensitive phase separation process. This time, the ferrite to HCl ratio was 0.8 g ferrite to 80 mL HCl 10% v/v. The sharpening in the corresponding XRD peaks suggested an improvement in crystallinity after the second treatment cycle.

The initially 9.2 kOe achieved in the second fraction generated during the first separation cycle was increased up to 11.9 kOe. This pretty large coercivity was obtained in the third fraction produced during the second acid washing/magnetic separation cycle. This dramatic enhancement in coercivity can be attributed to the larger crystal size in the corresponding fractions and a more efficient removal of any remaining superparamagnetic particles.

4.3. Effect of the ferrite composition

4.3.1. Variation of the Fe:Co mole ratio

The magnetic properties of cobalt ferrite depend on diverse factors such as crystal size [26], morphology [27], chemical composition [28, 29], and/or cation distribution [30]. The formation of nonstoichiometric cobalt ferrite nanocrystals will lead to atomic rearrangement between A- and B-sites and/or creation of vacancies [31]. Consequently, a change in magnetic properties will take place. On this basis, cobalt ferrite nanocrystals with different initial Fe:Co mole ratios (3:1, 2:1, 1.7:1, and 1.4:1) were synthesized by the traditional and modified coprecipitation routes. For this latter case, the flow rate at which the reactants were contacted was set to 0.67 mL/min. The reaction time was 1 hour in all our experiments. In addition to crystal growth, cation distribution, internal magnetic field, morphology, and chemical composition, the presence of a secondary phase (α-Fe_2O_3) by X-ray diffraction, see **Figure 6**, were identified responsible for the observed wide range of magnetic properties.

Average crystallite sizes in produced ferrite powders ranging between 11–12 nm and 15–19 nm were obtained using the conventional and the modified coprecipitation methods respectively. Excess of iron, (3:1 Fe:Co mole ratio), caused the formation of a secondary phase that was identified as hematite, α-Fe_2O_3. In turn, less crystalline powders were obtained when the Fe:Co mole ratio was decreased from 3:1 to 1.4:1 [32].

Table 2 summarizes the nominal and experimental (measured) Fe:Co mole ratios in synthesized ferrite nanocrystals. The experimental Fe:Co mole ratios represent an average from measurements at five different regions in the sample using energy dispersive X-ray spectroscopy. For all cases, approximately 78% of the Fe present in starting solutions was incorporated into the ferrite structure.

Average crystallite sizes of cobalt ferrite synthesized without control on the flow rate were quite similar (11–12 nm). Thus, any change in magnetic properties could be attributed to the

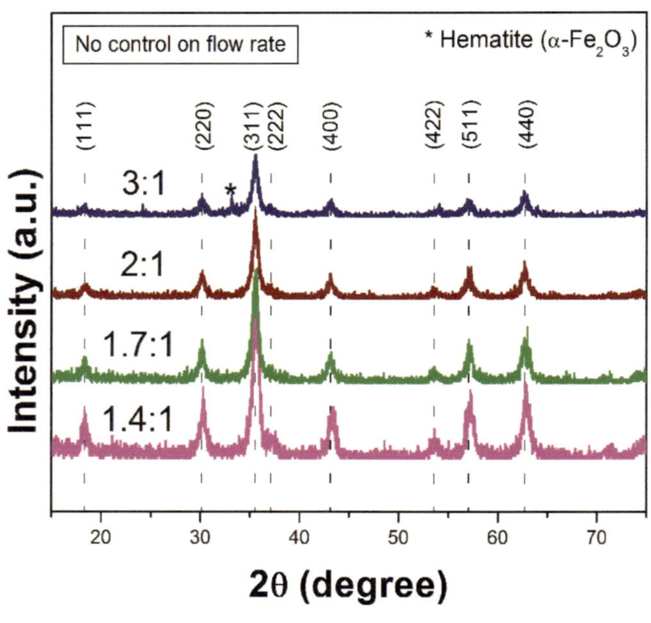

Figure 6. XRD patterns of cobalt ferrite nanocrystals synthesized without control on flow rate at different Fe:Co mole ratios. The peak with the asterisk corresponds to hematite, α-Fe_2O_3 [32].

http://dx.doi.org/10.5772/66842

	No control on flow rate	Flow rate: 0.67 mL/min
Nominal Fe:Co mole ratio	Experimental Fe:Co mole ratio	Experimental Fe:Co mole ratio
3:1	2.3:1 (±0.3)	2.3:1 (±0.2)
2:1	1.58:1 (±0.08)	1.54:1 (±0.08)
1.7:1	1.31:1 (±0.08)	1.30:1 (±0.06)
1.4:1	1.08:1 (±0.04)	1.0:1 (±0.1)

Table 2. Summary of Fe:Co mole ratios of cobalt ferrite synthesized with and without control on flow rate [32].

net effect of chemical composition. For this set of samples, the highest coercivity value (548 Oe) was obtained for the sample synthesized at a 3:1 Fe:Co mole ratio. This sample is a mixture of cobalt ferrite and hematite, so surface anisotropy and interparticle interactions that contribute to the net anisotropy [33] are different when compared to pure cobalt ferrite.

Since the flow rate was kept constant in all the experiments reported in this section (0.67 mL/min), any change in the average crystallite size of the ferrite could be attributed to the variation of the chemical composition of the powders. For instance, the sample synthesized at a starting Fe:Co mole ratio of 1.7:1 yielded an average crystallite size of 18 nm, whereas it was 19 nm for the sample produced with Fe:Co mole ratio of 3:1. The corresponding coercivity magnitudes were 4412 Oe and 4249 Oe, respectively. As reported in our earlier studies, the nonstoichiometry of cobalt ferrite powders (Fe/Co < 2) would influence the coercivity due to the distribution of Fe ions within A- and B-sites in the spinel structure [26].

Mössbauer spectroscopy measurements help to study the Fe cationic environment and how their distribution between octahedral and tetrahedral sites changes in addition to the internal magnetic field. This technique can also elucidate whether surface effects are present. These results are presented below.

4.3.2. Influence of the Fe:Co mole ratio on the ferrite magnetic structure: Mössbauer spectroscopy measurements

The fitting of the spectra corresponding to the ferrite synthesized at a Fe:Co mole ratio of 2 and 1.7 were carried out with the hyperfine field distribution (HFD) model, considering four sites. Samples were synthesized at a flow rate of 0.67 mL/min. The doublet peaks in **Figure 7** at the central part of the Mössbauer spectra is attributed to the phenomenon of superparamagnetic relaxation, arising from particles with single domain attributes. Given the broad aspects of this component, we can confirm some distribution of particle size within this superparamagnetic fraction, with some of them having their blocking temperature close to 298 K (room temperature conditions).

The most remarkable discrepancy between the two spectra of **Figure 7** relies on both the relative abundances in each subsites, and the internal magnetic field of the octahedral and tetrahedral cationic positions. Based on the features of each Mössbauer spectrum, the superparamagnetic components in each sample was estimated to be 17% of the total material

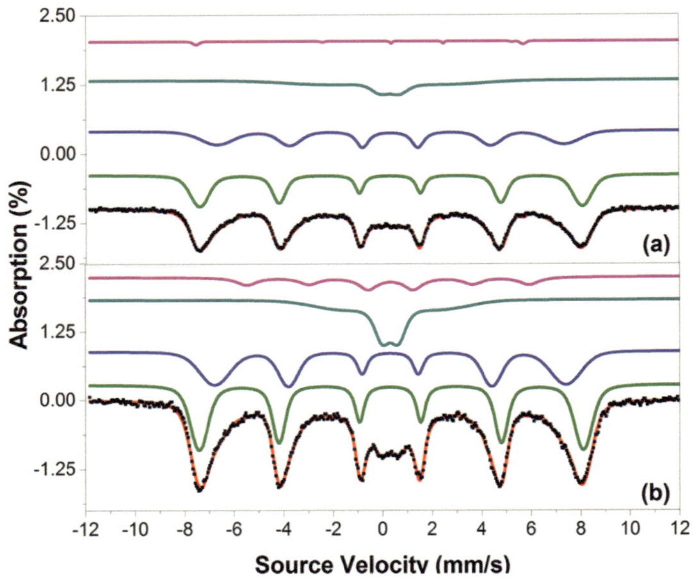

Figure 7. Fitted Mössbauer spectra corresponding to cobalt ferrite nanocrystals synthesized at 0.67 mL/min and nominal Fe:Co mole ratios of 2:1 (a) and 1.7:1 (b). The green, blue, cyan, and magenta fittings correspond to sites 1, 2, 3, and 4, respectively [32].

suggesting that the synthesis process did not affect the particle distribution. The corresponding internal magnetic field of the octahedral and tetrahedral Fe is found to be very similar in both samples. A noticeable difference between both spectra arises from the fourth component. The fourth component of the spectrum "b" (Fe:Co 1.7:1), accounted for 10% of the sample, which was very high when compared to the component but for sample "a" (Fe:Co 2:1) where it represented 0.9% of the sample. The surface cations (site-4) exhibit smaller internal magnetic field and broadened absorption lines indicating a more disordered state at the surface (spin canting) being responsible for the lack of magnetization saturation and higher coercivity value due to surface anisotropy [34].

4.3.3. Effect of the flow rate on the ferrite magnetic structure: Mössbauer spectroscopy measurements

Mössbauer spectra of cobalt ferrite nanocrystals produced at a 1.4:1 Fe:Co mole ratio with, and without control of the flow rate, were analyzed. Site-1 is attributed to Fe cations in the octahedral position of the spinel structure, while site-2 corresponds to the Fe located in the tetrahedral position.

Regarding the analysis of the spectrum corresponding to the sample synthesized without control of flow rate, the internal magnetic field values corresponding to the Fe-sites were higher than those for the powders synthesized at 0.67 mL/min. Furthermore, the relative abundance of sites-1 and 2 in the samples synthesized without control of flow rate (42.32 and 41.26%, respectively) followed an antagonistic trend when compared to the same parameters for the samples produced at 0.67 mL/min (36.80 and 42.75%); i.e., there are more Fe cations in site-1 when the powders are synthesized with no control of the flow rate. In

http://dx.doi.org/10.5772/66842

other words, the control of the flow rate of the addition of reactants during the synthesis of the cobalt ferrite nanocrystals affects the average crystallite size and the distribution of Fe ions within A and B sites in the spinel structure. This unexpected change in the atomic distribution should also be involved with the magnetic properties observed in those high-coercivity samples.

4.4. Inhibition of crystal growth: use of oleic acid sodium salt during the cobalt ferrite crystal formation

The use of surfactants or polymers comes from the necessity to produce extremely small particles which could help us to determine the actual superparamagnetic limit in the cobalt ferrite. Oleic acid sodium salt (Na-oleate) was used to inhibit crystal growth, because of expected surface interaction between polar groups and nanoparticles, as well as to prevent nanoparticle aggregation through strong steric interactions promoted by the adsorption of the surfactant ("capping ligand"). This surfactant has long chains that confer a physical impediment preventing particle growth and/or aggregation.

As discussed above, a reaction time as short as 5 minutes was conducive to the formation of ferrite nanocrystals in the 10–11 nm range, which suggested a very fast formation and growth of earlier ferrite nuclei. Therefore, as an attempt to avoid crystal growth, the cobalt ferrite powder was removed right after the metal solution contacted the boiling NaOH/Na-oleate solution.

The corresponding average crystallite sizes were estimated at 4 and 9 nm when the synthesis took place with and without addition of the surfactant, respectively. This result suggests that the ferrite formation was interrupted in the nucleation stage with practically no time for the nuclei to grow any further.

The extremely small crystal size was also confirmed by TEM analyses. TEM images from **Figure 8** confirmed the formation of nanocrystals with a diameter around 4 nm.

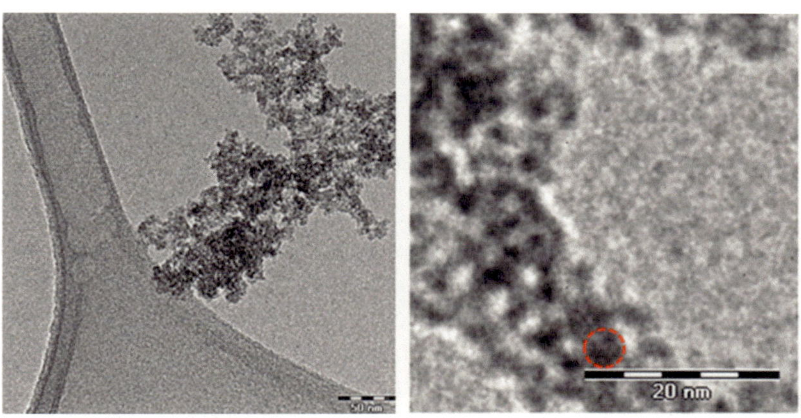

Figure 8. TEM images of cobalt ferrite nanocrystals synthesized in presence of 0.011 M Na-oleate.

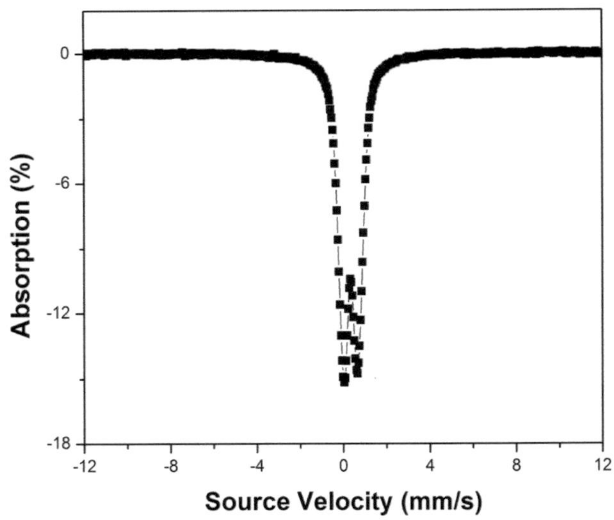

Figure 9. Mössbauer spectrum of cobalt ferrite synthesized in presence of Na-oleate.

The 9 nm nanoparticles exhibited a coercivity of 189 Oe whereas the 4 nm ones reported coercivity as low as 3 Oe. The latter value falls within the experimental error of the instrument; i.e., these particles practically did not show any coercivity.

The presence of the quadrupole splitting in the Mössbauer spectrum of the 4 nm cobalt ferrite nanocrystals (**Figure 9**) may suggest the superparamagnetic nature of the synthesized crystals. Electric quadrupole interaction occurs if at least one of the nuclear states involved possesses a quadrupole moment (which is the case for nuclear states with the spin $I > 1/2$) and if the electric field at the nucleus is inhomogeneous. In the case of ^{57}Fe, the first excited state (14.4 keV state) has a spin $I = 3/2$ and therefore also an electric quadrupole moment. It can be visualized by the precession of the quadrupole moment vector about the field gradient axis sets in and splits the degenerate $I = 3/2$ level into two substates with magnetic spin quantum numbers $mI = \pm 3/2$ and $\pm 1/2$. The energy difference between the two substates (ΔEQ) is observed in the spectrum as the separation between the two resonance lines [35]. On this basis, the superparamagnetic nature of the 4 nm nanocrystals can be proposed.

4.5. ZFC/FC measurements as a function of ferrite crystal size

In order to study the magnetic properties as a function of crystal size in more detail, cobalt ferrite nanocrystals with different crystal sizes in the 10–23 nm range were selected from previous sections. The magnetization curves of the superparamagnetic sample (4 nm in average size) were investigated in the zero-field-cooled/field-cooled (ZFC/FC) protocols using a Quantum Design SQUID unit, under an external magnetic field of 100 Oe and in the temperature range from 300 to 2 K. The powders were fixed in a solid matrix of poly(styrene-divinylbenzene) as described by Calero-Diaz del Castillo and Rinaldi [36]. The ZFC/FC profile for the

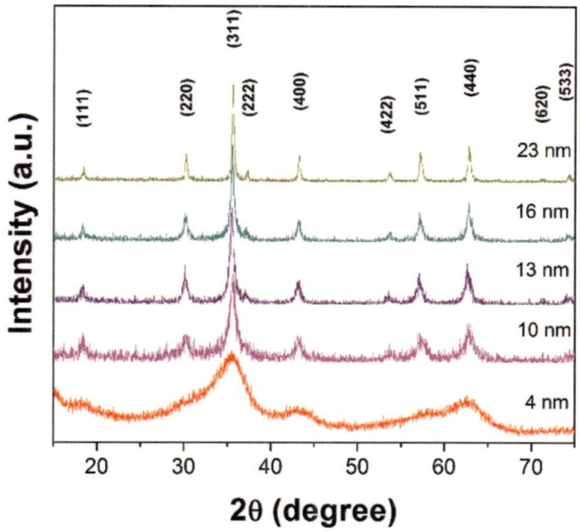

Figure 10. XRD patterns of cobalt ferrite with different average crystallite sizes.

4 nm crystals indicated they were blocked below 115 K, i.e., these nanoparticles would behave ferromagnetically below that blocking temperature [17]. As evident in **Figure 10**, the sharpening and intensity of the XRD peaks of cobalt ferrite powders were enhanced in those powders exhibiting larger average crystallite sizes.

Finally, the M-H loops of the different samples were also measured in the SQUID unit at 2K and 300K. As the M-H loop of **Figure 11** shows, the coercivity of the 23 nm samples

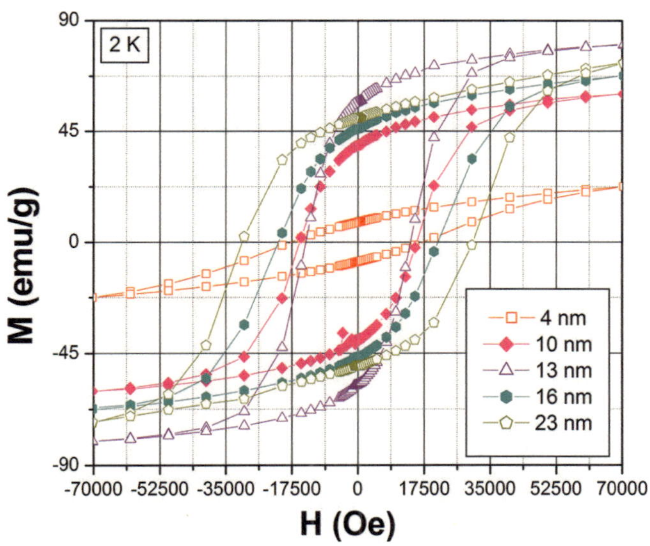

Figure 11. M-H loops at 2 K of cobalt ferrite powders with different average crystallite sizes.

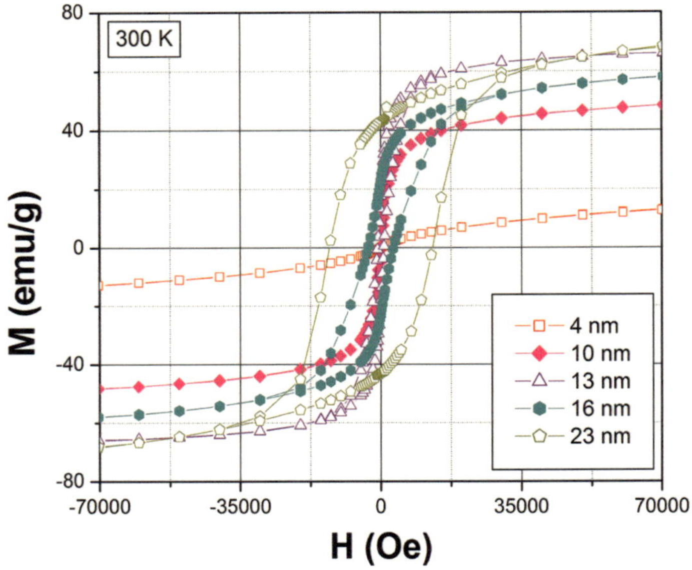

Figure 12. M-H loops at 300 K of cobalt ferrite powders with different average crystallite size.

at 2 K reached a value as high as 30 kOe. To our knowledge, this is the highest coercivity value reported for cobalt ferrite nanocrystals at 2 K. Sun et al. [37] and Meron et al. [38] reported 20 and 15 kOe, respectively, for highly monodisperse cobalt ferrite, measured at 10 K. The maximum magnetization also increases at low temperatures due to the absence of thermal energy responsible for spin rotation and randomization. The M-H loops at 2 and 300 K are shown in **Figures 11** and **12**. An increase in coercivity as a function of crystal size was observed as expected. Coercivity values ranged between 17–30 kOe at 2 K and 0.2–12.9 kOe at 300 K.

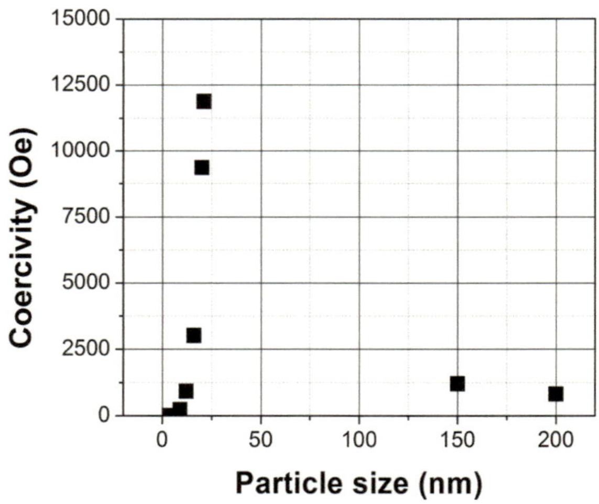

Figure 13. Variation of coercivity with particle size in cobalt ferrite particles synthesized in the present study.

http://dx.doi.org/10.5772/66842

4.6. Particle size dependence of coercivity revisited

In the above, the particle size dependence of the coercivity in cobalt ferrite nanocrystals has been demonstrated and discussed. Different approaches such as experimental design, size-sensitive phase separation method, and surfactants discussed above were the essential tools to obtain particles with a wide range of sizes. They helped us to achieve the main objective of this research: to determine the limits of the single-domain and multidomain regions as a function of the composition, structure, and crystal size in ferrimagnetic ferrites. It is important to notice that the theoretical coercivity values have been greatly surpassed. An experimental plot of coercivity versus particle size summarizes some of the values obtained and it is shown in **Figure 13**. Data corresponding to 150 and 200 nm were obtained using a postsynthesis thermal treatment [39].

5. Conclusions

Tuning the magnetic properties of cobalt ferrite at the nanoscale was achieved by controlling crystal size, chemical composition, and cation distribution. Average crystallite sizes ranging between 4 and 23 nm were successfully synthesized. The corresponding room temperature coercivity values varied between 3 Oe and 11.9 kOe. The 23 nm nanocrystals exhibited coercivity as high as 11.9 kOe, which is reported here for the first time. The corresponding magnetization and squareness ratio were 48 emu/g and 0.72, respectively.

The suitable combination of flow rate-controlled synthesis complemented by acid washing and magnetically assisted size-sensitive separation processes provided favorable conditions for the enhancement of nanoparticles monodispersity and tailoring of the corresponding magnetic properties. Smaller and more soluble nanocrystals that contribute negatively to coercivity (superparamagnetic particles) were selectively dissolved during the acid washing step. This selective dissolution of the superparamagnetic fraction took place in addition to the removal of the poorly crystalline layer on the surface of the nanocrystals.

The effect of the nominal Fe:Co mole ratio in starting solutions on the structural and magnetic properties of the nanocrystals was also investigated. Findings revealed that the cationic distribution in the ferrite was dependent on the nominal Fe:Co mole ratio whereas the contribution of the surface cations, and hence the surface anisotropy, became remarkable under flow rate-controlled synthesis conditions. Accordingly, the colossal coercivity attained in nanometric cobalt ferrite crystals became possible not exclusively by control of crystal size, but also through the promotion of the mentioned effects viz. surface anisotropy and atomic rearrangements of Fe species within the ferrite lattice.

About 4 nm cobalt ferrite crystals were produced using Na-oleate as a surfactant that inhibited crystal growth. The superparamagnetic behavior of these nanocrystals was confirmed by M-H measurements and Mössbauer spectroscopy techniques.

The temperature dependence of the magnetic properties of cobalt ferrite was studied as a function of crystal size. The 23 nm sample (room temperature coercivity of 11.9 kOe) reported a 2 K coercivity as high as 30 kOe.

Acknowledgements

This material is based upon work supported by The National Science Foundation under Grant No. HRD 13455156 (CREST II program).

Author details

Oscar Perales-Pérez[1]* and Yarilyn Cedeño-Mattei[2, 3]

*Address all correspondence to: oscarjuan.perales@upr.edu

1 Department of Engineering Sciences and Materials, University of Puerto Rico, Mayagüez, Puerto Rico

2 Department of Biology, Chemistry, and Environmental Sciences, Interamerican University of Puerto Rico, San Germán, Puerto Rico

3 Department of Chemistry and Physics, University of Puerto Rico, Ponce, Puerto Rico

References

[1] Sun C, Lee JS, Zhang M. Adv. Drug Delivery Rev. 2008;**60**:1252–1265.

[2] Arulmurugan R, Vaidyanathan G, Sendhilnathan S, Jeyadevan BJ. Magn. Magn. Mater. 2006;**298**:83–94.

[3] Terzzoli MC, Duhalde S, Jacobo S, Steren L, Moina CJ. Alloys Compd. 2004;**369**:209–212.

[4] Song Q, Zhang ZJ. J. Am. Chem. Soc. 2004;**126**:6164–6168.

[5] Hyeon T, Chung Y, Park J, Lee SS, Kim Y, Park BH. J. Phys. Chem B. 2002;**106**:6831–6833.

[6] Khedr MH, Omar AA, Abdel-Moaty SA. Colloids Surf. A. 2006;**281**:8–14.

[7] Li S, John VT, O'Connor C, Harris V, Carpenter EJ. Appl. Phys. 2000;**87**:6223–6225.

[8] Berkowitz AE, Schuele WJ. J. Appl. Phys. 1959;**30**:S134.

[9] Neel L. C.R. Acad. Sci. Paris. 1947;**224**:1488–1490.

[10] Goldman A. Modern Ferrite Technology. New York: Van Nostrand Reinhold; 1990.

[11] Callister Jr. W. Materials Science and Engineering an Introduction. 6th ed. New Jersey: John Wiley & Sons Inc.; 2003.

[12] Luborsky FE. J. Appl. Phys. 1961;**32**:S171.

[13] O'Handley RC. Modern Magnetic Materials: Principles and Applications. 1st ed. New York: Wiley Interscience; 1999.

[14] Bedanta S, Kleemann WJ. Phys. D: Appl. Phys. 2009;**42**:013001.

[15] Stoner EC, Wohlfarth EP. Trans. R. Soc. Lond. A. 1948;**240**:599–642.

[16] Chinnasamy C, Jeyadevan B, Perales-Perez O, Shinoda K, Tohji K, Kasuya A. IEEE Trans. Magn. 2002;**38**:2640–2642.

[17] Perales-Perez O, Sasaki H, Kasuya A, Jeyadevan B, Tohji K, Hihara T, Sumiyama K. J. Appl. Phys. 2002;**91**:6958–6960.

[18] Moumen N, Bonville P, Pileni MP. J. Phys. Chem. 1996;**100**:14410–14416.

[19] Cedeño-Mattei Y, Perales-Perez O, Tomar MS, Roman F, Voyles PM, Stratton WG. J. Appl. Phys. 2008;**103**:07E512.

[20] Sileo EE, Garcia-Rodenas L, Paiva-Santos CO, Stephens PW, Morando PJ, Blesa MA. J. Solid State Chem. 2006;**179**:2237–2244.

[21] Letellier P, Mayaffre A, Turmine M. J. Phys. Condens. Matter. 2007;**19**:436229.

[22] Figueroa CA, Sileo EE, Morando PJ, Blesa MA. J. Colloid Interface Sci. 2000;**225**:403–410.

[23] Sawatzky GA. J. Appl. Phys. 1968;**39**:1204–1205.

[24] Manova E, Kunev B, Paneva D, Mitov I, Petrov L, Estournès C, D'Orléan C, Rehspringer J, Kurmoo M. Chem. Mater. 2004;**16**:5689–5696.

[25] Zhao L, Zhang H, Xing Y, Song S, Yu S, Shi W, Guo X, Yang J, Lei Y, Cao F. J. Solid State Chem. 2008;**181**:245–252.

[26] Cedeño-Mattei Y, Perales-Perez O, Uwakweh ONC, Xin Y. J. Appl. Phys. 2010;**107**:09A741.

[27] Olsson RT, Salazar-Alvarez G, Hedenqvist MS, Gedde UW, Lindberg F, Savage S. J. Chem. Mater. 2005;**17**:5109–5118.

[28] Globus A, Monjaras R. IEEE Trans. Magn. 1975;**11**:1300–1302.

[29] Salazar-Alvarez G, Olsson RT, Sort J, Macedo WAA, Ardisson JD, Baró MD, Gedde UW, Nogués J. Chem. Mater. 2007;**19**:4957–4963.

[30] Henderson CMB, Charnock JM, Plant DA. J. Phys. Condens. Matter. 2007;**19**:076214.

[31] Töpfer J, Dieckmann R. J. Eur. Ceram. Soc. 2004;**24**:603–612.

[32] Cedeño-Mattei Y, Perales-Perez O, Uwakweh ONC. Mat. Chem. Phys. 2012; **132**: 999–1006.

[33] Xu P, Han X, Zhao H, Liang Z, Wang J. Mater Lett. 2008;**62**:1305–1308.

[34] Papaefthymiou GC. Nano Today. 2009;**4**:438–447.

[35] Miller JS, Drillon M. Magnetism: Molecules to Materials. New York: Wiley-VCH; 2003.

[36] Calero-Diaz del Castillo VL, Rinaldi C. IEEE Trans. Magn. 2010;**46**:852–859.

[37] Sun S, Zeng H, Robinson DB, Raoux S, Rice PM, Wang SX Li G. J. Am. Chem. Soc. 2004;**126**:273–279.

[38] Meron T, Rosenberg Y, Lereah Y, Markovich G. J. Magn. Magn. Mater. 2005;**292**:11–16.

[39] Cedeño-Mattei Y, PhD Dissertation, University of Puerto Rico: Mayaguez Campus, 2011.

Section 2

Electronic and Magnetic Properties

Chapter 4

Nature of Magnetic Ordering in Cobalt-Based Spinels

Subhash Thota and Sobhit Singh

Additional information is available at the end of the chapter

http://dx.doi.org/10.5772/65913

Abstract

In this chapter, the nature of magnetic ordering in cobalt-based spinels Co_3O_4, Co_2SnO_4, Co_2TiO_4, and Co_2MnO_4 is reviewed, and some new results that have not been reported before are presented. A systematic comparative analysis of various results available in the literature is presented with a focus on how occupation of the different cations on the A- and B-sites and their electronic states affect the magnetic properties. This chapter specifically focuses on the issues related to (i) surface and finite-size effects in pure Co_3O_4, (ii) magnetic-compensation effect, (iii) co-existence of ferrimagnetism and spin-glass-like ordering, (iv) giant coercivity (H_C) and exchange bias (H_{EB}) below the glassy state, and (v) sign-reversal behavior of H_{EB} across the ferri/antiferromagnetic Néel temperature (T_N) in Co_2TiO_4 and Co_2SnO_4. Finally, some results on the low-temperature anomalous magnetic behavior of Co_2MnO_4 spinels are presented.

Keywords: ferrimagnetic materials, Néel temperature, spin-glass, exchange bias

1. Introduction

The atomic arrangement of the spinel compounds is interpreted as a pseudo-close-packed arrangement of the oxygen anions with divalent cations occupying tetrahedral A-sites and trivalent cations residing at the octahedral B-sites of the cubic unit cell of space group Fd-3m(227). A spinel with the configuration (A^{2+}) $[B_2^{3+}]$ O_4 is termed as "normal spinel" whereas the other possible configuration (B^{3+}) $[A^{2+}B^{3+}]$ O_4 is called "inverse spinel." The continuum of possible atomic distribution between these two extremes is quantified by a parameter denoted as x (inversion parameter), which describes the fraction of "B" cations on tetrahedral sites. Thus, x = 0 for a normal spinel, 2/3 for a spinel with entirely arbitrary configuration, and 1 for a fully inverse spinel. Among many varieties of spinel compounds, ferrites and cobaltites are widely used in the high-frequency electronic circuit components such as transformers, noise filters, and magnetic recording heads [1, 2]. The key property of these spinels is that at high frequencies (>1 MHz), their dielectric permittivity (ε) and magnetic permeability (μ) are much higher than those of metals with low loss-factor (tanδ). These properties make them very advantageous for the development of magnetic components used in the power electronics industry. Also, the nanostructures of these spinels continue to receive large attention because of their potential applications in solid-oxide fuel cells, Li-ion batteries, thermistors, magnetic recording, microwave, and RF devices [1, 2].

In this review, we mainly focus on the nature of magnetic ordering of several insulating cobalt spinels of type Co_2MO_4 (where "M" is the tetravalent or trivalent metal cation such as Sn, Ti, Mn, Si, etc.) which are not yet well studied in literature. This review will primarily illustrate how the magnetic ordering changes when we substitute the above-listed metal cations at the tetrahedral B-sites. It is well known that the dilution of magnetic elements significantly disrupts the long-range magnetic ordering and leads to more exotic properties like magnetic frustration, polarity reversal exchange bias, and reentrant spin-glass state near the magnetic-phase transition [3, 4]. The dilution essentially alters the super-exchange interactions of J_{AB}, J_{BB}, and J_{AA} between the magnetic ions, which is the main source of anomalous magnetic behavior. In this review, we first start with the simplest case of antiferromagnetic (AFM) normal-spinel Co_3O_4 with configuration $(Co^{2+})[Co_2^{3+}]O_4$ and discuss the role of surface and finite-size effects on antiferromagnetic (AFM) ordering. In the second section, we focus on the coexistence of ferrimagnetism and low-temperature spin-glass behavior of cobalt orthostannate (Co_2SnO_4) and cobalt orthotitanate (Co_2TiO_4). A detailed comparative analysis of some recent experimental results dealing with the temperature and frequency dependence of ac-magnetic susceptibility is presented. In the subsequent section, some unusual magnetic properties of Co_2MnO_4 are discussed.

2. Magnetic properties of bulk versus nanocrystalline Co_3O_4

Cobalt forms two binary compounds with oxygen: CoO and Co_3O_4. While CoO has face-centered cubic (NaCl-type) structure, Co_3O_4 shows a normal-spinel structure with a cubic close packed arrangement of oxygen ions and Co^{2+} and Co^{3+} ions occupying the tetrahedral "8a" and the octahedral "16d" sites, respectively [5]. The magnetic properties of Co_3O_4 were first investigated over 58 years ago; however, its magnetic behavior under reduced dimensions still attracted immense scientific interest mainly because of its distinctly different magnetic ordering under nanoscale as compared to its bulk counterpart [6]. Co_3O_4 can be synthesized in various nanostructural forms such as nanorods, nanosheets, and ordered nanoflowers with ultrafine porosity [7–10]. Such engineered nanostructures play vital roles as catalysts, gas sensors, magneto-electronics, electrochromic devices, and high-temperature solar selective absorbers [11–18]. At first glance, the normal-spinel structure of Co_3O_4 may look similar to that of Fe_3O_4 (inverse spinel) but Co_3O_4 exhibits strikingly different magnetic ordering as compared to Fe_3O_4. In particular, Co_3O_4 does not exhibit ferrimagnetic ordering of the type observed in Fe_3O_4 because Co^{3+} ions on the octahedral B-sites are in the low spin $S = 0$ state [5].

http://dx.doi.org/10.5772/65913

Instead, it exhibits antiferromagnetic ordering with each Co^{2+} ion at the A-site having four neighboring Co^{2+} ions of opposite spins (with an effective magnetic moment of $\mu_{eff} \sim 4.14\ \mu_B$) [5]. Earlier studies by Roth reported that below the Néel temperature $T_N \sim 40$ K, Co_3O_4 becomes antiferromagnetic in which the uncorrelated spins of the $8Co^{2+}$ (in 8(a), F.C. +000, $\frac{1}{4}\frac{1}{4}\frac{1}{4}$) cations in the paramagnetic state (space group $O^7{}_h$—Fd-3m) are split in the antiferromagnetic state (space group $T^2{}_d$—F$\bar{4}$3m) into the two sublattices with oppositely directed spins of $4Co^{2+}\uparrow$ (4(a), +000) and $4Co^{2+}\downarrow$ (4(c) $\frac{1}{4}\frac{1}{4}\frac{1}{4}$) [5–7]. For $T < T_N$, the neutron diffraction studies did not show any evidence of a structural phase transition.

As shown in **Figure 1**, the recent magnetic studies by Dutta et al. [6] have reported a significant difference in the antiferromagnetic ordering temperature $T_N \sim 30$ K of Co_3O_4 as compared to the earlier data (40 K); this new result however is in excellent agreement with $T_N = 29.92 \pm 0.03$ K obtained by the heat capacity C_p versus T measurements reported by Khriplovich et al. [19]. It is well known that the peak in the magnetic susceptibility data of antiferromagnets usually occurs at a temperature few percent higher than T_N because the magnetic specific heat of a simple antiferromagnet (in particular, the singular behavior in the region of the transition) should be closely similar to the behavior of the function $d(\chi_p T)/dT$ [14]. Therefore, T_N is better

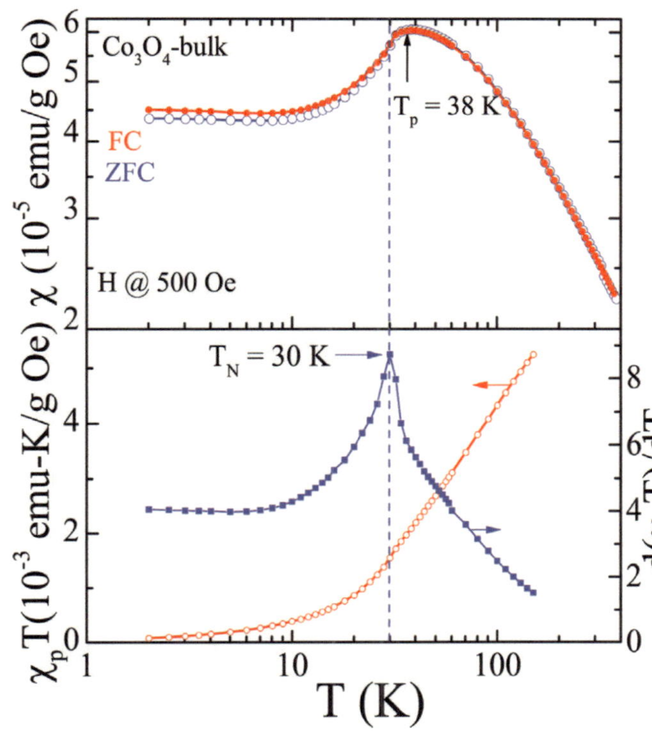

Figure 1. (a) Temperature dependence of the dc-magnetic susceptibility $\chi(T)$ for bulk Co_3O_4 under the zero-field-cooled (ZFC) and field-cooled (FC) conditions. Here, T_p denotes the peak position in χ versus T plots. (b) Plots of $(\chi_p T)$ versus T (LHS scale) and $d(\chi_p T)/dT$ versus T plots (RHS scale) for the bulk Co_3O_4. Here, the paramagnetic susceptibility $\chi_p = \chi - \chi_0$ with $\chi_0 = 3.06 \times 10^{-6}$ emu/g Oe being the temperature-independent contribution [6, 7].

defined by the peak in $\partial(\chi T)/\partial T$ versus T plot [20]. **Figure 1** shows the temperature dependence of paramagnetic susceptibility $\chi_p(T)$ (LHS scale) and $d(\chi_p T)/dT$ versus T (RHS scale). For bulk Co_3O_4 the peak temperature value (30 K) in the $d(\chi_p T)/dT$ versus T plots is lower than $T_N \simeq 40$ K often quoted for Co_3O_4 [5–7, 9–10]. Thus, $T_N = 30$ K determined from two independent techniques (i.e., χ_p and C_p measurements) is consistent with each other and is the accurate characteristic value for bulk Co_3O_4. On the other hand, the nanoparticles of Co_3O_4 exhibit lower T_N values and reduced magnetic moment than the bulk value (30 K, 4.14 μ_B) which is a consequence of finite-size and surface effects [6]. Salabas et al. first reported Co_3O_4 nanowires of diameter 8 nm and lengths of up to 100 nm by the nanocasting route and observed $T_B \simeq 30$ K and exchange bias (H_e) for $T < T_B$ [10].

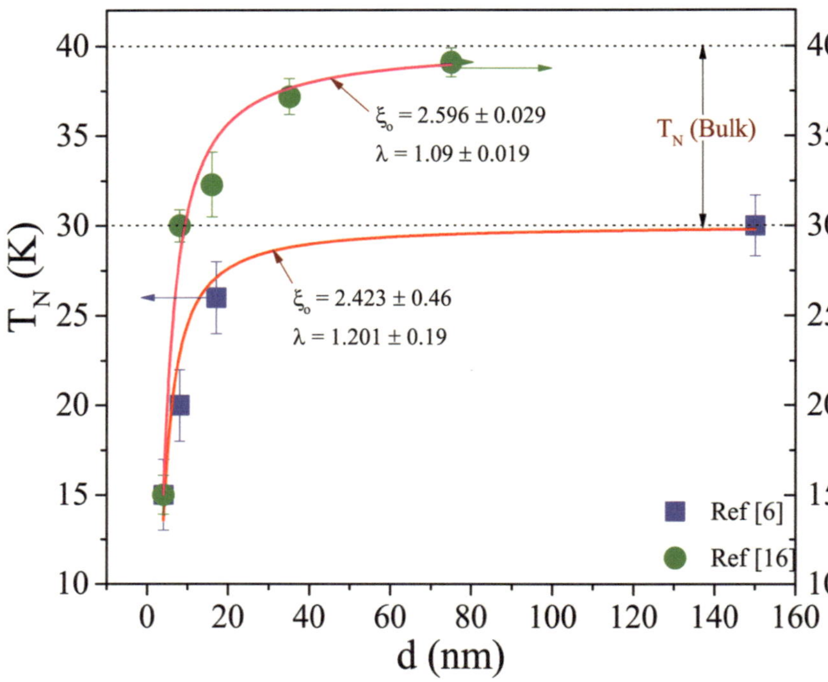

Figure 2. Variation of the Néel temperature T_N of Co_3O_4 as a function of particle size (d). The data pertaining to solid-square symbols are taken from Refs. [6–8] in which T_N is considered as a peak point in"$d(\chi_p T)/dT$" versus"T" data. However, the data related to solid green circles are taken from Refs. [8, 9] in which T_N is considered as just the peak point in the "χ"versus"T" plot, not from the derivative plots. The scattered symbols are the raw data corresponding to T_N and the solid lines are the best fit to Eq. (1).

A plot of T_N values versus particle size d reported by several authors for various crystallite sizes of Co_3O_4 is shown in **Figure 2**. The lowest T_N value reported till now is about 15 ± 2 K for 4.34 nm size Co_3O_4 particles [8]. These nanoparticles were synthesized using biological containers of Listeria innocua Dps proteins and LDps as constraining vessels. Lin and Chen studied the magnetic properties of various sizes (diameter d = 16, 35, and 75 nm) of Co_3O_4 nanoparticles prepared by chemical methods using $CoSO_4$ and $CoCl_2$ as precursors [9]. These authors reported that the variation of T_N follows the finite-size scaling relation:

http://dx.doi.org/10.5772/65913

$$T_N(D) = T_N(\infty)\left[1 - \left(\frac{\xi_0}{d}\right)^{\lambda}\right] \quad (1)$$

for various sizes of Co_3O_4 nanoparticles. Accordingly, they obtained the shift exponent $\lambda = 1.1 \pm 0.2$ and the correlation length $\xi_o = 2.8 \pm 0.3$ nm from the fitting analysis of T_N versus d (**Figure 2**). However, these authors considered T_N values as the direct peak temperature values from χ versus T instead of the peak point in $d(\chi_p T)/dT$. Also, for the bulk grain sizes $T_N(\infty) = 40$ K was considered instead of 30 K obtained from $d(\chi_p T)/dT$ analysis as discussed above. Therefore, we repeated the analysis but considering T_N values obtained from $d(\chi_p T)/dT$ versus T and the $T_N(\infty) = 30$ K for various sizes of the Co_3O_4 nanoparticles obtained by sol-gel process (these values were obtained from our earlier works [6, 7]). Accordingly, we obtained $\lambda = 1.201 \pm 0.2$ and the correlation length $\xi_o = 2.423 \pm 0.46$ nm, which are slightly different from the earlier reported values $\lambda = 1.1 \pm 0.2$ and the correlation length $\xi_o = 2.8 \pm 0.3$ nm [9]. Nonetheless, in both the cases T_N follows the finite-size scaling relation Eq. (1).

For $T > T_N$, the data of χ versus T (**Figure 3**) are fitted to the modified Curie-Weiss law $\chi_P = \chi_0 + [C/(T+\theta)]$ with $C = N\mu^2/3k_B$, $\mu^2 = g^2 J(J+1)\mu_B^2$, θ is the Curie-Weiss temperature and χ_0 contains two contributions: the temperature-independent orbital contribution mentioned earlier and

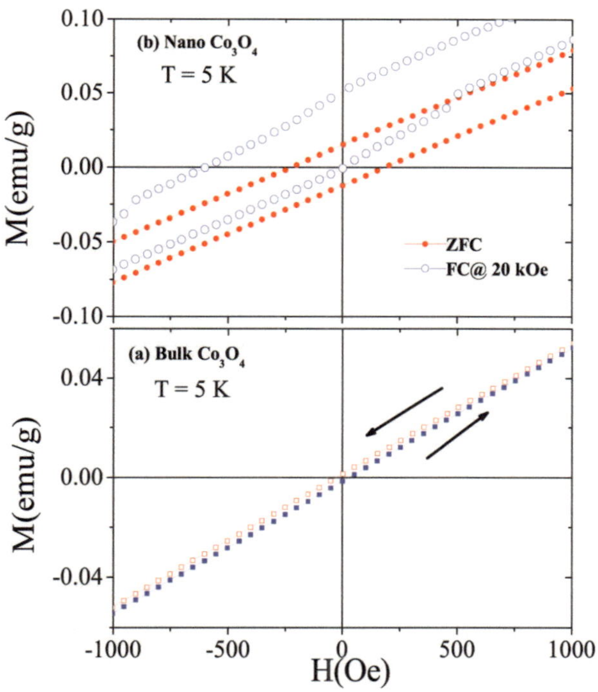

Figure 3. $1/\chi_p$ versus T plots for the bulk and nanocrystalline (~17 nm) Co_3O_4 with $\chi_0 = 3.06 \times 10^{-6}$ emu/g Oe (LHS scale). The solid lines represent linear fit to the Curie-Weiss law: $\chi_p = C/(T + \theta)$. On the RHS scale same figures are plotted except for $\chi_0 = 9 \times 10^{-6}$ and 7.5×10^{-6} emu/g Oe for the bulk and Co_3O_4 nanoparticles of size ~17 nm, respectively [6, 7].

the diamagnetic component $\chi_d = -3.3 \times 10^{-7}$ emu/g Oe [6]. Usually χ_0 is estimated from the plot of χ versus $1/T$ in the limit of $1/T \rightarrow 0$ using the high-temperature data. The value of χ_0 was estimated as 3.06×10^{-6} emu/g Oe for bulk Co_3O_4 using the inverse paramagnetic susceptibility ($1/\chi_P$) versus temperature (T) data (shown in the left-hand-side scale of **Figure 3**) [5, 6]. A similar procedure for experimental data, shown in the right-hand side scale of **Figure 3**, yields $\chi_0 = 9 \times 10^{-6}$ and 7.5×10^{-6} emu/g Oe for the bulk and nanoparticles (size ~17 nm) of Co_3O_4, respectively.

It is well known that the origin of the antiferromagnetic ordering in transition metal oxides can be explained by the super-exchange interaction between the magnetic elements via oxygen ion. In the present case, there are two possible paths for super-exchange interaction between magnetic ions in Co_3O_4, i.e., Co^{2+} ions: (tetrahedral site) A – O (oxygen) – A (tetrahedral site) and A – O – B – O – A with the number of nearest-neighbors $z_1 = 4$ and next-nearest-neighbors $z_2 = 12$, respectively. If the corresponding exchange constants are represented by J_{1ex} and J_{2ex}, the expressions for T_N and θ, using the molecular-field theory, can be written as [5, 6]

$$T_N = \frac{J(J+1)}{3k_B}\left(J_{1ex}z_1 - J_{2ex}z_2\right) \tag{2}$$

$$\theta = \frac{J(J+1)}{3k_B}\left(J_{1ex}z_1 + J_{2ex}z_2\right) \tag{3}$$

In order to determine J_{1ex} and J_{2ex}, the magnitude of effective $J(J + 1)$ for Co^{2+} is required. Since the Curie constant C is equivalent to $N\mu^2/3k_B$ with $\mu = g\,[\,J(J + 1)]^{1/2}\mu_B$ where g is the Landé g-factor and J is the total angular momentum. Using the magnitude of $g = 2$ and C from **Figure 3**, one can estimate the effective magnetic moment $\mu_{eff} = 4.27\mu_B$ for bulk Co_3O_4 and $\mu = 4.09\mu_B$ for Co_3O_4 nanoparticles of size ~17 nm. The spin contribution to the above magnitudes of μ is $3.87\mu_B$ for Co^{2+} with spin $S = 3/2$. Obviously, there is some additional contribution resulting from the partially restored orbital angular moment for the $^4F_{9/2}$ ground state of Co^{2+} [5, 6]. Using Eqs. (2) and (3) and the values of "θ," "T_N," and "μ" for the two cases yields $J_{1ex} = 11.7$ and $J_{2ex} = 2.3$ K for bulk, and $J_{1ex} = 11.5$ and $J_{2ex} = 2.3$ K for the Co_3O_4 nanoparticles (d ~ 17 nm). Thus, both the exchange constants J_{1ex} and J_{2ex} correspond to antiferromagnetic coupling. From the magnitudes of C in **Figure 3**, the value of μ is obtained as 3.28 and 3.43 μ_B for bulk and Co_3O_4 nanoparticles (d ~17 nm), respectively. These magnitudes of μ are lower than the spin contribution (3.87 μ_B) of Co^{2+} ion itself. Consequently, the magnitudes of "θ" in **Figure 3** seem to be questionable. This may be due to the fact that the use of molecular-field theory in determining the exchange constants has its own limitations since higher order spin correlations are neglected in this model [6].

For a typical bulk antiferromagnet, below T_N, the magnetization is expected to vary linearly with applied external magnetic field H below the spin-flop field. Therefore, the corresponding coercive field H_c and exchange-bias field H_e must become zero. This was indeed observed in bulk Co_3O_4 (**Figure 4a**) [5, 6]. Conversely, for the Co_3O_4 nanoparticles (d ~ 17 nm), the data at

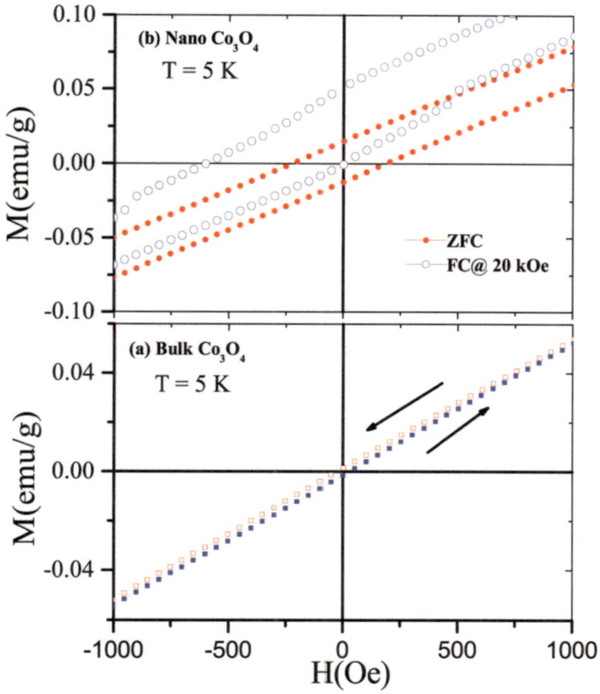

Figure 4. The magnetization (M) versus external applied field (H) plots recorded in standard five-cycle hysteresis mode for bulk and nanoparticles (d ~17 nm) of Co_3O_4 measured at 5 K in the lower field region of ±1 kOe. (a) Irreversibility observed for the direct and reverse field scans for bulk Co_3O_4, however, (b) a asymmetric shift in the hysteresis loop with enhanced coercivity can be clearly noticed in the case of nanoparticles (d ~ 17 nm) of Co_3O_4 measured under field-cooled protocol (FC) of H = 20 kOe [6, 7].

5 K show a symmetric hysteresis loop with H_c = 200 Oe for the zero-field-cooled sample and asymmetric (shifted) hysteresis loop with H_c = 250 Oe and H_e = −350 Oe for the sample cooled in magnetic field H = 20 kOe from 300 to 5 K as shown in **Figure 4b**. Thus, cooling the sample in a magnetic field produces an exchange bias and leads to the enhancement of H_c as well. The temperature dependence of H_c and H_e for the nanoparticles of Co_3O_4 cooled under H = 20 kOe from 300 K to the measuring temperature is shown in **Figure 5**. Both H_c and H_e approach to zero above T_N. The inset of **Figure 5** depicts the training effect, i.e., change in the magnitude of H_e for the sample cycled through several successive hysteresis loops (designated by "n" at 5 K) [6, 10]. A similar effect has been recently reported by Salabas et al. [10] in the Co_3O_4 nanowires of 8 nm diameter although the magnitudes of H_e and H_c in their case are somewhat smaller. The existence of the exchange bias suggests the presence of a ferromagnetic (shell)/antiferromagnetic (core) interface with FM-like surface spins covering the core of the antiferromagnetically ordered spins in the nanoparticles of Co_3O_4. Salabas et al. reported that H_e falls by ~25% measured between the first and the second loops. The observation of the training effect and open loops of up to 55 kOe suggests that the surface spins are in an unstable spin-glass-like state [10]. Such a spin-glass ordering results from the weaker exchange coupling experienced by the surface spins due to reduced coordination at the surface. These effects however disappear above T_N when the spins in the core become disordered. The observation

of somewhat lower magnetic moment per Co^{2+} ion, smaller values of exchange constants J_{1ex} and J_{2ex}, and lower T_N was noticed for the nanoparticles of Co_3O_4 in relation to the bulk Co_3O_4. This could be due to the weak exchange coupling and reduced coordination of the surface spins.

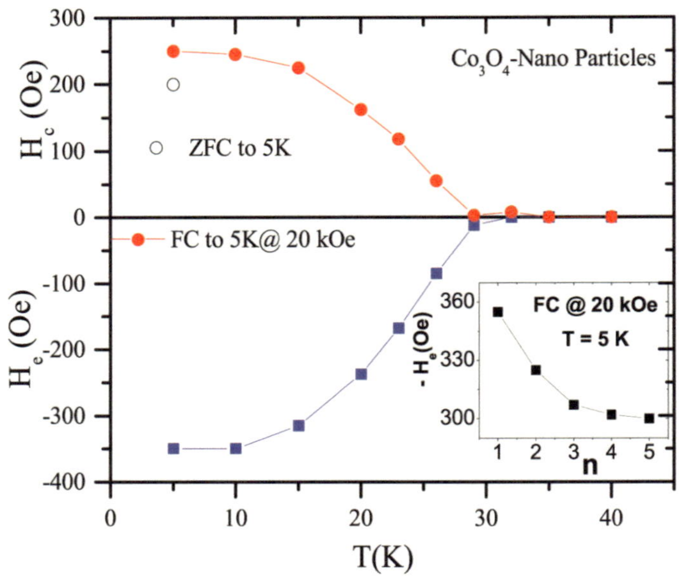

Figure 5. The temperature dependence of H_c and H_e for the nanoparticles of Co_3O_4 (d ~ 17nm) measured from T = 5 to 40 K under field-cooled (FC) mode at 20 kOe and at 5 K under zero-field-cooled (ZFC) condition [6]. One can clearly notice $H_c \rightarrow 0$, $H_e \rightarrow 0$ as T approaches T_N. The inset shows progressive decrease of the magnitude of H_e after successive scan (number of cycles"n") at 5 K [6].

3. Co-existence of ferrimagnetism and spin-glass states in some inverted spinels

In 1975, Sherrington and Kirkpatrick (SK) first predicted the reentrant behavior of spinels using mean-field approach for certain relative values of the temperature and exchange interaction [21, 22]. Later, Gabay and Toulouse extended the SK Ising model calculation to the vector spin glasses and showed that it is possible to have multiple phase transitions such as ferro/ferri/antiferromagnetic state $\xleftrightarrow{T_C \text{ or } T_N}$ paramagnetic state $\xleftrightarrow{T_{M1}}$ Mixed phase-1 $\xleftrightarrow{T_{M2}}$ Mixed phase-2 [22, 23]. The present section deals with such kind of systems in which the longitudinal ferrimagnetic ordering coexists with the transverse spin-glass state below the Néel temperature T_N [23–28]. In this connection, we mainly focus on the magnetic ordering in two inverse-spinel systems, namely (i) cobalt orthostannate (Co_2SnO_4) and (ii) cobalt orthotitanate (Co_2TiO_4) which exhibits the reentrant spin-glass behavior [24–40]. At first glimpse, both systems Co_2SnO_4 and Co_2TiO_4 are expected to show similar magnetic properties because of the fact that

http://dx.doi.org/10.5772/65913

nonmagnetic ions Sn and Ti play identical role on the global magnetic ordering of the system. But in reality, they exhibit markedly different magnetic structures below their ferrimagnetic Néel temperatures [24–40], which is discussed in detail below.

Usually, the ferrimagnetic ordering in Co_2SnO_4 and Co_2TiO_4 arises from the unequal magnetic moments of Co^{2+} ions at the tetrahedral A-sites and octahedral B-sites [24–40]. The corresponding magnetic moment at the tetrahedral A-sites $\mu(A)$ is equal to 3.87 μ_B and the magnetic moment at octahedral B-sites $\mu(B)$ is equal to 5.19 and 4.91 μ_B for Co_2TiO_4 and Co_2SnO_4, respectively [27–33]. In 1976, Harmon et al. first reported the low-temperature magnetic properties of polycrystalline Co_2SnO_4 system and showed the evidence for ferrimagnetic ordering with $T_N \sim 44 \pm 2$ K [24]. They also suggested that Co_2SnO_4 should contain two equally populated sublattices that align collinearly and couple antiferromagnetically [24]. On the basis of the Mössbauer spectroscopy results, these authors calculated the internal dipolar fields at the Sn sites from the two Co^{2+} sublattices to be > 80 kOe [24]. They also reported very high values of coercive field $H_C > 50$ kOe below T_N [24]. The reported magnetization value at 16 K per Co^{2+} ion was about 2.2×10^{-3} μ_B with zero magnetization value at 12 K. The Curie-Weiss constant (C) = 4.3 ± 0.2 emu/mol, and the effective magnetic moment of the Co^{2+} ions = 5.0 ± 0.2 μ_B was close to the standard value of 4.13 emu/mol and 4.8 μ_B, respectively. A year later, Sagredo et al. reported that the zero-field-cooled and field-cooled magnetization curves of single crystal Co_2SnO_4 exhibit strong irreversibility below T_N [39].

Sagredo et al. reported that thermoremanent magnetization, magnetic training effects, and spin-glass phases present in this system are driven by the disordered-spin configurations [39]. Accordingly, they speculated that the random distribution of Sn^{4+} ions on the B-sites might break the octahedral symmetry of the crystal field and result in the frustrated magnetic behavior [39]. In 1987, Srivastava et al. reported multiple peaks in the temperature dependence of ac-magnetic susceptibility $\chi_{ac}(T)$ for both Co_2SnO_4 and Co_2TiO_4 below their T_N providing the evidence of Gabay and Tolouse mixed-phase transitions [25, 28]. Nevertheless, some recent studies proved the existence of transverse spin-glass state T_{SG} (~39 K) just below the T_N (= 41 K) in Co_2SnO_4 [27, 38, 40]. Similar type of results in Co_2TiO_4 was reported by Hubsch et al. and Srivastava et al. but with different T_{SG} (~46 K) and T_N (=55 K) [25, 28, 31, 34].

In order to get a precise understanding of the magnetic properties of these systems, we have plotted the temperature dependence of dc-magnetic susceptibility $\chi_{dc}(T)$ for both Co_2SnO_4 and Co_2TiO_4 measured under ZFC and FC (H@100 Oe) conditions in **Figure 6**. These $\chi_{dc}(T)$ plots show typical characteristics of ferrimagnetic ordering with peaks across the Néel temperatures T_N = 47 K (for Co_2TiO_4) and 39 K (for Co_2SnO_4). However, for $T \leq 31.7$ K an opposite trend in the $\chi_{dc}(T)$ values was noticed for Co_2TiO_4 with $\chi_{dc} \sim 0$ at magnetic-compensation temperature T_{CMP} = 31.7 K at which the two-bulk sublattices magnetizations completely balances with each other [31–33]. Such compensation behavior in Co_2SnO_4 system is expected to appear at very low temperatures ($T < 10$ K) as $\chi_{dc}(T)$ approaches to zero. Consequently, $\chi_{dc\text{-}ZFC}$ exhibits negative magnetization until $T_{SG.}$ It is expected that the different magnetic moments on the tetrahedral (A) and the octahedral (B) sites, and their different temperature dependence (i.e., $\mu_A(T)$, $\mu_B(T)$) play a major role on the global magnetic ordering of both systems.

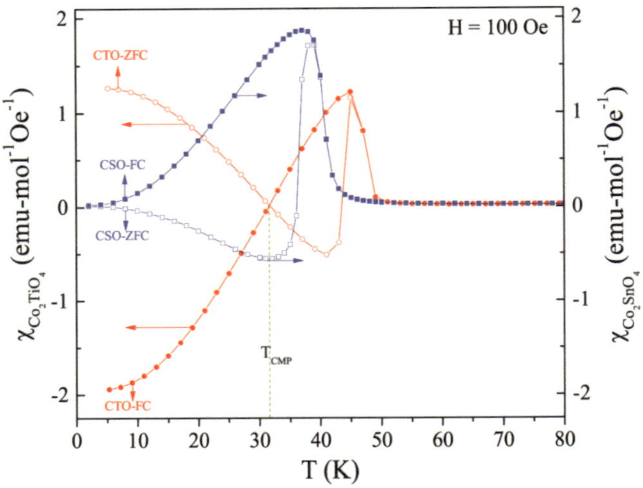

Figure 6. The temperature dependence of dc-magnetic susceptibility $\chi_{dc}(T)$ measured under ZFC and FC (H@100 Oe) conditions for both Co_2TiO_4 (LHS) and Co_2SnO_4 (RHS). The dotted line shows the T_{CMP}.

Recent X-ray photoelectron spectroscopic studies reveal that the crystal structure of Co_2TiO_4 consists of some fraction of trivalent cobalt and titanium ions at the octahedral sites, i.e., $[Co^{2+}][Co^{3+}Ti^{3+}]O_4$ instead of $[Co^{2+}][Co^{2+}Ti^{4+}]O_4$ [34]. On the contrary, the Co_2SnO_4 shows the perfect tetravalent nature of stannous ions without any trivalent signatures of Co^{3+} $[Co^{2+}][Co^{2+}Sn^{4+}]O_4$ [27]. Such distinctly different electronic structure of the ions on the B-sites of cobalt orthotitanate plays a significant role on the anomalous magnetic ordering below T_N, for example, exhibiting the magnetic-compensation behavior, sign reversible zero-field exchange-bias, and negative slopes in the Arrott plots (H/M versus M^2) [27, 34].

At high temperatures (for all $T > T_N$), the experimental data of inverse dc-magnetic susceptibility (χ^{-1}) for both the systems Co_2TiO_4 and Co_2SnO_4 fit well with the modified Néel expression for ferrimagnets $(1/\chi) = (T/C) + (1/\chi_0) - [\sigma_0/(T - \theta)]$. **Table 1** summarizes various fitting parameters obtained from the Néel expression for both Co_2SnO_4 and Co_2TiO_4. The fit for Co_2TiO_4 yields the following parameters: $\chi_0 = 41.92 \times 10^{-3}$ emu/mol-Oe, $\sigma_0 = 31.55$ mol-Oe-K/emu, $C = 5.245$ emu K/mol Oe, $\theta = 49.85$ K. The ratio $C/\chi_0 = T_a$ =125.1 K represents the strength of the antiferromagnetic exchange coupling between the spins on the A- and B-sites and is often termed as the asymptotic Curie temperature T_a. For both the systems, the effective magnetic moment μ_{eff} is determined from the formula $C = N\mu_{eff}^2/3k_B$. The experimentally observed magnetic moments at the B-sites $\mu(B) = 5.19\ \mu_B$ obtained from the temperature dependence of magnetization values are in line with the above-discussed spectroscopic properties of Co_2TiO_4, i.e., the total moment $\mu(B)$ is perfectly matching with the contribution from the magnetic moments due to Co^{3+} (4.89 μ_B) and Ti^{3+} (1.73 μ_B), $\mu(B) = \sqrt{\left(\mu_{Co^{+3}}\right)^2 + \left(\mu_{Ti^{+3}}\right)^2}$ [34]. Also, the analysis of the dc and ac susceptibilities combined with the weak anomalies observed in the C_p versus T data has shown the existence of a quasi-long-range ferrimagnetic state below $T_N \sim 47.8$ K and a compensation temperature of $T_{CMP} \sim 32$ K [34].

http://dx.doi.org/10.5772/65913

System	C (emu K/mol/Oe)	χ_o (emu/mol/Oe)	σ_o (Oe mol K emu^{-1})	θ (K)	μ_{eff} (μ_B)	μ (A) (μ_B)	μ(B) (μ_B)
Co_2TiO_4	5.245	0.0419	31.55	49.85	6.5	3.87	5.19
		N_{AA}	N_{AB}	N_{BB}	J_{AA}	J_{AB}	J_{BB}
		17.319	35.700	12.720	3.25 k_B	4.47 k_B	3.18 k_B
Co_2SnO_4	4.889	0.0436	102.370	39.5	6.25	3.87	4.91
		N_{AA}	N_{AB}	N_{BB}	J_{AA}	J_{AB}	J_{BB}
		21.564	33.201	10.68	4.05 k_B	5.26 k_B	4.28 k_B

Table 1. The list of various parameters obtained from the Néel fit of χ^{-1} versus T curve recorded under zero-field-cooled condition.

The real and imaginary components of the temperature dependence of ac-susceptibility data $\chi_{ac}(T)$ (= $\chi'(T) + i\,\chi''(T)$) recorded at different frequencies for both the polycrystalline samples Co_2SnO_4 and Co_2TiO_4 show the dispersion in their peak positions ($T_P(f)$) similar to the compounds exhibiting spin-glass-like ordering [27, 33]. **Figure 8** shows the $\chi'(T)$ and $\chi''(T)$ data of Co_2SnO_4 and Co_2TiO_4 recorded at different measuring frequencies ranging from 0.17 to 1202 Hz with peak-to-peak field amplitude $H_{ac} = 4$ Oe under zero dc-bias field. It is clear from these figures that the peaks seen in both cases show pronounced frequency dependence, which suggests the dynamical features analogous to that of observed in spin-glass systems. A detailed analysis of such frequency dependence of $\chi'(T)$ and $\chi''(T)$ using two scaling laws described below provides the evidence for spin-glass-like characteristics below T_N. For example, applying the Vogel-Fulcher law (below equation) for interacting particles

$$\tau = \tau_0 \exp\left(\frac{E_a}{k_B\left(T - T_0\right)}\right) \tag{4}$$

and the best fits of the experimental data (the logarithmic variation of relaxation time"τ" as a function of $1/(T_F - T_0)$ as shown in **Figure 8a** yields the following parameters: interparticle interaction strength $T_0 = 39.3$ K and relaxation time constant $\tau_0 = 7.3 \times 10^{-8}$ s for Co_2SnO_4. Here, we define the freezing temperature for each frequency is T_F, angular frequency ω as $2\pi f$ ($\omega = 1/\tau$), k_B is the Boltzmann constant, and E_a is an activation energy parameter. Such large value of τ_0 indicates the presence of interacting magnetic spin clusters of significant sizes in the polycrystalline Co_2SnO_4 system. The origin of such spin clusters may arise from a short-range magnetic order occurring due to the competition between ferrimagnetism and magnetic frustration. Another characteristic feature that the spin-glass systems follows is the power law (Eq. (5)) of critical slowing down in a spin-glass phase transition at T_{SG} (note that the $T_P(f)$ data represent a relatively small temperature interval):

$$\tau = \tau_0 \left(\frac{T}{T_g} - 1\right)^{-z\nu} \tag{5}$$

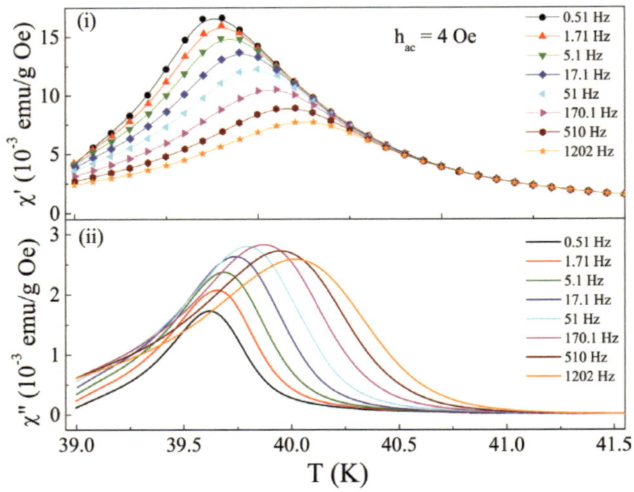

Figure 7(a). The temperature dependence of the ac-magnetic susceptibilities (i) real $\chi'(T)$ component and (ii) imaginary $\chi''(T)$ component of bulk polycrystalline Co_2SnO_4 system recorded at various frequencies under warming condition using dynamic magnetic field of amplitude $h_{ac} = 4$ Oe and zero static magnetic field $H_{dc} = 0$. The lines connecting the data points are visual guides [27].

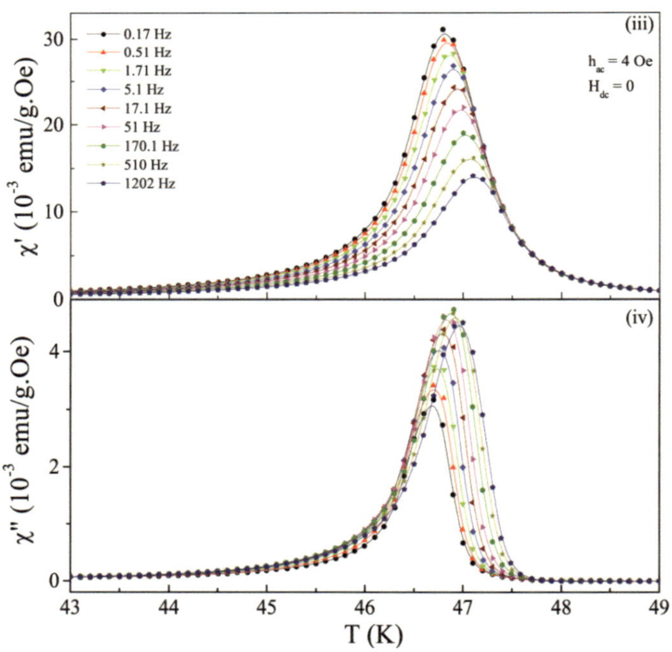

Figure 7(b). The temperature dependence of the ac-magnetic susceptibilities (iii) real $\chi'(T)$ component and (iv) imaginary $\chi''(T)$ component of bulk polycrystalline Co_2TiO_4 system recorded at various frequencies under warming condition using dynamic magnetic-field of amplitude $h_{ac} = 4$ Oe and zero static magnetic field $H_{dc} = 0$. The lines connecting the data points are visual guides [33].

The least-square fit using the power law of the data shown in **Figure 7** is depicted in **Figure 8b**. Here, T_{SG} is the freezing temperature, τ_0 is related to the relaxation of the individual cluster magnetic moment, and $z\nu$ is a critical exponent. The least-square fit analysis for Co_2SnO_4 gives $T_{SG} = 39.6$ K, $\tau_0 = 1.4 \times 10^{-15}$ s, and $z\nu = 6.4$. Since the value of $z\nu$ obtained in the present case lies well within the range (6–8) of a typical spin-glass systems; thus, one can conclude that Co_2SnO_4 exhibits spin-glass-like phase transition across 39 K just below the $T_N \sim 41$ K [20, 27]. In these studies, the difference in T_0 and T_{SG} is very small (~0.3 K) suggesting the close resemblance between the current Co_2SnO_4 system and the compounds exhibiting spin-glass-like transition. However, the situation for Co_2TiO_4 is bit different; in particular, the best fit to the Vogel-Fulcher law yields $T_0 = 45.8$ K and $\tau_0 = 3.2 \times 10^{-16}$ s and the power law yielded fairly unphysical values of the fitting parameters: for example, $\tau_0 \sim 10^{-33}$ s with $z\nu > 16$, indicating the lack of spin-glass-like phase transition [33]. Although, the magnitude and shift of the ac-susceptibility values both $\chi'(T)$ and $\chi''(T)$ strongly suppressed in the presence of dc-magnetic field (H_{DC}) in a similar way as it occurs in a typical spin-glass system perfectly following the linear behavior of $H^{2/3}$ versus T_P (AT-line analysis). Under such tricky situation, it is very difficult to conclude that Co_2TiO_4 is a perfect spin-glass or not (of course one can call it as a pseudo-spin-glass system). Nevertheless, the ac-magnetic susceptibility data and its analysis suggested that the both Co_2SnO_4 and Co_2TiO_4 systems consist of interacting magnetic clusters close to a spin-glass state.

Another interesting feature of both Co_2SnO_4 and Co_2TiO_4 is that they show asymmetry in M-H hysteresis loops unveiling giant coercivities and bipolar exchange bias under both ZFC and FC cases below their T_N [27, 31, 33]. Earlier studies by Hubsch et al. have shown unusual temperature dependence of coercive field $H_C(T)$ in polycrystalline Co_2TiO_4 sample where the M-H loops were measured in 20-kOe field at different temperatures below 60 K [31]. It is well known that the discovery of exchange-bias (H_{EB}) effect in the structurally single-phase materials with mixed magnetic phases has recently gained tremendous attention because of its technological applications in the development of Read/Write heads of the magnetic

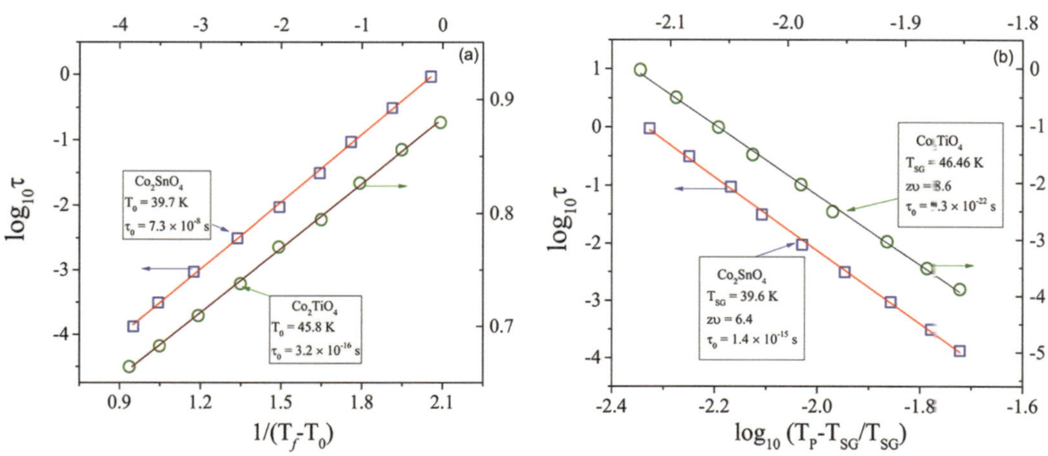

Figure 8. The best fit of the relaxation times to the (a) Vogel-Fulcher law and the (b) power law for the spinels Co_2SnO_4 and Co_2TiO_4.

recording devices [41]. Generally, H_{EB} has been experimentally observed only in the systems cooled in the presence of external magnetic field (FC mode) from above the Néel temperature or spin-glass freezing point. Such systems usually comprise of variety of interfaces such as ferromagnetic (FM)-antiferromagnetic (AFM), FM-SG, FM-ferrimagnetic, AFM-ferrimagnetic, and AFM-SG [41–49]. However, few recent papers have reported significant H_{EB} even under the zero-field-cooled samples of bulk Ni-Mn-In alloys and in bulk Mn_2PtGa [48, 49]. The source of such unusual H_{EB} under zero-field-cooled sample was attributed to the presence of complex magnetic interfaces such as ferrimagnetic/spin-glass or AFM/spin-glass phases [48–51]. Some recent reports have suggested that large exchange anisotropy can originate from the exchange interaction between the compensated host and ferromagnetic clusters [48–51]. Strikingly, Hubsch et al. observed the $H_C \rightarrow 0$ anomalies across T_{CMP}, T_{SG}, and T_N in the temperature-dependent data of H_C for Co_2TiO_4 samples [31]. Slightly different results were reported by Nayak et al. in Ref. [33], where H_C values drops monotonically on approaching T_N. However, the behavior of temperature dependence of exchange-bias field $H_{EB}(T)$ and remanent magnetization $M_R(T)$ in Ref. [33] closely resembles with the trend of $H_C(T)$ reported by Hubsch et. al. in polycrystalline Co_2TiO_4 samples. On the other hand, the temperature behavior of $H_{EB}(T)$, $H_C(T)$, and $M_R(T)$ in Co_2SnO_4 is way different from that of Co_2TiO_4 though they are isostructural with each other. It is likely that the different magnitudes and different temperature dependences of the moments on the Co^{2+} ions on the "A"- and "B"-sites in Co_2SnO_4 are responsible for the anomalous behavior in the $H_{EB}(T)$, $H_C(T)$, and $M_R(T)$ observed below T_N. Moreover, in Co_2SnO_4 below about 15 K, the data suggest that there is nearly complete effective balance of the antiferromagnetically coupled Co^{2+} moments at the"A"- and"B"-sites leading to negligible values of H_C and M_R.

Results from neutron diffraction in Co_2TiO_4 suggested the presence of canted-spins, likely resulting from magnetic frustration caused by the presence of nonmagnetic Ti^{4+} ions on the"B"-sites. A similar canting of the spins might be present in Co_2SnO_4 although neutron diffraction studies are needed to verify this suggestion [27, 31, 33]. Other Co-based spinel compounds that display the reversal in the orientation of the magnetic moments along with negative magnetization due to the magnetic-compensation phenomena are $CoCr_2O_4$ and $Co(Cr_{0.95}Fe_{0.05})_2O_4$ [3, 52, 53].

4. Magnetic properties of bulk and nanocrystalline Co_2MnO_4

Among various Co_2XO_4 (X = Mn, Ni, Co, Zn, etc.) spinels, Co_2MnO_4 has retained a unique place. In particular, Mn- and Co-based spinel oxides have gained considerable interest in the recent past due to their numerous applications in the Li-ion batteries [54, 55], sensors [56–58], thermistors [59], energy-conversion devices [60], and as a catalyst for the reduction of nitrogen oxides [61]. Moreover, Co_2MnO_4 nanocrystals have demonstrated outstanding catalytic properties for oxygen reduction reaction (ORR) and oxygen evolution reaction (OER) [60]. ORR and OER are the essential reactions in the electrochemistry-based energy-storage and energy-conversion devices. Co_2MnO_4 has shown superior catalytic activities compared to the commercial 30 wt% platinum supported on carbon black (Pt/C). Due to the special surface morphology [62], Co_2MnO_4 spinel is a very promising pseudo-capacitor material [63, 64]. Co_2MnO_4 has also demonstrated potential applications for protective coating on ferrite stainless steel interconnects in solid-oxide fuel cells (SOFCs) [65, 66]. Furthermore, colossal magnetoresistance (CMR) has been observed in the Mn- and Co-based spinel oxides [67].

In addition to the novel catalytic properties, Co_2MnO_4 spinel exhibits intriguing magnetic properties. Lotgering first observed the existence of ferromagnetic ordering in Co_2MnO_4 spinels [68]. Ríos et al. systematically studied the effect of Mn concentration on the magnetic properties of $Co_{3-x}Mn_xO_4$ solid solutions prepared by spray pyrolysis [69]. As we have discussed in Section 2 that Co_3O_4 has an antiferromagnetic order with $T_N = 30$ K. When we replace Co in Co_3O_4 by Mn cation, large ferromagnetic ordering appears, which ultimately dominates the antiferromagnetic ordering at $x = 1$. Therefore, pure Co_2MnO_4 shows ferromagnetic behavior and this has been confirmed by detailed magnetic measurements [70, 71]. However, for value of x other than 0 (Co_3O_4) and 1 (Co_2MnO_4), both ferromagnetic and antiferromagnetic magnetic exchange interactions coexist in the system, which yields a ferrimagnetic ordering in the $Co_{3-x}Mn_xO_4$ ($0 < x < 1$) solid solutions [72]. Tamura performed pressure-dependent study of Curie temperature (T_C) and found that T_C decreases with increase in pressure [71].

Pure Co_2MnO_4 possess a cubic inverse-spinel structure: $(B^{3+})[A^{2+}B^{3+}]O_4$ (shown in **Figure 9**). In an ideal case, the octahedral [B] sites are occupied by divalent cations together with half of the trivalent cations and the rest of the trivalent cations occupy the tetrahedral (A) sites in the spinel structure. However, due to the presence of different oxidation states of Mn and Co cations at (A) and [B] sites, the actual cationic distribution of Co_2MnO_4 is very complex and it has been a topic of considerable debate [59, 70–84]. Different sample preparation conditions also play an important role in the cationic distribution of Co_2MnO_4. On the basis of X-ray diffraction, electrical conductivity, magnetic, physiochemical, and neutron diffraction measurements, several different cationic distributions for nonstoichiometric $Co_{3-x}Mn_xO_4$ ($0 < x < 1$) have been reported in literature [59, 70–84]. From the magnetic measurements, Wickham and Croft proposed the following cationic distribution: $Co^{2+}[Co^{3+}_{2-x}Mn^{3+}_x]O_4 (0 < x < 2)$ for solid solutions of $Co_{3-x}Mn_xO_4$ systems obtained after the thermal decomposition of co-precipitated manganese and cobalt salts [72]. Later studies by Blasse suggested a different cationic distribution: $Co^{2+}[Co^{2+}_{2-x}Mn^{4+}_x]O_4$ [73]. Based on the physicochemical properties, Ríos et al. proposed a more complicated cationic distribution: $Co^{2+}_{0.88}Mn^{2+}_{0.12}[Co^{3+}_{0.87}Co^{2+}_{0.22}Mn^{3+}_{0.09}Mn^{4+}_{0.76}\blacksquare_{0.06}]O_4$ for Co_2MnO_4 powder samples prepared by thermal decomposition of nitrate salts [74]. From the neutron diffraction and magnetic measurements, Boucher et al. reported the cationic distribution $Co^{2+}[Co^{3+}_{2-x}Mn^{3+}_x]O_4$ similar to the one reported by Wickham and Croft [72, 75]. Yamamoto et al. [76] performed neutron diffraction measurements on Co_2MnO_4 oxides prepared by chemical methods at low temperatures, and reported $Mn_{1-n}Co_n[Mn_xCo_{2-n}]O_4$ as atomic distribution. Here, n is the inversion parameter of the inverse-spinel structure. Gautier et al. suggested two different possible cationic distributions: $Co^{2+}[Co^{3+}Mn^{2+}_{0.35}Mn^{3+}_{0.29}Mn^{4+}_{0.36}]O_4$ and

$Co^{2+}[Co^{3+}_{0.95}Mn^{2+}_{0.015}Mn^{3+}_{0.50}Mn^{4+}_{0.485}\blacksquare_{0.05}]O_4$ [77, 78]. On the contrary, the electrical conductivity measurements suggest a different cationic distribution: $Co^{2+}[Co^{2+}_{x}Co^{3+}_{2(1-x)}Mn^{4+}_{x}]O_4$, and $Co^{2+}[Co^{2+}Mn^{4+}]O_4$ or $Co^{3+}[Co^{3+}Mn^{2+}]O_4$ [59, 79] for Co_2MnO_4 spinel. Aoki studied the phase diagram and cationic distribution of various compositions of manganese and cobalt mixed spinel oxides [80]. He further investigated the effect of temperature and Mn concentration on the structure of manganese-cobalt spinel oxide systems. Control over morphology, crystallite site, grain size, and specific surface area of Co_2MnO_4 powders can be achieved by thermal decomposition of precursors in a controlled atmosphere [81, 82].

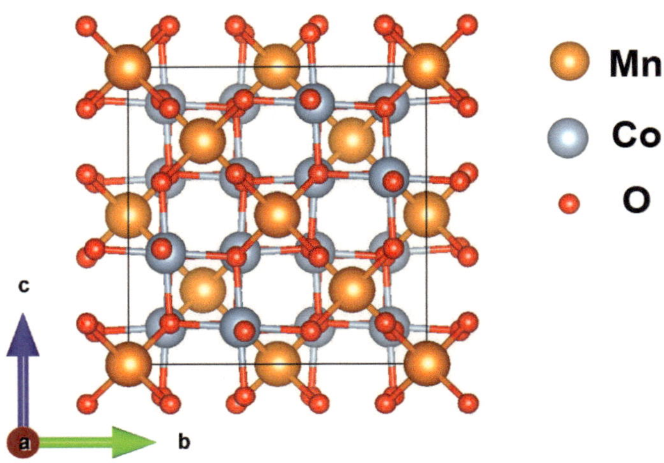

Figure 9. Cubic inverse-spinel structure of Co_2MnO_4 in Fd-3m (227) space group.

In 2008, Bazuev et al. investigated the effect of oxygen stoichiometry on the magnetic properties of $Co_2MnO_{4+\delta}$ [83]. Accordingly, two oxygen-rich compositions, (i) $Co_2MnO_{4.62}$ and (ii) $Co_2MnO_{4.275}$, were prepared by the thermal decomposition of presynthesized Co and Mn binary oxalates ($Mn_{1/3}Co_{2/3}C_2O_4 \cdot 2H_2O$). Above studies also report existence of anomalous behavior in the magnetic properties of $Co_2MnO_{4.275}$ spinel at low temperatures and high magnetocrystalline anisotropy in $Co_2MnO_{4.62}$. Bazuev et al. also noticed that T_N of $Co_2MnO_{4+\delta}$ is highly sensitive to the oxygen stoichiometry, imperfections in the cationic sublattice and variation in the Mn oxidation states. The imperfect $Co_2MnO_{4+\delta}$ inherits ferrimagnetic ordering that arises due to the antiferromagnetic exchange between Co^{2+} ($e_g^4 t_{2g}^3$) and Mn^{4+} ($t_{2g}^3 e_g^0$)) cations located at the tetrahedral and octahedral sites, respectively. To further investigate the electronic states of Mn and Co cations in the Co_2MnO_4 lattice, Bazuev at al. employed the X-ray absorption near-edge spectroscopy (XANES) to probe the electronic states of the absorbing atoms and their local neighborhood [83]. XANES spectra of both $Co_2MnO_{4.62}$ and $Co_2MnO_{4.275}$ compositions revealed that Co is present in both Co^{2+} and Co^{3+} oxidation states while Mn is present as Mn^{4+} and Mn^{3+}. Additionally, it was found that Mn is located in a higher symmetry octahedral crystal field environment [83]. Bazuev at al. further proposed following cationic distributions for $Co_2MnO_{4.275}$ and $Co_2MnO_{4.62}$ compositions: $Co^{2+}_{0.936}[Co^{III}_{0.936}Mn^{3+}_{0.421}Mn^{4+}_{0.515}]O_4$ and $Co^{2+}_{0.66}[Mn^{4+}_{0.866}Co^{III}_{1.072}]O_4$, respectively. Here, Co^{III} is a low-spin cation while Co^{2+} and Mn^{3+} are high-spin cations.

Co_2MnO_4 spinels have gained peculiar interest of researchers due to their unusual magnetic hysteresis behavior at low temperatures. Joy and Date [70] first observed the unusual magnetic hysteresis behavior in Co_2MnO_4 nanoparticles below 120 K. By means of the magnetic hysteresis loop measurements at low temperatures, they realized that the initial magnetization curve (virgin curve) lies outside the main hysteresis loop at 120 K. However, for $T > 120$ K, they observed normal hysteresis behavior in Co_2MnO_4. This unusual behavior of hysteresis loop at low temperatures can be associated with the irreversible domain wall motion. At low magnetic fields ($H < H_C$), domain walls experience substantial resistance during their motion with increasing magnetic field. Similar behavior has been observed for some other alloys [85–87]. Such irreversible movement of domain walls was ascribed to the rearrangement of valence electrons at low temperatures. Generally, in a system with mixed oxidation states of Mn at high temperatures, short-range diffusion of Co ions, $Co^{3+} \leftrightarrows Co^{2+}$ associated with $Mn^{3+} \leftrightarrows Mn^{4+}$, gets activated at low temperatures. This may cause change in the local ordering of the ions at the octahedral sites. Consequently, the resistance for domain wall motion increases and this slows down the motion of domain walls. Soon after Joy and Date, Borges et al. [84] confirmed that the unusual hysteresis of Co_2MnO_4 compound indeed arises due to the irreversible domain wall motion. Borges et al. prepared various different size nanoparticles of Co_2MnO_4 by Pechini method and performed the magnetic hysteresis measurements at low temperatures [84]. They observed that the samples below a critical diameter ($d < 39$ nm) exhibit normal hysteresis behavior, while the bulk grain size samples ($d \sim 200$ nm) show unusual hysteresis behavior at low temperatures. Using the spherical particle model, Borges et al. further calculated the critical diameter (d_{cr}) of a single-domain wall in Co_2MnO_4 and obtained that $d_{cr} = 39$ nm [88]. Therefore, all particles with diameter $d < d_{cr}$ can be considered as a single-domain particle. Since particles with $d < d_{cr}$ show normal hysteresis while particles with $d > d_{cr}$ show unusual hysteresis behavior, one can conclude that the unusual hysteresis behavior is indeed due to the irreversible motion of the domain walls.

The dynamic magnetic properties of Co_2MnO_4 nanoparticles of average diameter 28 nm were reported by Thota et al. [89]. A detailed study of dc- and ac-magnetic susceptibility measurements of these nanoparticles reveals the low temperature spin-glass-like characteristics together with the memory and aging effects (**Figure 10**). **Figure 10** shows the temperature dependence of real and imaginary part of the ac-susceptibility ($\chi'(T)$ and $\chi''(T)$) of Co_2MnO_4 nanoparticles recorded at different values of frequencies (f) between 1 Hz and 1.48 kHz at a peak-to-peak amplitude of 1 Oe. Both $\chi'(T)$ and $\chi''(T)$ exhibit a sharp peak at the onset of ferromagnetic ordering ($T_C = T_1 = 176.4$ K) and a broad cusp centered at T_2 ($<T_C$).

The temperature at which both $\chi'(T)$ and $\chi''(T)$ attain the maximum value shifts toward high-temperature side as the frequency increases from 1 Hz and 1.48 kHz similar to spin-glass behavior. Such frequency dependence of $\chi'(T)$ and $\chi''(T)$ follows the Vogel-Fulcher law (Eq. (4) and insets of **Figures 10a(ii)** and **b(ii)**) and power law of critical slowing down (Eq. (5) and insets of **Figures 10a(i)** and **b(i)**). Least-square fit to these equation yields the following parameters (insets of **Figure 10**): interparticle interaction strength (T_0) = 162 K for T_1 (118 K for T_2) and relaxation time constant $\tau_0 = 6.18 \times 10^{-15}$ s for T_1 (4.4×10^{-15} s for T_2), critical exponent ($z\nu$) = 6.01 for T_1 (7.14 for T_2), and spin-glass transition temperature (T_{SG}) = 162.6 K for T_1 (119.85 K for T_2). Since the values of $z\nu$ for both the peaks T_1 and T_2 of Co_2MnO_4 nanoparticles lie well within the range (6–8) of a typical spin-glass systems, one can conclude that Co_2MnO_4 exhibits spin-glass-like phase transition across 162.6 K just below the $T_C \sim 176.4$ K [89]. **Figure 11a** shows the magnetization relaxation of Co_2MnO_4 nanoparticles under ZFC and FC protocols with a temperature quench to 70 K at H = 250 Oe [89]. The magnetization relaxation during the third cycle appears as a continuation of first cycle (**Figures 11b** and **c**). Relaxation of ZFC magnetization with temperature quenching confirms the existence of memory effects in Co_2MnO_4 nanoparticles. A noticeable wait-time (t_{wt}) dependence of magnetization relaxation (*aging*) at 50 K in both M_{ZFC} and M_{FC} was noticed in these Co_2MnO_4 nanoparticles, which further supports the presence of the spin-glass behavior observed in this system.

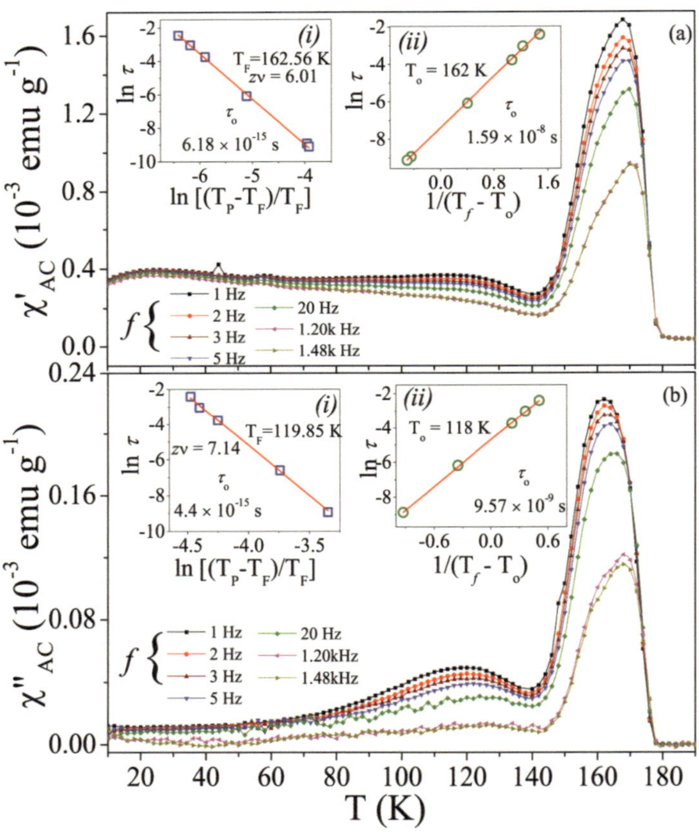

Figure 10. The real and imaginary components of the ac-magnetic susceptibilities ($\chi'(T)$ and $\chi''(T)$) of Co_2MnO_4 nanoparticles measured at various frequencies. The insets a(i) and b(i) represent the Vogel-Fulcher law whereas the insets a(ii) and b(ii) represent the power law of critical slowing down for both the peaks T_1 and T_2.

http://dx.doi.org/10.5772/65913

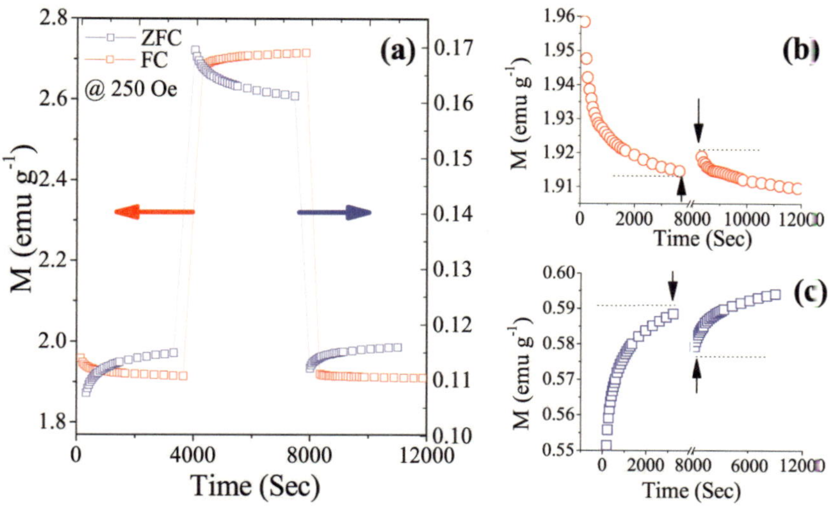

Figure 11. Magnetization relaxation $M(t)$ under ZFC and FC protocols with a temperature quench to 70 K at H = 250 Oe. The continuation of the first and third relaxation process during (b) ZFC and (c) FC cycles.

The high-temperature inverse magnetic susceptibility ($1/\chi_{ZFC}$ versus T) data of Co_2MnO_4 nanoparticles (**Figure 12**) fit well with the Néel's expression for ferrimagnets:

$$\frac{1}{\chi}=\frac{T}{C}+\frac{1}{\chi_0}-\left[\frac{\sigma_0}{T-\theta}\right] \tag{6}$$

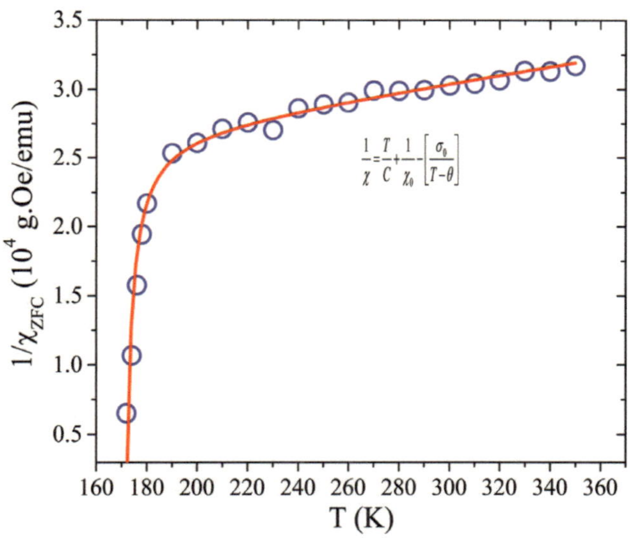

Figure 12. The scattered data represent the high-temperature inverse magnetic susceptibility ($1/\chi_{ZFC}$ versus T) of Co_2MnO_4 nanoparticles and the solid line represents the best fit to the Néel's expression for ferrimagnets Eq (6).

The fit (red line in **Figure 12**) yields: $C = 0.0349$ emu K/g Oe, $\chi_0 = 4.5 \times 10^{-5}$ emu/g Oe, $\sigma_0 = 5.76 \times 10^4$ g Oe K/emu, $\theta = 169.8$ K, and the asymptotic Curie temperature $T_a = C/\chi_0 = 775.6$ K [90, 91]. T_a gives us information about the strength of antiferromagnetic exchange coupling between Mn^{3+} and Mn^{4+} at octahedral sites. Moreover, the effective magnetic moment μ_{eff} calculated using expression: $C = \frac{N\mu_{eff}^2}{3k_B}$ turnouts to be 8.13 μ_B.

5. Concluding remarks

In this review, magnetic properties of bulk and nanoparticles of the Co-based spinels Co_3O_4, Co_2SnO_4, Co_2TiO_4, and Co_2MnO_4 have been summarized. The fact that the observed magnetic properties of these spinels are so different is shown to result from the different occupation of the cations on the A- and B-sites and their different electronic states at these sites. The richness of the properties of spinels thus results from these differences.

Acknowledgements

The authors thank Prof. M. S. Seehra for his suggestions and guidance in organizing this chapter. Sobhit Singh acknowledges the support from the West Virginia University Libraries under OAAF program.

Author details

Subhash Thota[1*] and Sobhit Singh[2]

*Address all correspondence to: subhashiitk@gmail.com

1 Department of Physics, Indian Institute of Technology, Assam, India

2 Department of Physics & Astronomy, West Virginia University, Morgantown, USA

References

[1] A. Goldman, Magnetic Components for Power Electronics, Springer (Nov. 2001).

[2] D. W. Bruce, Functional Oxides, Wiley (2010).

[3] R. Padam, S. Pandya, S. Ravi, A. K. Nigam, S. Ramakrishnan, A. K. Grover, and D. Pal, Appl. Phys. Lett. 102, 112412 (2013).

[4] K. De Stropper, C. Henriet-Iserentant, G. Robbrecht, and V. Brabers, C. R. Acad. Sci. Ser. B 277, 75 (1973).

[5] W. L. Roth, J. Phys. Chem. Sol. 25, 1 (1964).

[6] P. Dutta, M.S. Seehra, S. Thota and J. Kumar, J. Phys. Condens. Matter, 20, 015218 (2008).

[7] S. Gangopadhyay, G. C. Hadjipanayis, C. M. Sorensen, and K. J. Klabunde, J. Appl. Phys. 73, 6964 (1993).

[8] D. A. Resnick, K. Gilmore, Y. U. Idzerda, M. T. Klem, M. Allen, T. Douglas, E. Arenholz, and M. Young, J. Appl. Phys. 99, 08Q501 (2006).

[9] L. He, C. Chen, N. Wang, W. Zhou, and L. Guo, J. Appl. Phys. 102, 103911 (2007).

[10] E. L. Salabas, A. Rumplecker, F. Kleitz, F. Radu and F. Schuth, Nano. Lett. 6, 2977 (2006).

[11] M. Andok, T. Kobayashi, S. Iijima and M. Haruta, J. Mater. Chem. 7, 1779 (1997).

[12] H. Yamaura, J. Tamaki, K. Moriya, N. Miura and N. Yamazoe, J. Electrochem. Soc. 144, L158 (1997).

[13] P. Nkeng, J. Koening, J. Gautier, P. Chartier and G. Poillerat, J. Electroanal. Chem. 402, 81 (1996).

[14] S. G. Chrisoskova, M. Stoyanova, M. Georgieva and D. Mehandjiev, Mater. Chem. Phys. 60, 39 (1999).

[15] M. M. Natile and A. Glisenti, Chem. Mater. 14, 3090 (2002).

[16] T. Maruyama and S. Arai, J. Electrochem. Soc. 143, 1383 (1996).

[17] G. B. Smith, A. Ignatiev and G. Zajac, J. Appl. Phys. 51, 4186 (1980).

[18] K. Chidambaram, K. L. Malhotra and K. L. Chopra, Thin Solid Films 87, 365 (1982).

[19] L. M. Khriplovich, E. V. Kholopov and I. E. Paukov, J. Chem. Thermodyn. 14, 207 (1982).

[20] M. E. Fisher, Philos. Mag. 7, 1731 (1962).

[21] D. Sherrington and S. Kirkpatrick, Phys. Rev. Lett. 35, 1792 (1975).

[22] A. J. Bray and G. J. Rodgers, Phys. Rev. B, 38, 9252 (1988); A. J. Bray, Phys. Rev. Lett. 60, 720 (1988).

[23] M. Gabay and G. Toulouse. Phys. Rev. Lett. 47, 201 (1981); J. Villain, Z. Phys. B 33, 31 (1979).

[24] E. Harmon, D. J. Simkin, R. J. Haddad, and W. B. Muir, J. Phys. Colloq. 38, CI–131 (1977); AIP Conf. Proc. 29, 576 (1976).

[25] J. Srivastava and J. Kulkarni, J. Phys. C: Solid State Phys. 20, 2139 (1987).

[26] B. Antic, G. Goya, H. Rechenberg, V. Kusigerski, N. Jovic, and M. Mitric, J. Phys. Condens. Matter 16, 651 (2004).

[27] S. Thota, V. Narang, S. Nayak, S. Sambasivam, B.C. Choi, T. Sarkar, M. S. Andersson, R. Mathieu, and M. S. Seehra, J. Phys.Condens. Matter, 27, 166001 (2015)

[28] J. K. Srivastava et al., J. Appl. Phys. 61, 3622 (1987).

[29] K. De Strooper, Phys. Stat. Sol. A 39, 431 (1977).

[30] I. Mihailova, L. Radev, and D. Mehandjiev, Oxid. Commun. 35, 58 (2012).

[31] J. Hubsch and G. Gavoille, Phys. Rev. B 26, 3815 (1982).

[32] N. Sakamoto and Y. Yamaguchi, J. Phys. Soc. Jpn. 17, 276 (1962).

[33] S. Nayak, S. Thota, D. C. Joshi, M. Krautz, A. Waske, A. Behler, J. Eckert, T. Sarkar, M. S. Andersson, R. Mathieu, V. Narang, and M. S. Seehra, Phys. Rev. B 92, 214434 (2015).

[34] G. Gavoille, J. Hubsch, and S. Koutani, J. Magn. Magn. Mater. 102, 283 (1991).

[35] K. De Strooper and G. Robbrecht, Physica B 86, 934 (1977).

[36] S. Ogawa and S. Waki, J. Phys. Soc. Jpn. 20, 540 (1965).

[37] K. De Strooper, A. Govaert, C. Dauwe, and G. Robbrecht, Phys. Stat. Sol. A 37, 127 (1976).

[38] S. Thota and M. S. Seehra, J. Appl. Phys. 118, 129901 (2015).

[39] V. Sagredo, B. Watts and B. Wanklyn, J. Phys. IV France 7, C1-279 (1997).

[40] S. Thota and M. S. Seehra, J. Appl. Phys. 113, 203905 (2013).

[41] R. L. Stamps, J. Phys. D: Appl. Phys. 33, R247 (2000).

[42] W. H. Meiklejohn and C. P. Bean, Phys. Rev. 102, 1413 (1956).

[43] M. Ali, P. Adie, C. H. Marrows, D. Greig, B. J. Hickey, and R. L. Stamps, Nat. Mater. 6, 70 (2007).

[44] W. C. Cain and M. H. Kryder, J. Appl. Phys. 67, 5722 (1990).

[45] G. S. Alvarez, J. Sort, S. Surinach, M. D. Baro, and J. Nogues, J. Am. Chem. Soc. 129, 9102 (2007).

[46] W. H. Meiklejohn, J. Appl. Phys. 33, 1328 (1962).

[47] A. Punnoose, M. S. Seehra, J. Van Tol, and L. C. Brunel, J. Magn. Magn. Mater. 288, 168 (2005).

[48] B. M. Wang, Y. Liu, P. Ren, B. Xia, K. B. Ruan, J. B. Yi, J. Ding, X. J. Li, and L. Wang, Phys. Rev. Lett. 106, 077203 (2011).

[49] A. K. Nayak, M. Nicklas, S. Chadov, C. Shekhar, Y. Skourski, J. Winterlik, and C. Felser, Phys. Rev. Lett. 110, 127204 (2013).

[50] A. K. Nayak, M. Nicklas, S. Chadov, P. Khuntia, C. Shekhar, A. Kalache, et al, Nat. Mater. 14, 679, (2015).

[51] K. Guduru, A. Puri, U. Zeitler, J. M. D. Coey and C. Felser, Nat. Mater. 14, 679 (2015).

[52] F. C. Romeijn, Philips. Res. Rep. 8, 304 (1953).

[53] S. Nayak, D. C. Joshi, M. Krautz, A. Waske, J. Eckert and S. Thota, J. Appl. Phys. 119, 043901 (2016).

[54] H. Wang, Y. Yang, Y. Liang, G. Zheng, Y. Li, Y. Cui and Hongji, Energy Environ. Sci. 5, 7931 (2012).

[55] R. Alcantara, M. Jaroba, P. Lavela and J. L. Tirado, Chim. Mater. 1, S108 (2002).

[56] P. Umadevi and C. L. Nagendra, Sens. Actuator A Phys. 96, 114 (2002).

[57] Y. Shimizu and M. Shiotsuka, Jpn. J. Appl. Phys. 41, 6243 (2003).

[58] N. Nagatomo, K. Eudo and T. Yonezawa, Key Eng. Mater. 159/160, 305 (1999).

[59] E. Jabrya, A. Roussetb and A. Lagrange, Phase Trans 13, 13 (1988).

[60] Xiaoming Ge, Yayuan Liu, F. W. Thomas Goh, T. S. Andy Hor, Yun Zong, Peng Xiao, et al. ACS Appl. Mater. Interfaces 6, 12684 (2014).

[61] G. N. Pirogova, N. M. Panich, R. J. Korosteleva, Y. V. Voronin and N. N. Popova, Russ. Chem. Bull. 49, 1536 (2000).

[62] T. Nissinen, Y. Kiros, M. Gasik and M. Lampinen, Mater. Res. Bull. 39, 1195 (2004).

[63] J. Li, S. Xiong, X. Li and Y. Qian, Nanoscale 5, 2045 (2013).

[64] T. Nguyen, M. Boudard, L. Rapenne, O. Chaix-Pluchery, M. João Carmezim and M. F. Montemor, RSC Adv. 5, 27844 (2015).

[65] A. D. Sharma, J. Mukhopadhaya and R. N. Basu, ECS Trans. 35, 2509 (2011).

[66] Y. Xu, Z. Wen, S. Wang and T. Wen, Solid State Ion. 192, 561 (2011).

[67] J. Philip and T. R. N Kutty, Mater. Lett. 39, 311 (1999).

[68] F. K. Lotgering, Philip Res. Rep. 11, 337 (1956).

[69] E. Ríos, O. Peña, T. Guizouarn and J.-L. Gautier, Phys. Stat. Sol. C 1, S108 (2004).

[70] P. A. Joy and S. K. Date, J. Magn. Magn. Mater. 210, 31 (2000).

[71] S. Tamura, J. Phys. Soc. Jpn. 61, 752 (1992).

[72] D. G. Wickham and W. J. Croft, J. Phy. Chem. Sol. 7, 351 (1958).

[73] G. Blasse, Philip Res. Rep. Suppl. 18, 383 (1963).

[74] E. Ríos, P. Lara, D. Serafini, A. Restovic and J. L. Gautier, J. Chil. Chem. Soc. 55, 261 (2010).

[75] B. Boucher, R. Bhul, R. Di Bella and M. Perrin, (Paris) J. Phys. 31, 113 (1970).

[76] N. Yamamoto, S. Higashi, S. Kawano and N. Achiwa, J. Mater. Sci. Lett. 2, 525 (1983).

[77] J. L. Gautier, S. Barbato and J. Brenet, C. R. Acad. Sci. Paris 294, 427 (1982).

[78] J. L. Gautier and C. Cabezas, J. Electroanal. Chem. 159, 137 (1983).

[79] A. Restovic, E. Ríos, S. Barbato, J. Ortiz and J. L. Gautier, J. Electroanal. Chem. 522, 141 (2002).

[80] I. Aoki, J. Phys. Soc. Jpn. 17, 1 (1962).

[81] J. L. Gautier, E. Ríos, M. Gracia, J. F. Marco and J. R. Gancedo, Thin Solid Films. 311, 51 (1997).

[82] S. Trasatti, Electrochim. Acta 36, 225 (1991).

[83] G.V. Bazuev and A.V. Korolyov, J. Magn. Magn. Mater. 320, 2262 (2008).

[84] F. M. M. Borges, D. M. A. Melo, M. S. A. Camera, A. E. Martinelli, J. M. Soares, J. H. De Araujo and F. A. O. Cabral, J. Magn. Magn. Mater. 302, 273 (2006).

[85] J. L. Snoek, Physica 8, 426 (1941).

[86] C. Heck, Magnetic Materials and their Applications, Butterworths, London, p. 147 (1974).

[87] S. Senoussi, J. Phys. 45, 315 (1984).

[88] J. Smit and H. P. J. Wijin, Les ferrites, Bibl. Tech. Philips, Paris, (1961).

[89] S. Thota, S. K. Das, A. Kumar, S. Sangaraju and B. C. Choi, IEEE Trans. Magn. 49, 3 (2013).

[90] G. Srinivasan and M. S. Seehra, Phys. Rev. B 28, 1 (1983).

[91] S. K. Singh, M. S. Seehra and S. Thota, APS March Meeting Abstracts 1, 1284 (2013).

Chapter 5

Electronic, Transport and Magnetic Properties of Cr-based Chalcogenide Spinels

Chuan-Chuan Gu, Xu-Liang Chen and Zhao-Rong Yang

Additional information is available at the end of the chapter

http://dx.doi.org/10.5772/110953

Abstract

The Cr-based chalcogenide spinels with general formula ACr_2X_4 (A = Cd, Zn, Hg, Fe; X = S, Se) host rich physical properties due to coexistence of frustration as well as strong coupling among spin, charge, orbital and lattice degrees of freedom. In this chapter, recent advances on the study of electronic transport and magnetic properties of ACr_2X_4 are reviewed. After a short introduction of the crystal structure and magnetic interactions, we focus on the colossal magnetoresistance (CMR) in $FeCr_2S_4$, colossal magnetocapacitance (CMC) in $CdCr_2S_4$, negative thermal expansion (NTE) in $ZnCr_2Se_4$ and complex orbital states in $FeCr_2S_4$. It is hoped that this chapter will be beneficial for the readers to explore the interplay among different degrees of freedom in the frustrated system.

Keywords: Cr-based spinel, colossal magnetoresistance, colossal magnetocapacitance, negative thermal expansion, frustrated magnet, complex orbital states

1. Introduction

Chromium-based chalcogenide spinels with general formula ACr_2X_4 (A = Cd, Zn, Hg, Fe; X = S, Se) have attracted special attention not only for pure academic interest, but also for potential applications [1–4]. On the one hand, the chromium-based chalcogenide spinels are typical strongly correlated electron materials. Due to strong coupling among spin, charge, orbital and lattice degrees of freedom, this system displays rich physical properties such as colossal magnetoresistance (CMR), colossal magnetocapacitance (CMC), gigantic Kerr rotation, and magnetoelectric effect [5–8]. On the other hand, the existence of frustration (magnetic or orbital frustration) increases complexity of the system, yielding complex behaviors such as spin glass, orbital glasses, and spin nematic [9–11]. Depending on external perturbation from magnetic field, electric field, chemical substitution, or disorder, different quantum states are subtle balanced, increasing difficulty for understanding the intrinsic physics of this system. For example, the polycrystalline (PC) sample of $FeCr_2S_4$ displays orbital ordering around 9K, while single crystal (SC) sample shows orbital glass [12, 13]. $CdCr_2S_4$ single crystal exhibits multiferroic behavior with the evidence of relaxor ferroelectricity and CMC [6]. However, the emergence of ferroelectricity and CMC effect in these thio-spinels was found to be highly sensitive to the detail of sample preparation and chemical doping. Both multiferroicity and CMC effect are absent in PC samples [14].

In this chapter, recent progresses in the studies of Cr-based chalcogenide spinels are described. We will focus on the origin of several physical effects including CMR, CMC and NTE. The origin of complex spin and orbital states will also be covered.

2. Crystal structure and magnetic interactions in ACr_2X_4 (A= Cd, Zn, Hg, Fe; X = S, Se)

The chromium-based spinels ACr_2X_4 (A = Cd, Zn, Hg, Fe; X = S, Se) have the cubic symmetry with space group $Fd3'm$ (O_h^7), as is shown in **Figure 1**. Each cubic cell has 8 molecular formulae, which possess 56 ions, including 8 A^{2+} ions, 16 Cr^{3+} ions and 32 X^{2-} ions. The X^{2-} anions form a close-packed face-centered cubic lattice with two-type coordinated interstice: tetrahedral and octahedral ones. In the normal type spinel, the A^{2+} ions occupy 1/8 of the tetrahedrally coordinated interstices and the B^{3+}ions occupy 1/2 of the octahedrally coordinated interstices. Otherwise, it is defined as an inverse spinel type, in which the A^{2+} and B^{3+} cations exchange the sites. In some situations, the compound with the same chemical elements could be formed by normal and inverse type spinel simultaneously [15, 16].

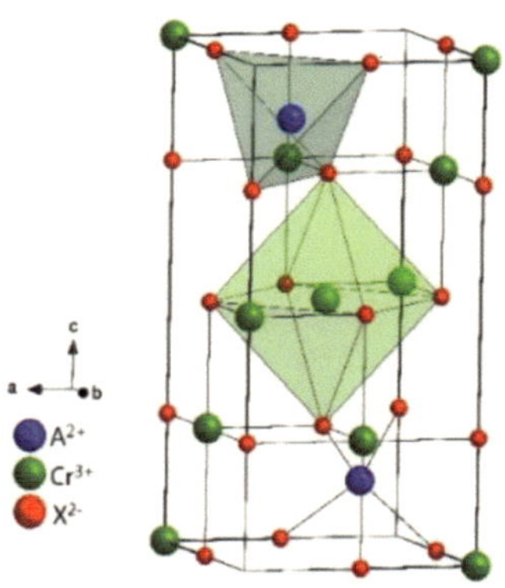

Figure 1. Schematic structure of spinel ACr_2X_4. A^{2+} ions are in tetrahedral and Cr^{3+} ions in octahedral site.

In ACr_2X_4, if magnetic cations (such as Mn^{2+}, Co^{2+}, Fe^{2+}) occupy the A site, the system usually displays a ferrimagnetic order with a net magnetization, due to the opposite spin direction of A and B sites. On the other hand, if A site is occupied by nonmagnetic ions (e.g., Zn^{2+}, Cd^{2+}, Hg^{2+}), the Cr^{3+} ions form a network of corner-sharing tetrahedral known as the pyrochlore lattice [17]. The dominant frustration in this network that hinders a simple antiparallel spin pattern is tightly associated with the way the Cr-Cr atoms are separated. The different ground states as a function of lattice constants are driven by the dominating exchange interactions: at small Cr-Cr separation, strong direct antiferromagnetic (AFM) exchange dominates, indicative of strong geometrical frustration [18]; at larger Cr-Cr separations, 90° Cr^{3+}-X^{2-}-Cr^{3+} ferromagnetic and Cr^{3+}-X^{2-}-A^{2+}-X^{2-}-Cr^{3+} or AFM Cr^{3+}-X^{2-}-X^{2-}-Cr^{3+} superexchange interactions come into play [1], with additional bond frustration dominanting in the sulfides and selenides.

3. Colossal magnetoresistance in $FeCr_2S_4$

The presence of the CMR effect in mixed valence manganese oxides $R^{3+}{}_{1-x}A^{2+}{}_xMnO_3$ (R is a rare earth and A is a divalent alkaline earth) in the vicinity of a paramagnetic (PM) to ferromagnetic (FM) transition has been the subject of much recent interest [19, 20]. This conductive behavior is described by the double-exchange (DE) model [21], which suggests that the conductivity is decided by the hopping of the electrons between hetrovalent Mn^{3+}/Mn^{4+} pairs [22]. Later, another CMR material—$FeCr_2S_4$ has been reported by Ramirez et al. [5]. Since the magneto-transport behaviors of $FeCr_2S_4$ are quite similar to manganite perovskites, one might expect to apply the DE theory and Jahn-Teller (JT) polaron mechanism to explain the CMR effects in $FeCr_2S_4$. However, the X-ray diffraction (XRD) pattern and Mössbauer spectrum give direct evidences that there is no DE mechanism in $FeCr_2S_4$ [23].

The magnetization (M) in the temperature range from 4.2 to 400 K at the field $H = 0.01$ T is shown in **Figure 2(a)**. With decreasing temperature, M first increases slowly, and then increases abruptly near T_C^{onset} (173 K) due to the PM-FM transition. The Curie temperature T_C is 168 K,

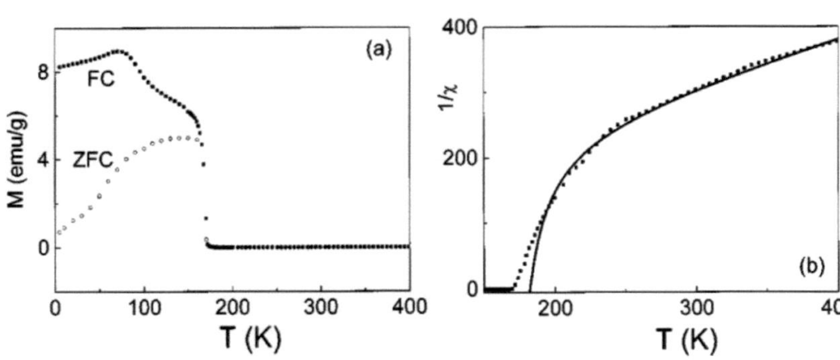

Figure 2. (a) Temperature dependence of the magnetization in field-cooled (FC) and zero-field-cooled (ZFC) sequences, respectively and (b) temperature dependence of the paramagnetic susceptibility χ as a plot of $1/\chi$ versus temperature [24].

which is defined as the temperature corresponding to maximum of |dM/dT|. To clarify the magnetism in $FeCr_2S_4$, the temperature dependence of the PM susceptibility χ is shown in **Figure 2(b)** as a plot of $1/\chi$ versus temperature. According to the physics of ferrimagnetism, the temperature dependence of the PM susceptibility χ can be described by Eq. (1) [25, 26]:

$$1/\chi = T/C + 1/\chi_0 - \sigma/(T-\theta). \quad (1)$$

Here χ_0, C, σ and θ are fitting parameters. The fit results of χ are shown as the solid line in **Figure 2(b)**, with $1/\chi_0 = 130(\pm2)$, $C = 1.50(\pm0.02)$, $\sigma = 3818(\pm50)$ and $\theta = 167(\pm3)$. Clearly, Eq. (1) fits the experimental result well except near T_C, indicating a ferrimagnet state. At temperatures near and above T_C, the system has strong short-range magnetic correlation, thus not in an ideal PM state.

The temperature-dependent resistivity ρ and magnetoresistance $MR_H = [\rho(0) - \rho(H)]/\rho(0)$ from 4.2 K to 300 K under three different applied fields (0, 3 and 5 T) are shown in **Figure 3**. Both ρ and MR_H display a peak near T_C^{onset}. The maximum of MR_5 is 16%. The temperature corre-

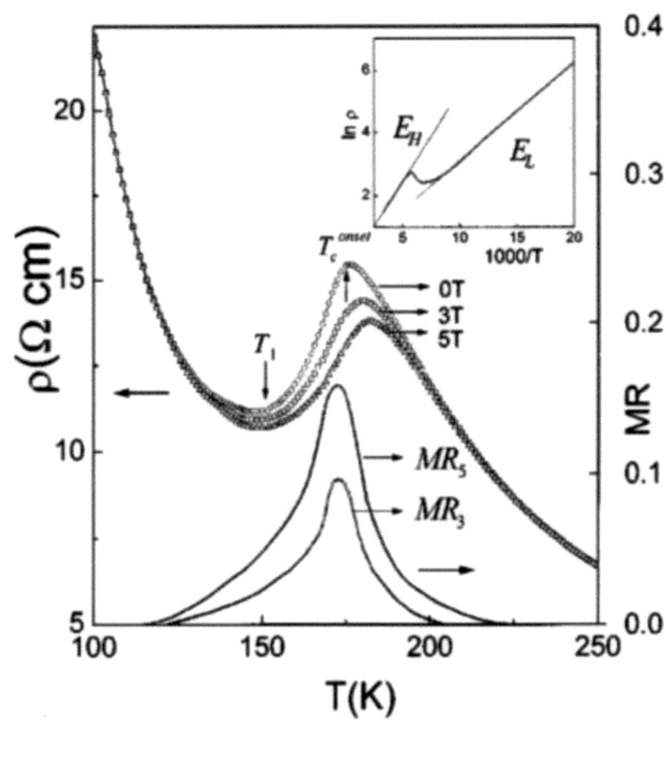

Figure 3. Resistivity ρ and magnetoresistance $MR_H = [\rho(0) - \rho(H)]/\rho(0)$, versus temperature for $FeCr_2S_4$. The inset shows the curve of ln ρ versus 1000/T [24].

sponding to the minimum resistivity in the ρ-T curve below T_C is defined as $T_1 = 153$ K. As seen from the inset of **Figure 3**, the curve of ln ρ versus 1000/T is linear at both $T > T_C^{onset}$ and $T < T_1$, indicating a semiconductor like behavior in two regions. The fits by $\rho = \rho_0 \exp(E/k_B T)$ give the activation energies E_L = 26 meV for $T < T_1$, and E_H = 47 meV for $T > T_C^{onset}$, respectively. This implies that the conduction mechanisms are different for the two temperature regions. $E_H > E_L$ indicates clearly that E_H, the activated energy at temperatures above T_C^{onset}, does not arise from a simple thermally activated effect. In order to understand the discrepancy between E_L and E_H, the thermoelectric power (S) is fitted by the experimental data through $S = (k_B/e)(a + E_S/k_B T)$ [24, 27]. E_S is obtained as 23 meV, which is approximately equal to 26 meV for E_L, and much lower than 47 meV for E_H. The discrepancy between E_S and E_H in manganite perovskites can be attributed to the presence of lattice or magnetic polarons. Since there is neither a structure transition nor a JT effect around T_C in $FeCr_2S_4$, the polarons mainly exist in the form of magnetic polarons [28].

The sample at temperatures above T_C^{onset} is in a typical PM state, in favor of the existence of magnetic polarons [24]. As is known, a magnetic polaron is a carrier coupled by short-range magnetic correlation within a magnetic cluster at temperatures above T_C^{onset}. Thus, the effective mass of a magnetic polaron increases greatly with respect to that of a naked carrier, which means it has a lower mobility and a higher activated energy [29, 30]. In an ideal FM order, magnetic polarons will be delocalized from the self-trapped state and turn into naked carriers completely [29]. In consequence, ρ decreases drastically at this point, accompanied with the system changing completely to one of thermal-activated transport of naked carriers. However, ρ decreases gradually from T_C^{onset} to T_1, and MR also expands to a broader temperature range below T_C^{onset}, indicative of the existence of magnetic polarons in this temperature range. Magnetic polarons will be delocalized gradually as the PM phase weakens. They then vanish completely at T_1, denoting that magnetic order destroys the environment for magnetic polarons. Hence, ρ in the temperature range from T_C^{onset} to T_1 can be described by a two-fluid model involving the coexistence of magnetic polarons and naked carriers. It could also explain that MR expands to a broader temperature range below T_C^{onset}, since the formation of magnetic polarons is inhibited by the enhancement of FM order at an external magnetic field.

To further investigate the evolution of magnetic polarons in $FeCr_2S_4$, nonmagnetic Cd is substituted for Fe. The XRD measurement for $Fe_{1-x}Cd_xCr_2S_4$ reveals that the substitution of Fe by Cd produces no structural change [31]. The resistivity ρ and magnetoresistance $MR_H = [\rho(0) - \rho(H)]/\rho(0)$, as a function of temperature are shown in **Figure 4**. Comparing with $FeCr_2S_4$, the Cd-containing sample has a higher zero-field resistivity. For $x = 0.2$, a peak is shown in both ρ and MR_H near T_C. Zero-field ρ reveals semiconductor-like behavior at temperatures for both $T > T_C$ and T far below T_C, which is similar to the results of $FeCr_2S_4$. Upon substitution of Cd for Fe, both the zero-field ρ and MR_H value increases monotonically, while the peak of the MR_H

curve shifts to lower temperatures. As mentioned above, it is believed that the conduction originates from Fe^{2+} narrow band at temperatures far below T_C, and is dominated by the magnetic polarons in the temperature region above T_C. For $Fe_{1-x}Cd_xCr_2S_4$, there are two sided influences of the introduced Cd^{2+} on the conduction behavior. On the one hand, since the conduction originates from Fe^{2+} narrow band, the increase of Cd ions will decrease carrier density, resulting in the increase of ρ. As is mentioned above, the magnetic polaron hopping conduction $\rho = \rho_0 T \exp(E/k_B T)$ should be considered for $T > T_C$. On the other hand, the introduced Cd ions will induce both Coulomb and magnetic potential fluctuations, as a sequence, leading to the formation of a mobility edge. Thus, thermally activated behavior $\rho = \rho_0 \exp(E/k_B T)$ due to a mobility edge may be more appropriate to describe the conduction above T_C [32, 33].

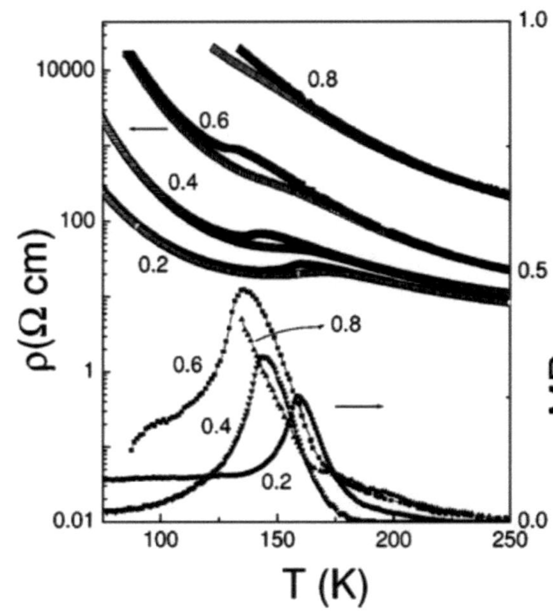

Figure 4. Resistivity ρ and magnetoresistance $MR_H = [\rho(0) - \rho(H)]/\rho(0)$, versus temperature for $Fe_{1-x}Cd_xCr_2S_4$ ($0 < x \leq 0.8$) [31].

Figure 5 shows the zero-field resistivity curves replotted as $\ln \rho - 1000/T$ and $\ln(\rho/T) - 1000/T$, respectively. For $x = 0.2$, the curve of $\ln \rho - 1000/T$ is linear both at $T > T_C$ and at temperatures far below T_C. The activation energy E, which is obtained from the fits to $\rho = \rho_0 \exp(E/k_B T)$, is 39 meV for the temperatures far below T_C, and 50 meV for $T > T_C$. As is known, the discrepancy of the activation energy for $FeCr_2S_4$ is described by a two-fluid model concerning the coexistence of magnetic polarons and naked carriers. As is shown in **Figure 5(a)**, magnetic polaron hopping conduction [$\rho = \rho_0 T \exp(E/k_B T)$] is more appropriate for the resistivity behavior at $T > T_C$. Moreover, it gives more convincible description as compared to thermal-activated mechanism for higher Cd concentration, which implies that the magnetic polarons may be more stable in the high-Cd samples.

http://dx.doi.org/10.5772/110953

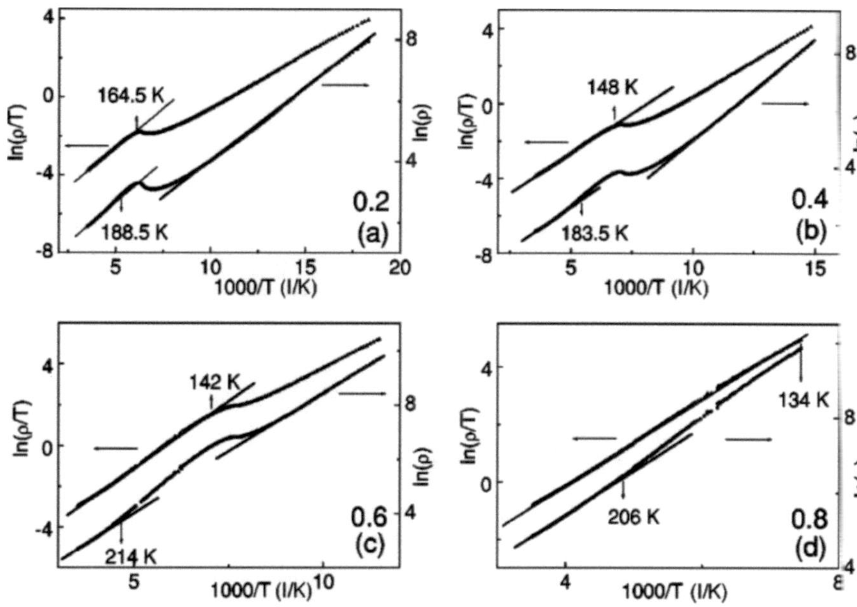

Figure 5. Zero-field resistivity curves for $Fe_{1-x}Cd_xCr_2S_4$ ($0 < x \leq 0.8$) replotted as $\ln \rho - 1000/T$ (closed circle •) and l $\ln(\rho/T) - 1000/T$ (solid square ■), respectively [31].

Direct evidence of the magnetic polarons could be best sought with electron spin-resonance spectroscopy (ESR) measurements [31]. Above $T^* \sim 113$ K, each spectrum consists of a single line with a Lorentzian line shape, and further decreasing temperature, some distortions occur. A similar behavior has been observed in manganese perovskites, which is attributed to the presence of the FM clusters embedded in PM matrix [34, 35]. In addition, as T is increased, the peak-to-peak line-width ΔH_{PP} decreases in a broad temperature range [31]. T_{min} (corresponding to the temperature where ΔH_{PP} is minimum) is much higher than T^*, suggesting that the relaxation mechanism dominating in $CdCr_2S_4$ is different with that in the manganese. The analysis of ESR experiments in CMR materials has pointed out that line-width ΔH_{PP} will decrease exponentially upon warming if the magnetic polaron acts as a part in the relaxation mechanism, as a result of the motionally narrowed relaxation mechanism [36]. Thus in $CdCr_2S_4$, since magnetic polaron is more stable upon Cd substitution, it is plausible that the hopping of magnetic polarons will make contribution to the relaxation process, and certainly decrease the line-width ΔH_{PP}. The decrease of line-width with increasing temperature obviously gives a direct evidence for the presence of magnetic polarons in $CdCr_2S_4$.

For a pure magnetic polaron system, the magnetic polaron forms by self-trapping in a ferromagnetically aligned cluster of spins at $T < T_P$ (T_P denotes the temperature where polarons start forming). As the temperature is lowered toward T_C, the polaron grows in size. According to the ML model [37], for a low carrier density ferromagnet, when $n\xi \approx 1$ is accomplished, magnetic polarons overlap, which leads to the carrier delocalization. Here n is the carrier density, and ξ is FM correlation length. It provides the temperature ratio of about $T/T_C \leq 1.05$–1.1 [37]. Therefore, for a pure magnetic polaron system, the magnetic polarons are periodically spaced before overlapping, similar to a gas of magnetic polarons. When the system is in a PM state, no magnetic correlation occurs among the polarons to be considered as superparamagnetic. On the other hand, if the system is inhomogeneous, there may be magnetic correlation, which means magnetic polarons themselves are no longer isolated. Due to the correlation, the system could not stay in the superparamagnetic state, which results in the existence of some distortions in the resonance line [34]. Hence, it can be concluded that magnetic polarons first exist as a gas of polarons at temperatures $T < T_P$, and then transform into correlated polarons at temperatures $T < T^*$ ($T_P > T^* > T_C$).

4. Colossal magnetocapacitance in $CdCr_2S_4$

Multiferroic materials that exhibit simultaneous magnetic and ferroelectric order as well as concomitant magnetoelectric coupling have attracted special interest in recent years [38–40]. $CdCr_2S_4$ was originally investigated as a FM semiconductor more than 40 years ago [41]. Recently, it was reported that SC $CdCr_2S_4$ exhibits multiferroic behavior with the evidence of relaxor ferroelectricity and CMC [6]. Soon after, similar magnetoelectric effect has also been revealed in other related thio-spinel compounds, that is, $CdCr_2Se_4$ and $HgCr_2S_4$ [42, 43]. However, the emergence of ferroelectricity and CMC effect in these thio-spinels was found to be highly sensitive to the detail of sample preparation and chemical doping. Annealing SC samples in vacuum or sulfur atmosphere led to a suppression of relaxation features, and no remanent electric polarization could be found at low temperatures [14]. In addition, multiferroicity and CMC effect are absent in undoped PC samples, but present in indium-doped PCs [14]. Accordingly, the question of whether the magnetoelectric effect in $CdCr_2S_4$ is intrinsic or not has been raised [44, 45]. In this section, the origin of CMC effect in the $CdCr_2S_4$ system would be revealed systematically through the magnetic, dielectric and electric transport measurements of $CdCr_2S_4$ and $Cd_{0.97}In_{0.03}Cr_2S_4$ PC samples before and after annealing in cadmium vapor.

Upon doping and annealing, no impurity phases are detected [46]. The *M-T* curves under an external magnetic field of 1 T for all samples are shown in the inset of **Figures 6(b)** and **7(b)**. After annealing in cadmium vapor, the magnetic properties of PC $CdCr_2S_4$ and $Cd_{0.97}In_{0.03}Cr_2S_4$ are slightly changed. The Curie temperature T_C is about 92.1 K for $CdCr_2S_4$ and 92.9 K for $Cd_{0.97}In_{0.03}Cr_2S_4$, respectively. After annealing, the increase of T_C is less than 4 K. **Figures 6** and **7** display the temperature dependence of dielectric constant ε' and *ac*-conductivity σ' at two different frequencies in magnetic field of 0 T and 4.5 T for the as-prepared and annealed samples. Both ε' and σ' increase with temperature monotonously and do not show anomaly and magnetic field dependence in the whole measured temperature range, as seen from **Figures 6(a, b)** and **7(a, b)**. After annealing $CdCr_2S_4$ at 800°C (**Figure 6(c, d)**), ε' displays a hump in the low temperature region and the hump temperature increases with frequency, implying a relaxor-like behavior of the dielectric property [14]. However, the magnetic field of 4.5 T still has no obvious effect on either ε' or σ'.

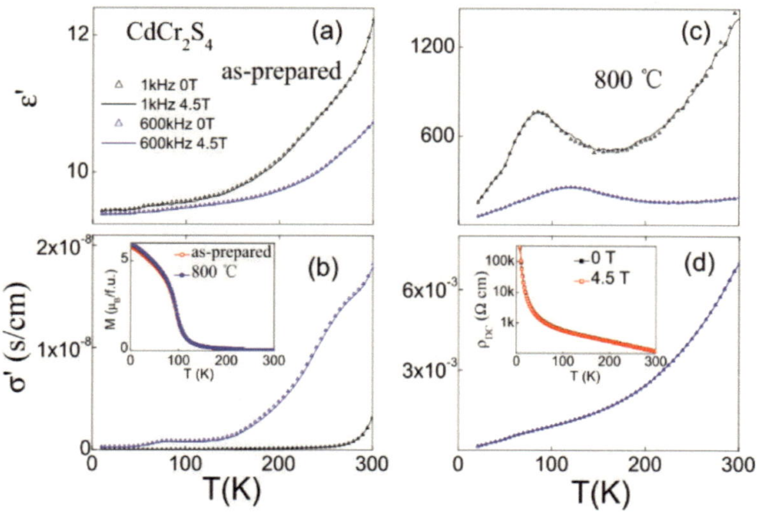

Figure 6. Temperature dependence of dielectric constant (upper frames) and *ac*-conductivity (lower frames) for the as-prepared and annealed $CdCr_2S_4$ samples. Inset of (b) and (d) show temperature dependence of magnetization and resistivity in magnetic field of 0 and 4.5 T for annealed $CdCr_2S_4$, respectively [46].

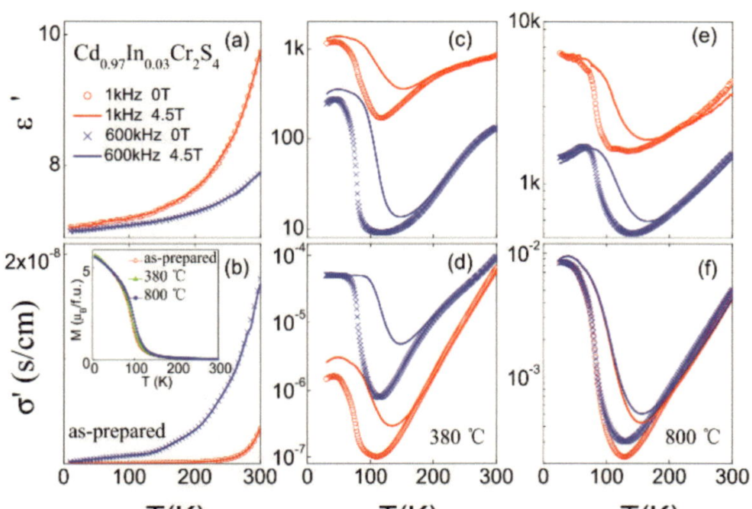

Figure 7. Temperature dependence of dielectric constant (upper frames) and *ac*-conductivity (lower frames) for the as-prepared and annealed $Cd_{0.97}In_{0.03}Cr_2S_4$ samples. Inset of (b) shows the temperature dependence of magnetization [46].

In contrast, for the $Cd_{0.97}In_{0.03}Cr_2S_4$ sample annealed at 380°C (**Figure 7(c, d)**), a strong upturn of ε' is clearly observed with decreasing temperature near T_C. σ' has a similar temperature dependence as ε'. The external magnetic field of 4.5 T does not change the shape of ε'-T and σ'-T curves, but makes the upturn of ε' and σ' shifting toward higher temperatures. Magnetocapacitance, defined as MC = (ε'(4.5 T) − ε'(0 T))/ε'(0 T), reaches up to 290% at 1 kHz and

1950% at 600 kHz (**Figure 8(a)**). The sample annealed at 800°C has similar temperature dependence of ε' and σ' as the sample annealed at 380°C, while has a much higher ε' and σ', and a lower MC value (reaches up to 145% at 1 kHz and 104% at 600 kHz, see **Figure 8(c)**). A clear feature in **Figures 6(c, d)**, and **7(c, f)** is that the appearance of CMC effect is accompanied by the field-enhanced ac conductivity. The inset of **Figure 6(d)** shows the temperature dependence of DC-resistivity under 0 T and 4.5 T for the annealed $CdCr_2S_4$ sample. In accordance with the field independence of the *ac*-conductivity, the application of 4.5 T magnetic field has no evident influence on the DC-resistivity. Upon cooling, the DC-resistivity increases monotonously as a typical semiconductor behavior in the whole temperature range measured. However, as seen in **Figure 8(b, d)**, the temperature dependence of DC-resistivity for the annealed $Cd_{0.97}In_{0.03}Cr_2S_4$ samples is quite different from that of $CdCr_2S_4$. For annealed $Cd_{0.97}In_{0.03}Cr_2S_4$, the zero-field resistivity first increases with decreasing temperature. After reaching up to a maximum near T_C, the resistivity decreases abruptly, indicating the occurrence of insulator-metal transition. Upon further cooling to the low temperature region, the resistivity increases again. Being correlated to the insulator-metal transition around T_C, the external magnetic field of 4.5 T makes the resistivity peak moving to a higher temperature and dramatically depresses the peak value. Magnetoresistance, defined as $MR = (\rho(0\,T) - \rho(4.5\,T))/\rho(0\,T)$, reaches up to about 95% for the sample annealed at 380°C and 93% for the sample annealed at 800°C, much higher than the value of the most-investigated CMR material $FeCr_2S_4$, see the insets of **Figure 8(b, d)** [5, 31].

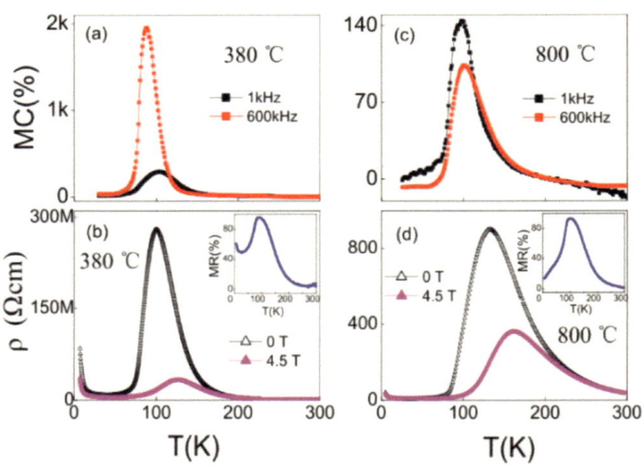

Figure 8. Temperature dependence of magnetocapacitance (upper frames) and resistivity (lower frames) for annealed $Cd_{0.97}In_{0.03}Cr_2S_4$. Insets of (b) and (d) display the magnetoresistance of the samples annealed at 380°C (left) and 800°C (right), respectively [46].

The appearance of CMC in the $CdCr_2S_4$ system is always accompanied by CMR. To confirm that the CMC can be described by a combination of CMR and Maxwell-Wagner effects, the impedance spectroscopy of $Cd_{0.97}In_{0.03}Cr_2S_4$ annealed at 380°C is fitted to a Maxwell-Wagner equivalent circuit. The impedance spectroscopy is converted from capacitance and dielectric loss tangent in the frequency range from 100 Hz to 600 kHz. The circuit consists of two

http://dx.doi.org/10.5772/110953

subcircuits in series, each containing a resistor (*R*) and a capacitor (*C*) in parallel, as depicted in the inset of **Figure 9(a)**. *Z* is better described by using the Cole-Cole equation: $Z = R/(1+(i\omega RC)^n)$, $(0 \le n \le 1)$ [47, 48]. The fitting was performed with all parameters unlocked. Take the data under zero magnetic field for the $Cd_{0.97}In_{0.03}Cr_2S_4$ sample annealed at 380°C for example. The impedance spectroscopy can be fitted well in the temperature range from 79 K to 189 K, see the solid lines in **Figure 9(a)**. The temperature-dependent total resistance $R_1 + R_2$ derived from the fitting coincides with the temperature dependence of the measured DC resistance, as shown in **Figure 9(b)**.

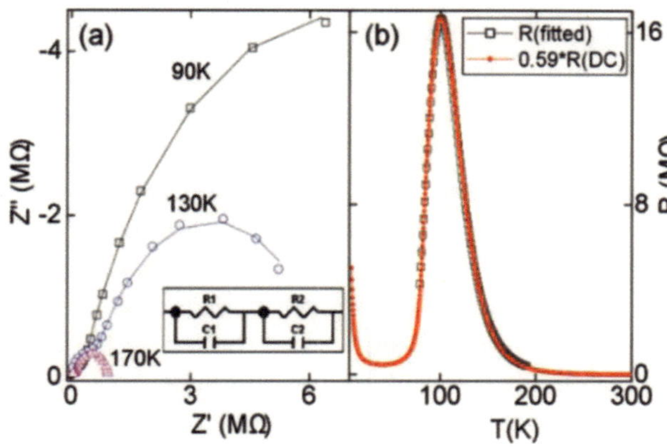

Figure 9. (a) The complex impedance plot at 90, 130 and 170 K for the $Cd_{0.97}In_{0.03}Cr_2S_4$ sample annealed at 380°C. The solid line is the fitting result based on the equivalent circuit displayed in the inset. (b) Temperature dependence of the fitted resistance and measured DC resistance (multiplied by 0.59) [49].

The perfect match between the resistance derived from the impedance spectroscopy and the DC resistance strongly suggests that the annealed $Cd_{0.97}In_{0.03}Cr_2S_4$sample contains two components with different electrical response. These two components can be a sample and an electrode, or a grain and a grain boundary. In these cases, an applied bias voltage can change the capacitance and the ac conductance [50–52]. Moreover, Hemberger et al. have experimentally excluded any electrode effects on the magnetodielectric response and excluded any nonhomogeneous impurity distribution in their SC $CdCr_2S_4$ [42, 45]. Thus, CMC effect could not be attributed to the extrinsic Maxwell-Wagner effect. In phase-separated manganite, strong magnetodielectric effect is resulted from the scenario of phase separation [53–56]. The thermoelectric-power and ESR measurements reveal the existence of magnetic polarons in the annealed $Cd_{0.97}In_{0.03}Cr_2S_4$ sample [49]. As is discussed in the above section, magnetic polaron is similar as the presence of the FM clusters embedded in PM matrix, indeed a kind of phase separation in nature [24, 57, 58]. For the exchange interaction with localized spins, the energy of the conduction electron in the FM cluster is smaller than that in the PM host, which results in a nanoscale charge and phase separation [57, 58]. Upon cooling, magnetic polarons grow in size, as a consequence, their overlap induces larger metallic clusters and finally a percolative metallic filament. After the insulator-metal transition, the coexistence of FM metallic and PM insulating phases is replaced by a complete FM order founded in all the regions. The interfacial polarization or space-charge yielded by the mixture of insulating and metallic regions would cause a dielectric response, which is intrinsic to the material. Therefore, the dielectric response can be attributed to a combination of magnetoresistance and an intrinsic Maxwell-Wagner effect, as observed in the phase-separated manganite [53–56].

5. Negative thermal expansion in $ZnCr_2Se_4$

Recently, $ZnCr_2Se_4$ has been observed to display NTE and a large magnetostriction [2]. To study the very origin of NTE, a set of experimental techniques is utilized to probe the spin-lattice correlation in $ZnCr_2Se_4$. **Figure 10(a)** shows the temperature dependence of lattice constant a. With decreasing temperature, a first decreases rapidly and then manifests a NTE behavior below about $T_E = 60$ K. Upon further cooling below 20 K, splitting of several peaks in XRD spectra is observed. **Figure 10(b)** presents the magnetization versus T at low-applied magnetic field of 100 Oe. A fitting of the inverse susceptibility $1/\chi$ according to the PM Curie-Weiss (CW) law $1/\chi = (T - \Theta_{CW0})/C$ is shown in **Figure 10(c)**. A large positive CW temperature $\Theta_{CW0} = 85$ K and the coefficient $C = 3.74$ are obtained. The large positive CW temperature implies a dominant FM exchange interaction; however, the compound shows an AFM ordering at low temperatures. On the other hand, both the constant g-factor and linear behavior of peak-to-peak linewidth (ΔH_{PP}) in the ESR spectrum evidence a well-defined PM state at least above 100 K [59]. Thus, the inverse susceptibility should be described by the PM CW law down to this temperature. However, it departs from the linear behavior at a temperature as high as about 180 K, see **Figure 10(c)**. Recalling the fitting process in **Figure 10(c)**, a constant CW temperature Θ_{CW0} has been assumed, that is, a constant magnetic exchange interaction J. In addition, the nearest neighbor FM Cr-Se-Cr and other neighbor AFM Cr-Se-Zn-Se-Cr super-exchange interactions depend strongly on the lattice constant [1]. It means that the total J may change since a decreases dramatically upon cooling. Accordingly, the traditional CW behavior should be modified within the present case. In specific, one should take a variable Θ_{CW} (or J) as a function of T or a into account.

In AFM spinel oxides, the CW temperature changes exponentially with the lattice parameter [60]. Naturally, an empirical description of $\Theta_{CW}(T) = \Theta_{CW0} - \Theta^* e^{-T/\beta}$ is postulated. The fitting of the inverse susceptibility above 100 K using the modified CW behavior $1/\chi = (T - \Theta_{CW}(T))/C$ is exhibited in **Figure 11(a)**. The parameters are $\Theta = 226$ and $\beta = 45$. Furthermore, a remarkable deviation below 100 K in blue short-dashed line indicates the appearance of the effective internal field originating from FM clusters. Next, based on the obtained Θ and β, Θ_{CW}(T) is extrapolated to low temperatures as exhibited in **Figure 11(b)**. It shows a derivation at about 180 K from the nearly constant value. With further lowering temperature, it decreases faster and faster and below $T \approx 45$ K, Θ_{CW}(T) even becomes negative. These features may interpret qualitatively the fact that $ZnCr_2Se_4$ is dominated by ferromagnetic exchange interaction but orders at low temperatures antiferromagnetically. Since the exchange integral and the CW temperature are linked by $J(T) \propto \Theta_{CW}(T)$ [$J(a) \propto \Theta_{CW}(a)$], we will use $J(T)$ [$J(a)$] instead in the following discussion. Given that J is changeable, magnetic exchange and lattice elastic energies can link

http://dx.doi.org/10.5772/110953

effectively with each other via magnetoelastic coupling. The free energy F in a magnetoelastic system is expressed as:

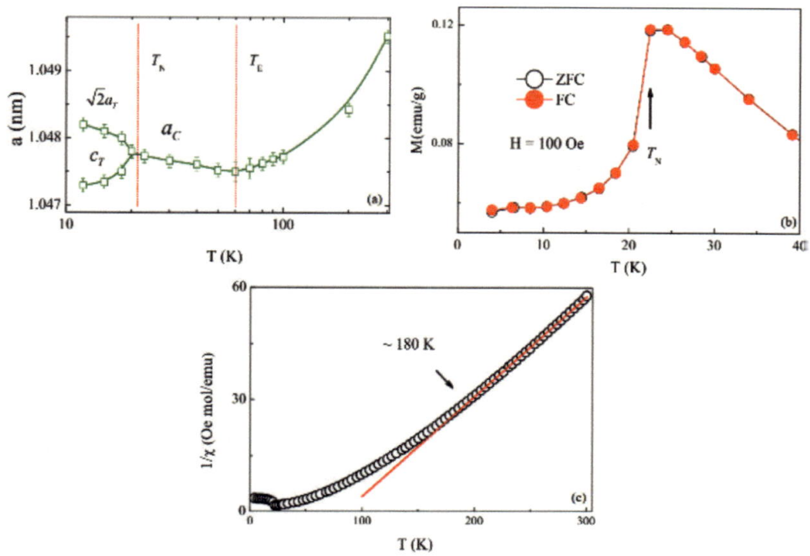

Figure 10. (a) Temperature dependent lattice parameter a versus T for $ZnCr_2Se_4$. (b) Low temperature dependence of the magnetization M at 100 Oe. (c) A Curie-Weiss fitting of the inverse susceptibility [59].

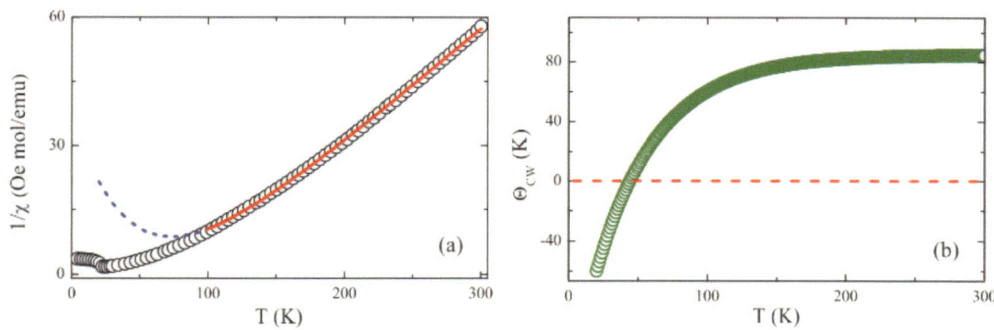

Figure 11. (a) A fitting of the inverse susceptibility above 100 K in red solid line using $1/\chi = (T - \Theta_{CW}(T))/C$ taking a temperature or lattice constant a dependent CW temperature into account. The short-dashed blue line is an extension of the fitting to low temperatures. (b) Θ_{CW} versus T and the red dashed line indicates $\Theta_{CW} = 0$ at 45 K [59].

$$F(T) = -J(T)\sum_{ij}\vec{S_i}\cdot\vec{S_j} + \frac{1}{2}N\omega^2\Delta^2(T) - T\cdot S(T) \tag{2}$$

where N is the number of the ion sites, ω is the averaged vibrational angular frequency, and Δ is the averaged strain relative to the equilibrium lattice constant. The first term is exchange energy (E_{ex}) as a function of J, the second is lattice elastic energy (E_{el}) related mainly to the lattice parameter a (or equivalent T), and the last is the entropy. From the above equation, if the system stays at an ideal PM state, then $E_{ex} = 0$. So E_{ex} and E_{el} will show no coupling.

When the system stays in FM state with a changeable J, there may exist a competition between E_{ex} and E_{el} since the former is negative while the latter always is positive. Indeed, it has been concluded above that J decreases exponentially **[Figure 11(a)]** and some FM clusters form gradually below 100 K. Therefore, the concomitant decrease of J and a causes an increasing E_{ex} but decreasing E_{el} in the FM clusters. At some critical point, a totally compensation between them may present. If a further decreases upon cooling, the variation of E_{ex} would gradually exceed that of E_{el} in magnitude. Especially when J drops sharply with respect to a, say here at T_E, a tiny decrement of a will give rise to a dramatic increment of E_{ex}, whereas E_{el} keeps nearly constant. The state in a system subjected to stimuli, such as cooling, always tends to develop toward one that can lower F. In this sense, it is favorable to lowering F by expanding the lattice parameter a to increase J ($J > 0$), and thereby to decrease E_{ex} due to its negative value, at the same time at a cost of small increases of E_{el} in magnitude. This means that a NTE of the lattice originating from the FM clusters with an exponentially changeable J is expected. It is noted that when applying a magnetic field to the system in the NTE temperature region, magnitude of NTE enhances [2]. This is because the size or population of the FM clusters increases when applying a magnetic field. On the contrary, when an AFM ordering appears, J becomes negative and the condition to stimulate NTE is not met any more. Normal thermal expansion upon cooling results in a simultaneous decreasing of E_{ex} and E_{el}, which is consistent to lowering F. In fact, a normal expansion feature is observed below T_N [2]. It should be noted that the existence of NTE evidences in turn that J is changeable [2, 61]. If J keeps constant in a FM cluster, E_{ex} will be almost constant and a normal thermal expansion of the lattice alone can give rise to a decrease of E_{el} and of F sufficiently.

The thermal expansion data for $ZnCr_2(Se_{1-x}S_x)_4$ ($0 \le x \le 0.1$) SC samples is represented as the function of $\delta = [L(T) - L_0]/L_0$ from 4 K to 300 K in **Figure 12(a)**, where L_0 is the length of the sample at room temperature. For $ZnCr_2Se_4$, as T is lowered, δ first decreases monotonically and reaches a minimum at $T_m \sim 50$ K. Then δ begins to increase, followed by a steep downturn upon further cooling across T_N. For the S-doped samples, the striction below T_N becomes gentler, while the temperature region of the NTE is expanded. As seen from the inset of **Figure 12(a)**, T_N moves to lower temperatures with increasing S content while T_m shifts oppositely. To further probe the nature of spin-lattice coupling in the complex magnetic state, the field dependence of both magnetization and magnetostriction δ at 5 K for all the $ZnCr_2(Se_{1-x}S_x)_4$ ($0 \le x \le 0.1$) samples are illustrated in **Figure 12(b, c)**. δ is defined as $\delta = [L(H) - L(0)]/L(0)$, and $L(0)$ is the length of the sample at zero field. Due to strong spin-lattice coupling, a one-to-one correlation between M-H and δ-H is clearly reflected in **Figure 12(b, c)**. At both of the characteristic fields of H_{C1} and H_{C2}, δ also displays an anomaly. Being associated to domain reorientation, hysteresis is also observed below H_{C1} as the magnetic field returns to zero, see **Figure 12(b, c)**. For $x = 0.05$ and 0.10, field dependence of magnetization is still inconsistent with the magnetostriction data. With increasing S content, the hysteresis behavior below H_{C1} wanes in both M-H and δ-H, and completely vanishes at $x =$

http://dx.doi.org/10.5772/110953

0.10 [9]. In addition, the magnetization and magnetostriction data are both suppressed by an order of magnitude.

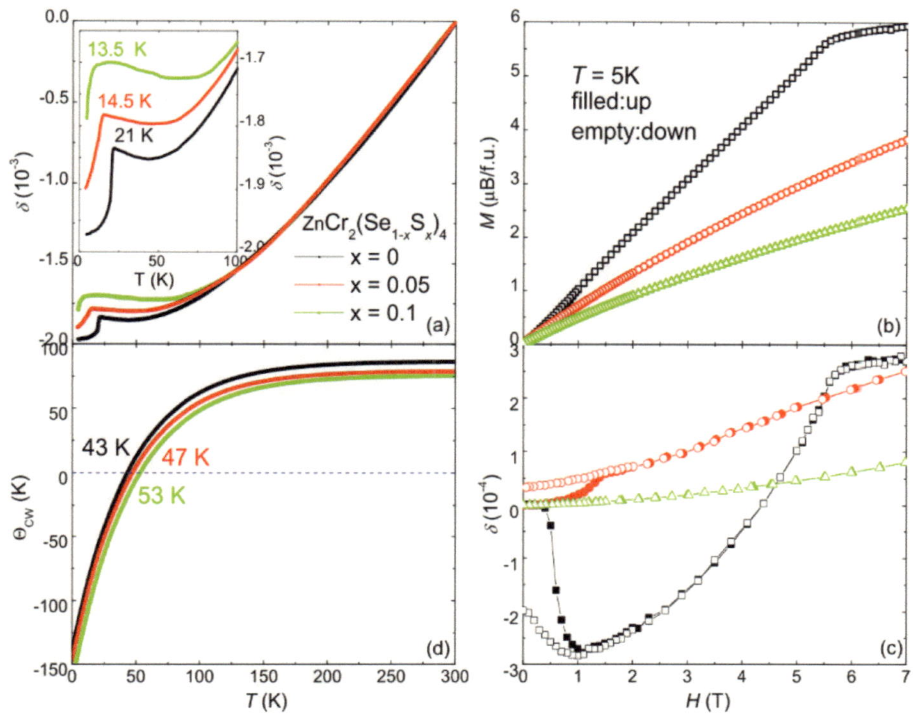

Figure 12. (a) The thermal expansion plotted as $\delta = [L(T) - L_0]/L_0$ for SC $ZnCr_2(Se_{1-x}S_x)_4$ ($0 \leq x \leq 0.1$) from 5 K to 300 K, where L_0 is the length of the sample in a paramagnetic state at 300 K. The inset shows the enlarged view of the low-temperature region. (b) Magnetic field dependence of magnetization; (c) Magnetostriction plotted as $\delta = [L(H) - L(0)]/L(0)$ at 5 K for SC $ZnCr_2(Se_{1-x}S_x)_4$ ($0 \leq x \leq 0.1$), where $L(0)$ is the length of each $ZnCr_2(Se_{1-x}S_x)_4$ ($0 \leq x \leq 0.1$) sample at zero field. (d) The variable CW temperature plotted as $\Theta_{CW}(T)$ versus T [9].

To investigate the NTE and magnetostriction in S-substituted samples, the inverse susceptibility was analyzed by the same method as is proposed above, which is shown in **Figure 12(d)** [59]. For $ZnCr_2Se_4$, $\Theta_{CW}(T)$ drops dramatically and changes sign at 43 K, close to the onset temperature of NTE. The substitution of S for Se drives Θ_{CW0} to lower temperature on account of an enhancement of the AFM exchange interaction, while the temperature corresponding to the sign change in $\Theta_{CW}(T)$ is increased. It could explain why the temperature region of the NTE is extended in $ZnCr_2(Se_{1-x}S_x)_4$ ($0 < x \leq 0.1$) samples. The substitution of sulfur for selenium not only enhances the AFM exchange interaction, but also leads to a Cr-Cr bond disorder [9]. The bond disorder frustrates the spins and finally leads to a spin-glass state. In a spin-glass state, since the spin arranges or freezes randomly upon cooling to lower temperature, the magnetic exchange and lattice elastic energy are decoupled. Note that, in the S-substituted samples, potential striction or shrinkage of the lattice can still be observed upon cooling across T_N, which implies that a partial long-range AFM ordering still manifests and coexists with the spin-glass state.

6. Complex orbital states in spinel $FeCr_2S_4$

Due to strong frustration, the spinel $FeCr_2S_4$ not only shows the fascinating CMR effect, but also displays complex orbital states [12, 13]. Recently, it was reported to display the orbital glass state in SC samples, while it displays orbital ordering in PC ones [12, 13]. In $FeCr_2S_4$, the Fe^{2+} ion is tetrahedrally coordinated by the sulfur ions, with an electronic configuration $e_g^3 t_{2g}^3$ and $S = 2$, is a typical JT active. Thus, it might induce long-range orbital order at low temperatures [62, 63]. In addition, the diamond lattice formed by Fe ions is geometrically frustrated for the orbital degrees of freedom [12, 13]. Thus, Fichtl et al. proposed two possible explanations to interpret the discrepancy of different orbital states in $FeCr_2S_4$ [12]. One is that the orbital glass state is attributed to geometric frustration in the SC, and broken in the PC by marginal disorder. Alternatively, the orbital ordering in the SC is suppressed by small disorder (such as chlorine defects from the growth process). To verify the role of disorder in these two possibilities, the physical properties of some SC and doped PC $FeCr_2S_4$ are presented in detail next.

Figures 13(a) and **14** display the magnetization as a function of temperature obtained in ZFC and FC processes with $H = 0.005$ T. All samples display clear anomalies around 65–75 K and irreversibility between ZFC and FC curves. The cusp-like anomaly was early correlated to an abrupt increase of magnetic anisotropy below $T_m \sim 70$ K, and recently attributed to a spin-reorientation transition associated with the onset of short-range orbital ordering [64, 65]. Moreover, the PC $FeCr_2S_4$ displays a step-like anomaly in the magnetization around $T_{OO} \sim 9$ K in **Figure13(a)**. For the other samples, the transition is replaced by a smooth temperature dependence, which is shown in the insets of **Figure 14**. The step-like anomaly for PC and smooth temperature dependence for other ones can be related to the orbital-ordering (OO) transition and orbital glass state, respectively [66]. A similar temperature-dependent magnetization between SC and doped PC samples suggests that the orbital order might also be frozen in these samples. Temperature dependence of specific heat for PC $FeCr_2S_4$ is shown in **Figure 13(b)**. It displays a well-defined λ-type anomaly around 9 K, indicative of the OO transition, which is consistent with the step-like transition in magnetization [12, 13]. For the other samples, the λ-type anomaly is entirely inhibited and becomes a broad hump, see **Figure 14(e)**. In addition, as T is lowered below 2 K, the heat capacity approaches zero following a strict T^2-dependence in the insets of **Figure 14(e)**. In contrast, the specific heat of PC deviates from the linear behavior in low temperatures below 2 K, shown in the inset of **Figure 13(b)**. The earlier reports pointed out that the phonon and magnon contribution to the specific heat of SC could be ignorable as compared to the orbital contribution at ultra-low temperatures [12, 13]. Thus, the existence of the orbital glass state in both SC and doped samples is plausible, since heat capacity of orbital order suppressed by random fields obeys the T^2-dependence predicted by the theoretical calculation [67]. Clearly, all the results reveal that the orbital moment has been frozen into the orbital glass state in the SC and doped PC samples.

For a different orbital state, the orientation of the orbital is coupled to the elastic response of the ionic lattice via electron-phonon interaction, and therefore a different orbital state is accompanied by a different charge distribution [68]. The resistivity curves in the function of $\ln \rho$-$1000/T$ and $\ln \rho$-$(1000/T)^{1/4}$ are shown in **Figures 13(c)** and **15**, respectively. The conduction mechanism for PC $FeCr_2S_4$ has been studied in detail in the previous section. In the low temperatures below 37 K, the resistivity could be better described by a semiconductor-like behavior. Upon warming, a curvature emerges, which might be related to the formation of short-range orbital order state [12]. For the other samples, the Mott's variable-range hopping is more appropriate to fit the feature of resistivity. As we know, for variable-range hopping, it is necessary that the system exists random potential fluctuations, being related to the degrees of disorder [32]. Hence, the presence of random potential could be the reason for the emergence of orbital glass state [12, 13].

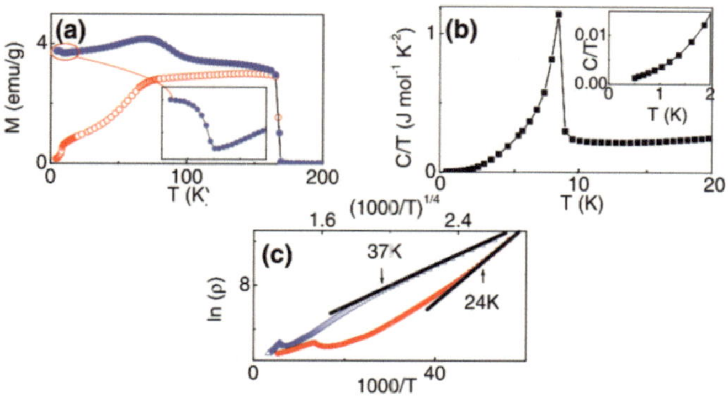

Figure 13. (a) Temperature dependence of ZFC (red circle) and FC (blue circle) magnetization in H = 0.005 T for PC $FeCr_2S_4$. The inset shows an enlarged view at low temperatures. (b) The specific heat plotted as C_P/T versus T for PC $FeCr_2S_4$. The inset shows an enlarged view at temperatures below 2 K. (c) Resistivity curves in the function of ln ρ – 1000/T (blue triangle) and ln ρ-$(1000/T)^{1/4}$ (red circle) for PC $FeCr_2S_4$. The solid lines are a guide to eyes [10].

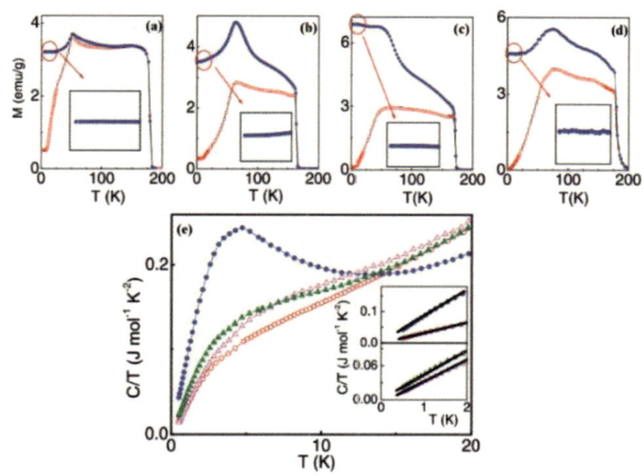

Figure 14. Temperature dependence of magnetization at 0.005 T for (a) SC $FeCr_2S_4$, (b) $FeCr_{1.9}Al_{0.1}S_4$, (c) $FeCr_{1.95}Ga_{0.05}S_4$ and (d) $Fe_{1.05}Cr_{1.95}S_4$. The insets show enlarged views of the magnetization at low temperatures. (e) The specific heat in the function of C_P/T versus T for SC $FeCr_2S_4$ (red circle), $FeCr_{1.9}Al_{0.1}S_4$ (blue circle), $FeCr_{1.95}Ga_{0.05}S_4$ (violet triangle) and $Fe_{1.05}Cr_{1.95}S_4$ (olive triangle). The insets show enlarged views below 2 K. The solid lines are a guide to eyes [10].

In conclusion, complex orbital states in $FeCr_2S_4$ are driven by the coexistence of strong electron-phonon coupling and geometrical frustration. The different orbital states in PC and SC $FeCr_2S_4$ reveal the exquisite balance between frustration and strong coupling among different degrees of freedom. The magnetism, heat capacity and resistivity properties of all the samples give clear evidence that the disorder in SC and doped samples raises random potential up, freezes the orbital order and finally results in orbital glass state.

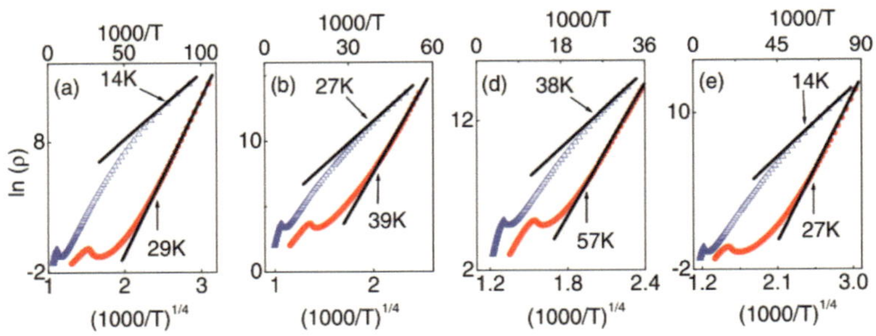

Figure 15. Resistivity curves in the function of ln ρ – 1000/T (blue triangle) and ln ρ – $(1000/T)^{1/4}$ (red circle) for (a) SC $FeCr_2S_4$, (b) $FeCr_{1.9}Al_{0.1}S_4$, (c) $FeCr_{1.95}Ga_{0.05}S_4$ and (d) $Fe_{1.05}Cr_{1.95}S_4$. The solid lines are a guide to eyes [10].

7. Conclusion

Due to the presence of frustration as well as strong coupling among spin, charge, orbital and lattice degrees of freedom, Cr-based chalcogenide spinels display rich physical effects and complex ground states. There is still some open questions such as the nature of spin nematic, correlation between orbital state and magnetic structure, which should be addressed in future under the conditions of high pressures and high magnetic fields.

Acknowledgements

This work is financially supported by the National Natural Science Foundation of China under Grant nos. U1332143 and 11574323.

Author details

Chuan-Chuan Gu, Xu-Liang Chen and Zhao-Rong Yang*

*Address all correspondence to: zryang@issp.ac.cn

High Magnetic Field Laboratory, Chinese Academy of Sciences, Hefei, China

References

[1] P. K. Baltzer, P. J. Wojtowicz, M. Robbins and E. Lopatin, *Phys. Rev.* 151, 367 (1966).

[2] J. Hemberger, H.-A. K. v. Nidda, V. Tsurkan and A. Loidl, *Phys. Rev. Lett.* 98, 147203 (2007).

[3] L. Q. Yan, J. Shen, Y. X. Li, F. W. Wang, Z. W. Jiang, F. X. Hu et al., *Appl. Phys. Lett.* 90, 262502 (2007).

[4] I. Kim, Y. S. Oh, Y. Liu, S. H. Chun, J.-S. Lee, K.-T. Ko et al., *Appl. Phys. Lett.* 94, 042505 (2009).

[5] A. P. Ramirez, R. J. Cava and J. Krajewski, *Nature* 386, 156 (1997).

[6] J. Hemberger, P. Lunkenheimer, R. Fichtl, H.-A. K. v. Nidda, V. Tsurkan and A. Loidl, *Nature* 434, 364 (2005).

[7] K. Ohgushi, T. Ogasawara, Y. Okimoto, S. Miyasaka and Y. Tokura, *Phys. Rev. B* 72, 155114 (2005).

[8] K. Siratori and E. Kita, J. Phys. Soc. Jpn. 48, 1443 (1980).

[9] C. C. Gu, Z. R. Yang, X. L. Chen, L. Pi and Y. H. Zhang, *J. Phys. Conden. Matt.* 28, 18LT01 (2016).

[10] R. Tong, Z. R. Yang, C. Shen, X. B. Zhu, Y. P. Sun, L. Li et al., *EPL* (Europhysics Letters) 89, 57002 (2010).

[11] V. Felea, S. Yasin, A. Günther, J. Deisenhofer, H.-A. Krug von Nidda, S. Zherlitsyn et al., *Phys. Rev. B* 86, 104420 (2012).

[12] R. Ficht, V. Tsurkan, P. Lunkenheimer, J. Hemberger, V. Fritsch, H.-A. Krug von Nidda et al., *Phys. Rev. Lett.* 94, 027601 (2005).

[13] N. Büttgen, J. Hemberger, V. Fritsch, A. Krimmel, M. Mücksch, H.-A. K. v. Nidda et al., *New J. Phys.* 6, 191 (2004).

[14] J. Hemberger, P. Lunkenheimer, R. Fichtl, S. Weber, V. Tsurkan and A. Loidl, *Phase Transit.* 79, 1065 (2006).

[15] Z. R. Yang, S. Tan and Y. H. Zhang, *Phys. Rev. B* 64, 024401 (2001).

[16] Z. R. Yang, S. Tan and Y. H. Zhang, *Phys. Rev. B* 65, 184404 (2002).

[17] S.-H. Lee, C. Broholm, W. Ratcliff, G. Gasparovic, Q. Huang, T. H. Kim et al., *Nature* 418, 856 (2002).

[18] S.-H. Lee, C. Broholm, T. H. Kim, W. Ratcliff II and S.-W. Cheong, *Phys. Rev. Lett.* 84, 3718 (2000).

[19] M. McCormack, S. Jin, T. H. Tiefel, R. M. Fleming, J. M. Phillips and R. Ramesh, *Appl. Phys. Lett*. 64, 3045 (1994).

[20] S. Jin, T. H. Tiefel, M. McCormack, R. A. Fastnacht, R. Ramesh and L. H. Chen, *Science* 264, 413 (1999).

[21] C. Zener, *Phys. Rev*. 82, 403 (1951).

[22] H. Y. Hwang, S-W. Cheong, N. P. Ong and B. Batlogg, *Phys. Rev. Lett*. 77, 2041 (1996).

[23] Z. W. Chen, S. Tan, Z. R. Yang and Y. H. Zhang, *Phys. Rev. B* 59, 11172 (1999).

[24] Z. R. Yang, S. Tan, Z. W. Chen and Y. H. Zhang, *Phys. Rev. B* 62, 13872 (2000).

[25] G. Srinivasan and M. S. Seehra, *Phys. Rev. B* 28, 1 (1983).

[26] L. Néel, *Proc. Phys. Soc. Sect. A* 65, 869 (1952).

[27] M. F. Hundley and J. J. Neumeier, *Phys. Rev. B* 55, 11511 (1997).

[28] A. P. Ramirez, *J. Phys. Conden. Matt*. 9, 8171 (1997).

[29] M. Rubinstein, D. J. Gillespie, J. E. Snyder and T. M. Tritt, *Phys. Rev. B* 56, 5412 (1997).

[30] Y. Shapira, S. Foner, N. F. Oliveira and T. B. Reed, *Phys. Rev. B* 10, 4765 (1974).

[31] Z. R. Yang, X. Y. Bao, S. Tan and Y. H. Zhang, *Phys. Rev. B* 69, 144407 (2004).

[32] M. Viret, L. Ranno and J. M. D. Coey, *Phys. Rev*. B 55, 8067 (1997).

[33] G. J. Snyder, R. Hiskes, S. DiCarolis, M. R. Beasley and T. H. Geballe, *Phys. Rev. B* 53, 14434 (1996).

[34] S. L. Yuan, W. Y. Zhao, G. Q. Zhang, F. Tu, G. Peng, J. Liu et al., *Appl. Phys. Lett*. 77, 4398 (2000).

[35] F. Rivadulla, M. Freita-Alvite, M. A. López-Quintela, L. E. Hueso, D. R. Miguéns, P. Sande et al., *J. Appl. Phys*. 91, 785 (2002).

[36] D. L. Huber, *J. Appl. Phys*. 83, 6949 (1998).

[37] P. Majumdar and P. Littlewood, *Phys. Rev. Lett*. 81, 1314 (1998).

[38] S.-W. Cheong and M. Mostovoy, *Nat. Mater*. 6, 13 (2007).

[39] M. Fiebig, *J. Phys. D: Appl. Phys*. 38, R123 (2005).

[40] W. Eerenstein, N. D. Mathur and J. F. Scott, *Nature* 442, 759 (2006).

[41] P. K. Baltzer, H. W. Lehmann and M. Robbins, *Phys. Rev. Lett*. 15, 493 (1965).

[42] J. Hemberger, P. Lunkenheimer, R. Fichtl, S. Weber, V. Tsurkan and A. Loidl, *Phys. B: Conden. Matt*. 378–380, 363 (2006).

http://dx.doi.org/10.5772/110953

[43] S. Weber, P. Lunkenheimer, R. Fichtl, J. Hemberger, V. Tsurkan and A. Loidl, *Phys. Rev. Lett*. 96, 157202 (2006).

[44] C. J. Fennie and K. M. Rabe, *Phys. Rev. B* 72, 214123 (2005).

[45] G. Catalan and J. F. Scott, *Nature* 448, E4 (2007).

[46] Y. M. Xie, Z. R. Yang, L. Li, L. H. Yin, X. B. Hu, Y. L. Huang et al., *J. Appl. Phys*. 112, 123912 (2012).

[47] J. R. Macdonald, *Ann. Biomed. Eng*. 20, 289 (1992).

[48] K. S. Cole and R. H. Cole, *J. Chem. Phys*. 9, 341 (1941).

[49] Y. M. Xie, Z. R. Yang, Z. T. Zhang, L. H. Yin, X. L. Chen, W. H. Songet al., *EPL* (Europhysics Letters) 104, 17005 (2013).

[50] W. Li and R. W. Schwartz, *J. Am. Ceram. Soc*. 90, 3536 (2007).

[51] C.-C. Wang, M.-N. Zhang, K.-B. Xu and G.-J. Wang, *J. Appl. Phys*. 112, 034109 (2012).

[52] N. Biškup, A. d. Andrés and J. L. Martinez, *Phys. Rev. B* 72, 024115 (2005).

[53] G. Catalan, *Appl. Phys. Lett*. 88, 102902 (2006).

[54] M. M. Parish and P. B. Littlewood, *Phys. Rev. Lett*. 101, 166602 (2008).

[55] J. Rivas, J. Mira, B. Rivas-Murias, A. Fondado, J. Dec, W. Kleemannet al., *Appl. Phys. Lett*. 88, 242906 (2006).

[56] J. A. Souza, R. F. Jardim, R. Muccillo, E. N. S. Muccillo, M. S. Torikachvili and J. J. Neumeier, *J. Appl. Phys*. 89, 6636 (2001).

[57] V. G. Storchak, J. H. Brewer, P. L. Russo, S. L. Stubbs, O. E. Parfenov, R. L. Lichti et al., *J. Phys. Conden. Matt*. 22, 495601 (2010).

[58] E. L. Nagaev, *Phys. Rep*. 346, 387 (2001).

[59] X. L. Chen, Z. R. Yang, W. Tong, Z. H. Huang, L. Zhang, S. L. Zhang et al., *J. Appl. Phys*. 115, 083916 (2014).

[60] T. Rudolf, C. Kant, F. Mayr, J. Hemberger, V. Tsurkan and A. Loidl, *New J. Phys*. 9, 76 (2007).

[61] F. Yokaichiya, A. Krimmel, V. Tsurkan, I. Margiolaki, P. Thompson, H. N. Bordallo et al., *Phys. Rev. B* 79, 064423 (2009).

[62] Y. Tokura and N. Nagaosa, *Science* 288, 462 (2000).

[63] P. G. Radaelli, *New J. Phys*. 7, 53 (2005).

[64] Z. R. Yang, S. Tan and Y. H. Zhang, *Appl. Phys. Lett*. 79, 3645 (2001).

[65] C. Shen, Z. R. Yang, R. Tong, G. Li, B. S. Wang, Y. P. Sun et al., *J. Magn. Magn. Mater.* 321, 3090 (2009).

[66] V. Tsurkan, V. Fritsch, J. Hemberger, H.-A. K. v. Nidda, N. Büttgen, D. Samusi et al., *J. Phys. Chem. Solids* 66, 2036 (2005).

[67] M. A. Ivanov, V. Y. Mitrofanov, L. D. Falkovskaya and A. Y. Fishman, *J. Magn. Magn. Mater.* 36, 26 (1983).

[68] R. Fichtl, P. Lunkenheimer, J. Hemberger, V. Tsurkan and A. Loidl, *J. Non-Cryst. Solids* 351, 2793 (2005).

Chapter 6

Structure-Property Correlations and Superconductivity in Spinels

Weiwei Xie and Huixia Luo

Additional information is available at the end of the chapter

http://dx.doi.org/10.5772/65943

Abstract

In this chapter, alternative views based on the structure have been presented in the spinel superconducting compounds, including the only oxide spinel superconductor, $LiTi_2O_4$, and non-oxide superconductors, $CuIr_2S_4$ and CuV_2S_4. Inspection of the atomic arrangements, electronic structures and bonding interactions of spinel superconductor, $LiTi_2O_4$ shows that $LiTi_2O_4$ can be interpreted as Li-doped TiO_2, which is similar with doping Cu into $TiSe_2$ to induce superconductivity. Different from $LiTi_2O_4$, the electronic structures of $CuIr_2S_4$ and CuV_2S_4 indicate a distinctive way to understand them in the structural viewpoint. The d^6 electron configuration and the octahedral coordination of Ir in $CuIr_2S_4$ can be analogous to the d^6 in perovskites, which sometimes host a metal-insulator transition. However, the superconductivity in CuV_2S_4 may be induced from the suppression of charge density waves. This kind of structural views will help chemists understand physical phenomena obviously more straightforward, though not sufficient, as clearly shown by the competition between each other, such as superconductivity and other structural phase transition (CDWs), oxidation fluctuation or magnsetism.

Keywords: superconductivity, chemical bonding, crystal structural analysis

1. Introduction

Superconducting phenomenon incorporates the exact zero electrical resistance and expulsion of magnetic flux fields occurring in many solid state materials when cooling below a certain critical temperature [1]. The expulsion of the magnetic flux fields, known as Meissner effect, and zero electric resistance has tremendous applications in the fields of transportation, electricity, and so on [2]. The "ideal" superconducting materials potentially could solve the most energy problems human being is facing. Back to the discovery of superconductivity in mercury in 1911, a century has passed by. However, the mechanisms of superconductivity are still undergoing extraordinary scrutiny. The conventional pictures arising from the Bardeen-Cooper-Scrieffer (BCS) theory merge the electron-phonon coupling to generate a pairing mechanism between electrons with the opposite crystal momenta that induce a superconducting state [3,4]. Derived from the BCS theory, the qualitative correlation between the superconducting critical temperature (T_c) and the density of states (DOS) at the Fermi level, $N(E_F)$, is $kT_c = 1.13\ \hbar\ \omega\ \exp(-1/N(E_F)V)$, where V is a merit of the electron-phonon interaction and ω is a characteristic phonon frequency, similar to the Debye frequency [5]. According to the expression, a large density of states at Fermi level $N(E_F)$ or electron-phonon interaction V or both leads to a higher T_c superconductor. Later, Eliashberg and McMillan extended the BCS theory and gave a better correspondence between experiment and prediction [6].

Until now, BCS theory is still in an even more dominant position to determine whether a superconductor will be classified as BCS-like or not. As more high temperature superconductors were discovered, more new "universal" mechanisms were sought for. However, neither BCS nor other exotic mechanisms established a relationship with the real chemical systems. Thus, the question appears whether the general statements of BCS theory can be associated with distinct chemical meanings, such as specific bonding situations, and whether the physical phenomenon of superconductivity can be interpreted from the viewpoint of chemistry.

2. Electron counting rules in chemistry and superconductors

Empirical observations of the range of electron counts to specific structural compounds are widely used in chemistry to help determine and find out the empirical rules to stabilize the compounds with specific structural frameworks, such as Wades-Mingo polyhedral skeletal rules for boron cluster compounds [7], $14e^-$ rules for DNA-like helix Chimney Ladders phases [8] and Hume-Rothery rules for multi-shelled clustering γ-brass phases [9, 10]. The electronic structure calculation and the bonding schemes allow us to determine a structure's preferred electron count for most compounds, for example, take Hume-Rothery rules in complex clustering compounds. The stability ranges of complex intermetallic alloys (CMAs) are frequently identified by specific valence electron-to-atom (*e*/*a*) ratios, such as 1.617 e-/a for transition-metal-free γ-brass systems, which are generally called Hume-Rothery rules and validated by the presence of pseudogaps at the corresponding Fermi level in the calculated electronic structures [10]. Moreover, a distinctive way to count electrons is applied for the transition-metal-rich systems, such as Chimney Ladders phases, endohedral gallides superconductors, and so on [11]. For example, Ga-cluster superconductor, $ReGa_5$, containing 11 bonding orbitals in the cluster would be fully occupied by 22e- (Re: 7e- from 5*d* and 6*s* orbitals + 5Ga: 3e- from 4*s* and 4*p* orbitals) and the Fermi level of $ReGa_5$ should be located in a gap or pseudo gap in the DOS [11]. Briefly, small values of density of states, $N(E_F)$, are corresponding to the stable electronic structures in the reciprocal space and chemical compounds in the real chemistry system [12]. As we mentioned in early section, $N(E_F)$ exists the

http://dx.doi.org/10.5772/65943

close relationship with the superconducting critical temperatures: the larger values of $N(E_F)$ in DOS, the higher T_c is likely to appear in the solid state materials. An empirical electron counting method, although this follows a different, less chemical-based electron counting process, was generated by Matthias [13]. It states that the number of valence electrons in a superconductor has lost nothing of its fundamental importance to the value of T_c. For the transition-metal-rich compounds with simple crystal structures, the maximum in T_c is seen to occur at approximately 4.7 and 6.5 valence electrons per atom [14]. The particular impressive example is the T_c dependence on the average number of valence electrons in A15 phases, for example, take Nb_3Ge. The valence electron concentration is calculated as follows: (5 e-/Nb × 3 Nb + 4 e-/Ge × 1 Ge)/4 = 4.75 e-/atom [15]. Similarly, 4.6 e-/atom work for $(Zr/Hf)_5Sb_{2.5}Ru_{0.5}$ [16, 17]. Therefore, to increase the T_c in superconductivity is at a high risk of destabilizing the compounds. From the chemistry viewpoints, BCS-like superconductors need to balance the structural stability and superconducting property and result in the limited T_c [18]. Circumventing the inherent conflict of structural stability and superconducting critical temperatures in BCS-like superconductors would be analogous to the thermoelectric materials with a phonon glass with electron-crystal properties [19].

In the past several decades, several new classes of high temperature superconductors were discovered, whose critical temperatures are way above the ones of conventional superconductors [20–23]. These discoveries give physicists hope to keep looking for the new mechanisms for superconductivity. Different from metallic superconductors, more chemistry terms can be applied for the high temperature superconductors, such as oxidation numbers, Zintl phases, valence-electron-precise systems, and so on [24, 25].

3. Inducing superconductivity by the suppression of charge density waves

In the semiconductor $BaBiO_3$ compound, the bonding interaction can be described by the formula $(Ba^{2+})(Bi^{4+})(O^{2-})_3$, Bi has the unusual oxidation state, +IV [26, 27]. At room temperature, it has the doubled perovskite unit cell and the structure distorted to monoclinic rather than being cubic. It contains two types of Bi atoms in different sized coordinated polyhedral, so the formula of $BaBiO_3$ can be modified as $(Ba^{2+})_2(Bi^{3+})(Bi^{5+})(O^{2-})_6$. Now the complex structural distortion can be interpreted as the relocalization of two electrons at the Bi^{3+} ion with the "long pair" configuration [28]. Contradictory, the two Bi atoms show slightly different in the oxidation states (+3.9 versa +4.1) from the band structure calculation [29]. Another argument was arisen that the structural distortion, as well as the non-equivalent Bi atoms, caused by the charge density waves (CDWs) [30]. Suppression of the charge density waves in Bi oxides may induce the superconductivity. It is achieved by doping Pb^{4+}, which has closed electron configuration and prefers a regularly coordinated environment to stabilize the structure. $BaPb_xBi_{1-x}O_{3-\delta}$ shows no CDWs but the superconducting transition when cooling to 13 K [31]. Another way to stabilize the regular structure is increasing the Bi^{5+} ions, which also has the closed electron configuration. To obtain this, K was used to partially replace Ba and $K_xBa_{1-x}BiO_3$ in cubic perovskite structure shows the superconducting transition around 30 K [26].

4. Superconductivity hosted in the specific structural frameworks

High temperature superconductivity in $ThCr_2Si_2$-type iron pnictides led to numerous investigations in these compounds in the past decade [32]. However, the structure of $ThCr_2Si_2$-type materials hosting superconductivity could be traced back to the quaternary superconductors, $LnNi_2B_2C$ (Ln = Ho, Er, Tm, Y and Lu) [33]. In $LnNi_2B_2C$, we could treat the B-C-B as a single chemical unit based on the short bonding distance and strong bonding interaction between B and C [34]. Therefore, the ionic formula of $LnNi_2B_2C$ can be treated as $Ln^{3+}(Ni^0)_2(B_2C)^{3-}$ [34]. In the viewpoint of chemistry, the large $N(E_F)$ in $LnNi_2B_2C$ was mainly arisen from the slight orbital distortion of B-C-B fragments. Moreover, the structure could be considered to represent the first member of a homologous series $(LnC)_n(Ni_2B_2)$, in which the LnC block adopts to a NaCl-type packing, which naturally drove us to investigate the $(LnC)_2(Ni_2B_2)$, written as (LnC)(NiB), in which the ionic formula could be written as $Ln^{3+}(Ni^0)(BC)^{3-}$ [33]. $(BC)^{3-}$ is isoelectronic with CO, and the B-C interaction rapidly changes from bonding to antibonding in addition to the dispersionless band from Ln orbital below Fermi level in LnNiBC may be the important factor to kill the superconductivity [34].

However, the "exotic" quantum mechanism for superconductivity is undergoing an unclear status even though the phenomenon has been discovered for more than a century. Superconductivity is still unpredictable currently. Condensed matter physicists try to predict superconductors based on analyzing the superconductivity through "*k*-space" pictures based on Fermi surfaces and particles interactions, that is, electron-phonon coupling [35]. Thus, there are few predictive rules from physics aspect, one of which, perhaps the most widely used, is that in intermetallic compounds of a known superconducting structure type, one can count electrons and expect to find the best superconductivity or the highest critical temperature (T_c) at ~4.7 or ~6.5 valence electrons per atom—Matthias rules mentioned above [14]. However, the chemists' viewpoint is from real space such as chemical compositions and atomic structures, which play critical roles in superconductivity, rather than reciprocal space [11]. One of the chemical views to increase the occurrence of new superconducting materials is to posit that it carries out in structural families. The well-known examples are found in $ThCr_2Si_2$-type such as $BaFe_2As_2$ and $LnNi_2B_2C$ systems and perovskites like bismuth oxides, which are fairly favored by superconductivity [32]. Laves phase compounds are previously well-investigated families for hosting superconductivity among alloys [36]. Here, we analyze the structural relationship between diamond framework and spinels from a molecular perspective, then apply this connection for interpretation and prediction of other possible new superconductors adopting to spinels and their derived structures.

5. Calculation details

5.1. Tight-binding, linear Muffin-Tin orbital-atomic spheres approximation (TB-LMTO-ASA)

Calculations of the electronic structures were performed by TB-LMTO-ASA using the Stuttgart code [37–39]. Exchange and correlation were treated by the local density approximation

(LDA) [40]. In the ASA method, space is filled with overlapping Wigner-Seitz (WS) spheres [41]. The symmetry of the potential is considered spherical inside each WS sphere, and a combined correction is used to take into account the overlapping part, and the overlap of WS spheresis limited to no larger than 16%. The empty spheres are necessary, and the overlap between empty spheres is limited to no larger than 40%.The convergence criterion was set to 0.1 meV.A mesh of ~100 *k* points [42] in the irreducible edge of the first Brillouin zone was used to obtain all integrated values, including the density of states (DOS) and Crystal Orbital Hamiltonian Population (COHP) curves [43].

5.2. WIEN2k

The electronic structures (density of states and band structure) of intermetallics were calculated using the WIEN2k code with spin orbital coupling, which has the full-potential linearized augmented plane wave method (FP-LAPW) with local orbitals implemented [44, 45]. For the treatment of the electron correlation within the generalized gradient approximation, the electron exchange-correlation potential was used with the parameterization by Perdew et al. (i.e. the PBE-GGA) [46]. For valence states, relativistic effects were included through a scalar relativistic treatment, and core states were treated fully relativistic [47]. The structure used to calculate the band structure was based on the single crystal data. The conjugate gradient algorithm was applied, and the energy cutoff was 500 eV. Reciprocal space integrations were completed over a 9 × 9 × 9 Monkhorst-Pack *k*-points mesh with the linear tetrahedron method. With these settings, the calculated total energy converged to less than 0.1 meV per atom.

5.3. Materials projects

The electronic structures of partial hypothetical compounds were predicted and calculated using the Materials Projects, which have been treated in the electron correlation within the generalized gradient approximation. The structure used to calculate the band structure was based on the single crystal data. The conjugate gradient algorithm was applied, and the energy cutoff was 520 eV. Reciprocal space integrations were completed over a 104 Monkhorst-Pack *k*-points mesh with the linear tetrahedron method.

6. Hierarchical structural interpretation of existing superconductors with spinels

Spinels, generally formulated as $A^{2+}(B^{3+})_2O_4$, crystallize in the cubic crystal system, with the oxide anions arranged in a cubic close-packed lattice and the cations A and B occupy the octahedral and tetrahedral sites in the lattice [48, 49]. An alternative tantalizing way to view the spinel structure is to treat spinels as void-filled cubic Laves phases, both of which exhibit some close relationships with the diamond structure. In the cubic Laves phase, $MgCu_2$, the Mg atom sites (Wyckoff designation 8*a*) arrange precisely into a three-dimensional (3D) diamond network. Within the voids, Cu atoms (Wyckoff designation 16*d*) form a 3D framework of vertex-sharing tetrahedra, as emphasized in **Figure 1** (Left). Thus, the Mg and Cu sites become A and B, respectively, in spinels. Furthermore, in spinels, the O atoms on 32*e* (x, x, x)

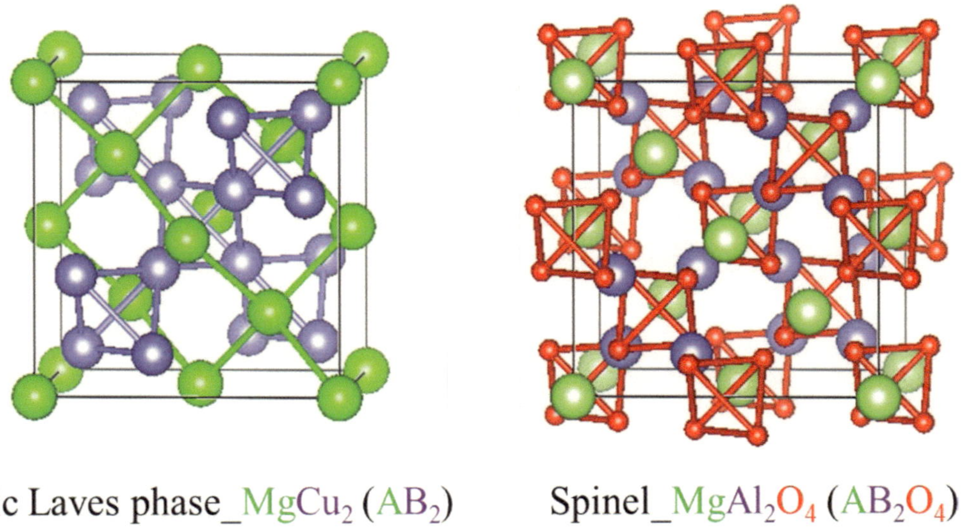

Figure 1. The structural relationship between the $MgCu_2$-type, cubic Laves phase structure and the spinel-type, $MgAl_2O_4$. *Left*: $MgCu_2$-type (Mg, green; Cu, purple); *right*: Spinel type $MgAl_2O_4$ (Mg, green; Al, purple and O, red).

sites forming isolated tetrahedral were inserted into the B_4 tetrahedra and center around 8*b* (½, ½, ½) sites in **Figure 1** (Right). The formation of the complete cubic unit cell from the cubic Laves phase to the spinels $A_8B_{16}O_{32}$ is, therefore, shown in **Figure 1**.

6.1. Li-doped "TiO_2": superconductivity in spinel $LiTi_2O_4$

Superconductivity in $Li_{1-x}Ti_{2+x}O_4$ was first reported in 1973, much earlier than the discovery of high-T_c cuprate superconductors. The superconducting transition temperature (T_c) of $LiTi_2O_4$ is around 11 K [50]. As the first oxide superconductor with a relatively high critical temperature, $LiTi_2O_4$ remains widely intriguing for scientists. The most frequent questions arose are why $LiTi_2O_4$ adopts to a unique structure type, which is different from other high temperature superconducting materials, such as perovskites or cuprates. However, Li-doped TiO_2 and $LiTi_2O_4$ can be treated as the analogy between Cu-doped $TiSe_2$ and $Cu_{0.08}TiSe_2$ in a certain way [51]. $TiSe_2$ adopts to the trigonal-layered structure (1T) (S.G. P-31*m*) with charge density waves observed around 220 K, and with doping Cu, the superconductivity in Cu_xTiSe_2 appears and the charge density wave was suppressed. Similarly, for both rutile- and anatase-TiO_2, Ti and O atoms form distorted Ti@O_6 octahedra; thus, there exist empty voids in the structure shown in **Figure 2a** and **b**. Based on the electronic structures of TiO_2in **Figure 3a** and **b**, both polymorphic TiO_2 compounds are well-known *n*-type semiconductors with ~2eV gaps above Fermi levels [52]. Through doping with Li atom, which can be considered as the nearly free electron in solid state chemistry, the empty voids in TiO_2 are occupied by the Li atoms, and the Fermi levels start lifting up and shifting to the peak in the DOS.

To confirm our assumptions, the electronic structures of $LiTi_2O_4$ are calculated using TB-LMTO-ASA with Crystal Orbital Hamilton Population (COHP) codes. In **Figure 4** (left), the DOS

http://dx.doi.org/10.5772/65943

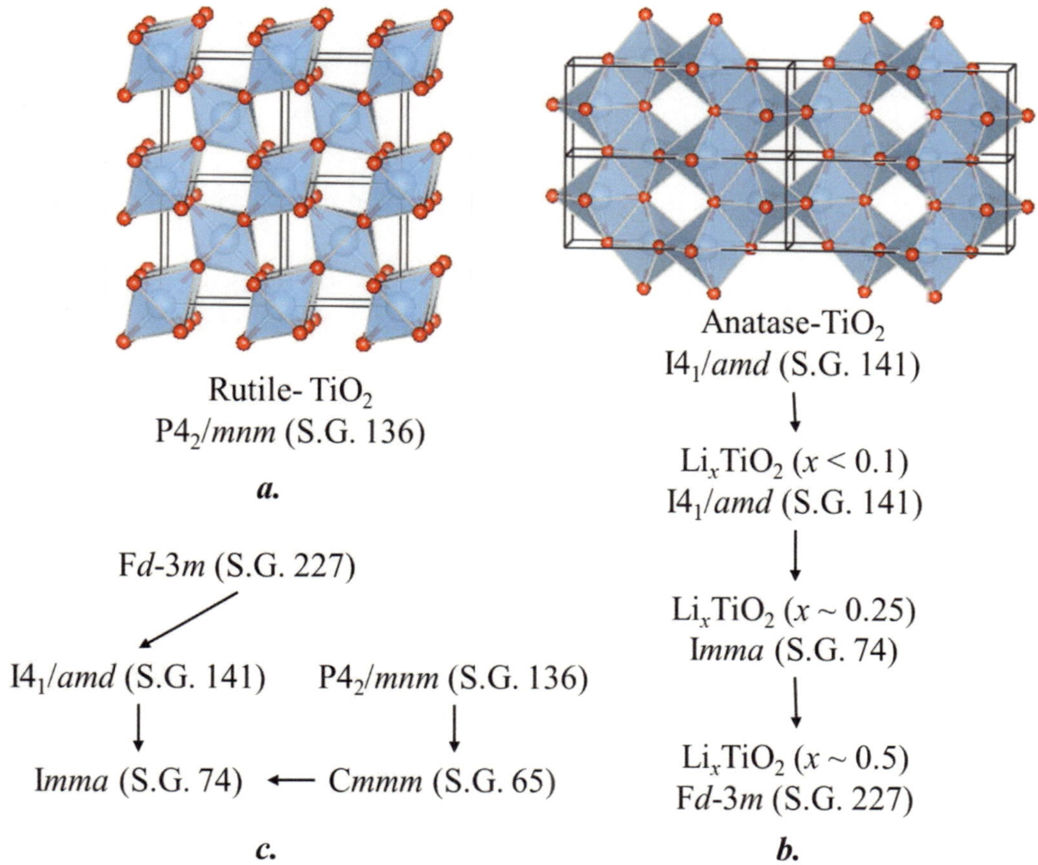

Figure 2. The structure and space group connections between two types of TiO_2 and Li_xTiO_2. (a) The crystal structure of rutile-TiO_2. The rutile-TiO2 adopts to the primitive tetragonal structure with space group $P4_2/mnm$. Each Ti atom surrounded by 6 O atoms forms the octahedral coordination. (b) The schematic picture showing the possible phase transitions when doping Li into TiO_2. (c) The space group and sub-space group relationship among TiO_2 and $LiTi_2O_4$.

qualitative features obtained by this calculation state that are 2–6 eV below the Fermi level (E_F) arise primarily from valence 4*s* and 3*d* orbitals from Ti and 2*p* orbitals from O, whereas the Li 2*s* band is broadly distributed from −6 to −2 eV. The contribution of Li 2*s* electrons to the DOS curve, as shaded in **Figure 4** in black, shows it just contributes one free electron to the system rather than making any change to the DOS features of TiO_2 in the diamond-like framework. The integrated DOS till the broad band gap around 1 eV below Fermi level results in the same electron counts as TiO_2. Comparing with the DOS of rutile- and anatase-TiO_2 in **Figure 3**, the difference between $LiTi_2O_4$ and TiO_2 is the up-lift 1 e- per formula of the Fermi level. The Fermi level for $LiTi_2O_4$ falls just above the topmost peak of the largely Ti 3*d* bands. Therefore, we employed Local Spin Density Approximation (LSDA) to see if a magnetic moment would spontaneously develop, but the converged result yielded zero magnetic moment. This result of the unstable electronic structure gives a strong indication of the occurrence of superconductivity. The "bond energy" term is evaluated by the crystal orbital Hamilton populations

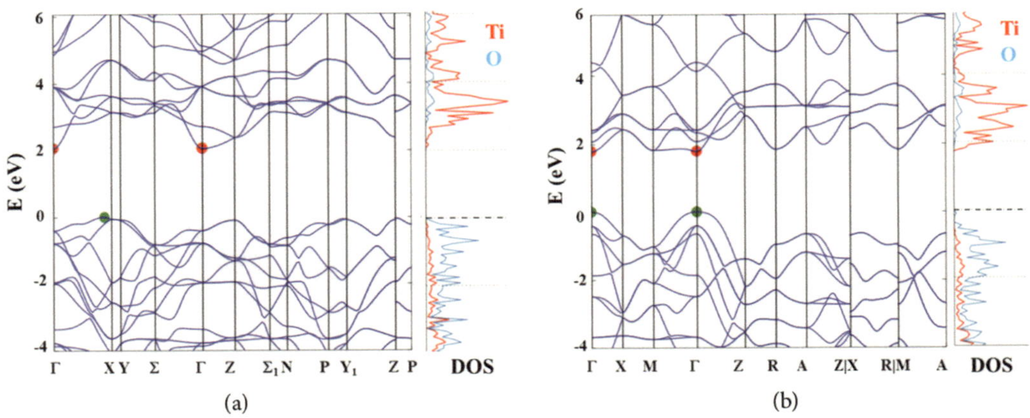

Figure 3. The band structures and density of states (DOS) of (a) anatase-TiO_2 with ~2eV indirect band gap and (b) rutile-TiO_2 with ~2eV direct band gap (generated from *Materials Projects*).

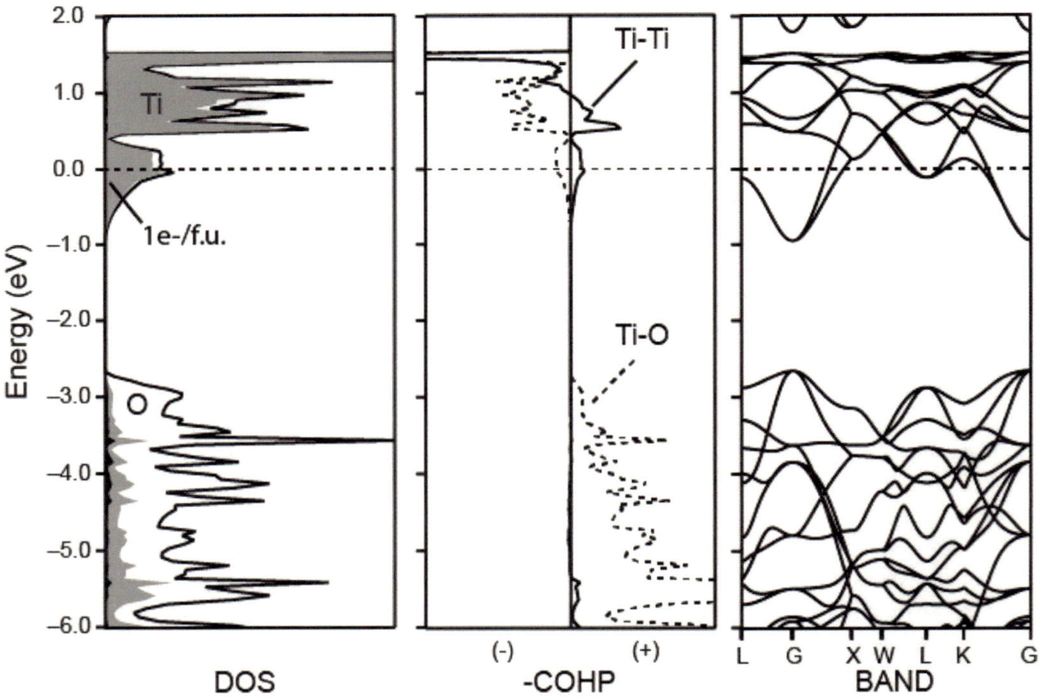

Figure 4. Electronic structure of spinel $LiTi_2O_4$. Partial DOS curves, –COHP curves and band structure of "$LiTi_2O_4$" obtained from non-spin-polarization (LDA). (+ is bonding/ – is anti-bonding).

(COHP) curves [53]. These curves illustrated in **Figure 4** (Middle) show that the band gap around 1eV below Fermi level corresponds to the non-bonding for the compound, similar in rutile- and anatase-TiO_2. Interestingly, there is no atomic interaction between Li and O or Li and Ti, which confirmed our claims discussed above—Li just acts as the electron-donator

http://dx.doi.org/10.5772/65943

to change the Fermi level as well as balance the charge, and empty sphere to fill the volume, rather than giving the impact on the electronic structure including changing the atomic interactions. In –COHP, the Fermi level is located on the Ti-O anti-bonding and Ti-Ti bonding interactions, the sum of the anti-bonding and bonding effects is close to zero, which indicates the possibility of the stabilization of the compound. The strong Ti–O antibonding interactions at the Fermi level, contributing the unstable factors in the electronic structure is significant relative with the superconductivity in spinel $LiTi_2O_4$. Meanwhile, in the band structure of spinel $LiTi_2O_4$, the Fermi level locates on the saddle points around U and W points in the Brillouin zone, which is a strong evidence for the unstable electronic structure.The structural connection might also be noted from the links between the space groups in **Figure 2**. If we divided the unit cell of the spinel structure into two along ½(***a*** + ***b***) and ½(***a*** - ***b***), the cubic structure will become tetragonal, and space group will decrease from F*d*-3m to $I4_1/amds$. Furthermore, the distortion of ***a*** and ***b*** decreases the symmetry from tetragonal to orthorhombic and the space group will become I*mma* instead of $I4_1/amds$. On the other side, I*mma* is also the direct subgroup of $P4_2/mnm$ (rutile-TiO_2). In summary, the structural transformation could be treated as the transition of a continuously doping Li process. When the amount of doped Li is small, the Li_xTiO_2 keeps in $I4_1/amds$. As x increases, the orthorhombic structure appears. When x is close to 0.5, the spinel phase is more favored than other phases [54].

Since the discovery of the superconductivity in spinel $LiTi_2O_4$, much effort has been put into finding more spinel oxide superconductors. The studies of spinel oxide superconductors endeavored for the physics community for many years. The alternative view on the spinel superconductor, $LiTi_2O_4$, could be considered as the electron-doping in transition metal dichalcogenides, similar with Cu-doped $TiSe_2$. Li-doped anatase-TiO_2 crystallizes in tetragonal structure with the space group of $I4_1/amds$. With doping more electrons into the system, the structural transitions happen from tetragonal to orthorhombic to cubic. The superconductivity was arisen when doping Li to ~1/2 per f.u. and the structure adopting to the cubic spinel. Similar structural transitions from tetragonal $I4_1/amd$ to F*d*-3*m* occur in another spinel superconductor, $CuIr_2S_4$ [55].

6.2. Superconductivity in non-oxide spinel $CuIr_2S_4$ and CuV_2S_4

$CuIr_2S_4$in the cubic structure with the space groupshows metallic properties at room temperature [56]. As the temperature decreases, $CuIr_2S_4$ undergoes a transition from a metal to an insulator around 230 K, which is also associated with a structural change from cubic to tetragonal [57]. Interestingly, a pseudogap is situated just above in the calculated density of states (DOS). In **Figure 5** (left), the DOS shows that ~6 eV range below the Fermi level (E_F) arises from all of the valence Cu, Ir and S orbitals. The contribution of Cu 4*s* and 3*d* electrons to the DOS curve, as shaded in **Figure 5** in black, states the filled-up *d* electrons are delocalized and hybridized with 5*d* electrons from Ir as well as 3*p* electrons from S, which is quite different from $LiTi_2O_4$. To further confirm our assumptions, the bonding/anti-bonding interactions (–COHP) in $CuIr_2S_4$ are calculated. Unlike the –COHP in $LiTi_2O_4$, which was dominated by Ti-O and Ti-Ti interactions, Ir-S and Cu-S interactions play the most important roles in the structural stabilization and superconducting properties in $CuIr_2S_4$. The band gap in $LiTi_2O_4$ corresponds to the non-bonding boundary in $LiTi_2O_4$; however, the 0–1 eV below the Fermi

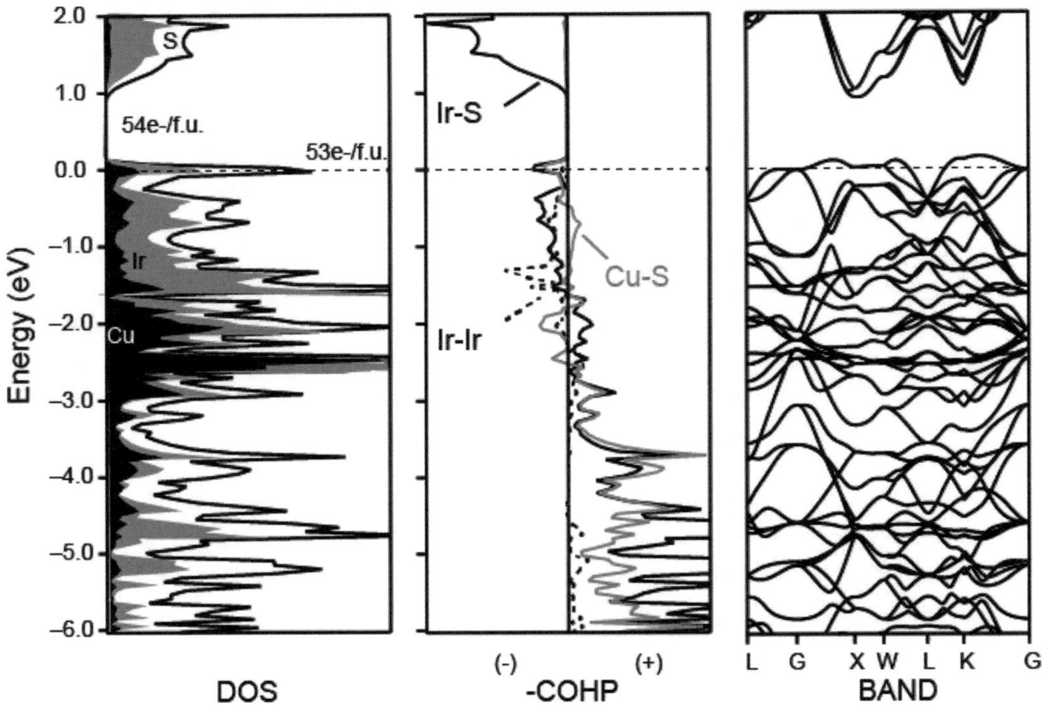

Figure 5. Electronic structure of spinel $CuIr_2S_4$. Partial DOS curves, –COHP curves and band structure of "$CuIr_2S_4$" obtained from non-spin-polarization (LDA). (+ is bonding/ – is anti-bonding, E_F for 53e⁻ is set to zero).

level in $CuIr_2S_4$ is on the mixed status of Ir-S anti-bonding and Cu-S bonding interactions. The Fermi level at anti-bonding interactions indicates the instability of electronic structure of $CuIr_2S_4$ and the possible occurrence of superconductivity. The integrated DOS of $CuIr_2S_4$ gives 53e- per f.u., whereas the band gap just above Fermi level corresponds to 54e- per f.u., which can be expressed with the hypothetical compound "$CuIr_2S_4$(1e-)" according to the Zintl-Klemm concept. The ionic formula of $CuIr_2S_4$ can be interpreted as $Cu^{2+}(Ir^{3+})_2(S^{2-})_4$(1e-), the electron configuration of Ir^{3+} becomes $5d^6$. Or $Cu^{+}(Ir^{4+})(Ir^{3+})(S^{2-})_4$(1e-) with two kinds of electron configurations of Ir, $5d^6$ and $5d^5$. The coordinated environment of Ir^{3+} is octahedral (O_h), thus, the d orbital will split into e_g and t_{2g}.

It has been well known even in textbooks that molecular transition metal complexes have a gap between the e_g and t_{2g} type in d orbitals, which is determined by the α and π boning of the coordinated ligands. However, the band gap between the e_g and t_{2g} in d bands in certain solids is dependent on more complex orbital considerations. Take perovskite $LaCoO_3$ for example, the band gap is very small, close to 0 eV, but the iso-electronic $LaRhO_3$ has ~1.6 eV band gap. Similarly, in $CuIr_2S_4$, the band gap between e_g and t_{2g} is so small that the $5d^6$ and $5d^5$ configurations can coexist [58]. Moreover, at higher temperatures, a whole series of transformations take place triggered by thermal excitation of electrons from the valence to the conduction band. In $CuIr_2S_4$ compound, the two possible oxidation state fluctuations of Cu^+/Cu^{2+} and Ir^{3+}/Ir^{4+} could be related to superconductivity. The presence of a metal-insulator

(M-I) transition on cooling or under pressure has been of particular interest in the heavy metal chalcogenide spinel systems to make superconductors. Based on the decreased lattice parameters, $CuRh_2S_4$ and $CuRh_2Se_4$ can be treated as the compressed and expanded format of $CuIr_2S_4$ [59].

Another representative non-oxide spinel superconductor is CuV_2S_4 [60]. Unlike $CuIr_2S_4$, the superconductivity in CuV_2S_4 is induced by suppressing the CDWs rather than the metal-insulator transition in $CuIr_2S_4$ [61]. Also, according to the Zintl-Klemm concept, the ionic formula of CuV_2S_4 can be written as $Cu^{2+}(V^{3+})(V^{3+})(S^{2-})_4$. From the electronic structural calculations of CuV_2S_4 in **Figure 6**, a ~0.3 eV band gap is located at 0.6 eV below the Fermi level. The integrated DOS shows the gap responds to the 42e- (45e- for Fermi level). The band gap above Fermi level corresponds to 54e-, just as "$CuIr_2S_4$ (1e)." The band structure indicates the similarity betweenCuV_2S_4 (early transition metal, V) and $LiTi_2O_4$ (early transition metal, Ti) and the difference between CuV_2S_4 (early transition metal, V) and $CuIr_2S_4$ (late transition metal, Ir). By analogy with Jahn-Teller distortion ideas, the partially occupied bands are subject to the geometrical distortions related to a lowering of the total energy and usually termed as the instability of the Fermi surface (CDWs). A band gap may open at the Fermi level to create a semiconductor or insulator as the structure changes. From the chemistry viewpoint, the highest occupied conduction band is filled to make insulators. For example, in MoS_2, the charge density waves were observed in the localized unit of S-Mo-S rather than a localized

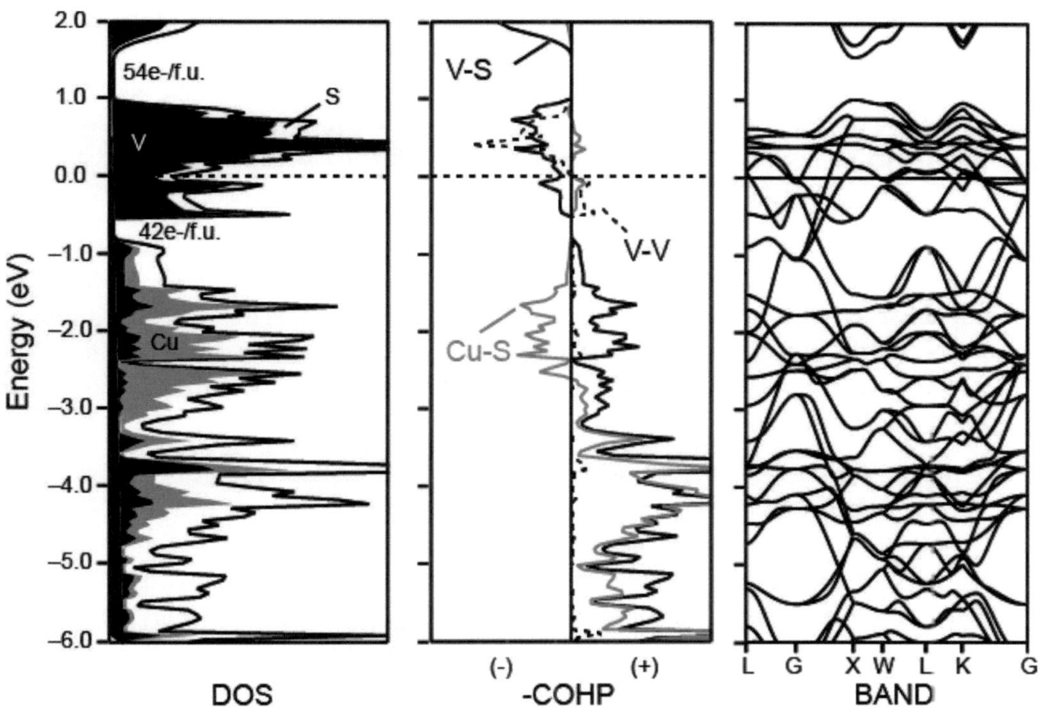

Figure 6. Electronic structure of spinel CuV_2S_4. Partial DOS curves,–COHP curves and band structure of "CuV_2S_4" obtained from non-spin-polarization (LDA). (+ is bonding/ – is anti-bonding, E_F for $45e^-$ is set to zero).

atom. A series of superconductors were reported by suppression of the charge density waves in MoS_2, just like CuV_2S_4 [58].

Based on the above considerations, one of the most interesting areas from both chemical and physical points of view is identification of the factors that determines whether a particular solid is a conductor of electricity or a specific structure type is favored to hold the conducting properties and how well they do it. Furthermore, how external events such as pressure and temperature may affect a system to transit from one regime to the other. Are there any surprises associated with the transition between metal and insulator? Indeed, one of the consequences of the discovery of this series of superconducting copper and bismuth oxides has been unraveling of the possible connection with the metal-insulator transition. But what are the rules associated with the generation of this state of affairs, and what are the factors which compete with them and which lead to the superconductivity and how can we use this to make new superconductors? Recently, the superconductivity was observed in the non-superconducting $CuIr_2Se_4$ spinel by partial substitution of Pt for Ir [62].

7. Concluding remarks

The understanding of superconductivity from the viewpoint of chemistry offers a relatively straightforward approach to the real space rather than thinking in reciprocal space from a physical viewpoint. This chemical thinking is obviously basic, though not sufficiently comprehensive, as clearly shown by the competition between superconductivity and other structural phase transition (CDWs), oxidation fluctuation or magnetism. In this work, the introduced ideas are coming from the chemistry and carried some way into physics, alternatively, using chemical concepts to explain some physical phenomenon. A few questions arise about chemical trivial materials, such as how to make an indirect band gap a direct one. Several empirical rules can be used for chemists to design new superconductors.

1. Matthias' rule to make diamond-related α-Mn type new superconductors: α-Mn framework can be treated as defected 2 × 2 × 2 diamond structure shown in Xie's yet unpublished work. The space group of α-Mn is I-43*m*, which is the direct subgroup of F*d*-3*m*. Re-rich binary compounds are favored by α-Mn structure. By tuning the electron counts to 6.5e- per atom, α-Mn type Re-rich compounds are highly likely to be superconductors.

2. Searching for the new pyrochlore-type superconductors: In a brief discussion of the structural chemistry of both cubic Laves phase and Ni_2In structures, it is suggested that spinels and pyrochlores structures show the similarities just like cubic Laves phases and Ni_2In. Pyrochlores can be treated as the superlattice of spinels according to the connection in the lattice parameters. The superconductor, $Cd_2Re_2O_7$, in the pyrochlore-type structure can be conducted the similar research to $LiTi_2O_4$. Moreover, more non-oxide pyrochlore compounds can be synthesized to examine the superconducting properties.

3. It is not straightforward to predict the metallic or insulating properties, even harder to predict the M-I transition including the accompanying superconductivity sometimes.

http://dx.doi.org/10.5772/65943

But many CDW instabilities are triggered by lowering the temperature and occur in a range of systems, which cover a wide range of chemical types, including metal oxides and sulfides and molecular metals. The surprise of superconductivity may be observed by suppressing the CDWs.

Acknowledgements

W. Xie thanks Louisiana State University for the start-up funding support and also acknowledges very helpful discussions with Professor Robert Cava (Princeton University) and Professor Gordon Miller (Iowa State University). W. Xie appreciates Yuze Gao for editing the references.

Author details

Weiwei Xie[1]* and Huixia Luo[2]

*Address all correspondence to: weiweix@lsu.edu

1 Department of Chemistry, Louisiana State University, Baton Rouge, USA

2 Department of Chemistry, Princeton University, Princeton, USA

References

[1] Meyers, H. P.; Myers, H. P. Introductory Solid State Physics, 2nd Edition; CRC Press, USA 1997.

[2] Larbalestier, D.; Gurevich, A.; Feldmann, D. M.; Polyanskii, A. Nature **2001**, 414 (6861), 368–377.

[3] Suhl, H.; Matthias, B. T.; Walker, L. R. Phys. Rev. Lett. **1959**, 3 (12), 552–554.

[4] Anderson, P. W. J.Phys.Chem.Solids **1959**, 11 (1), 26–30.

[5] Simon, A. Angew. Chem. Int. Ed. Engl. **1997**, 36 (17), 1788–1806.

[6] McMillan, W. L. Phys. Rev. **1968**, 167 (2), 331–344.

[7] Fässler, T. F.; Hoffmann, S. D. Angew. Chem. Int. Ed. Engl. **2004**, 43 (46), 6242–6247.

[8] Fredrickson, D. C.; Lee, S.; Hoffmann, R. Inorg. Chem. **2004**, 43 (20), 6159–6167.

[9] Xie, W.; Miller, G. J. Chem. Mater. **2014**, 26 (8), 2624–2634.

[10] Mizutani, U. Hume-Rothery Rules for Structurally Complex Alloy Phases; CRC Press, USA 2016.

[11] Xie, W.; Luo, H.; Phelan, B. F.; Klimczuk, T.; Cevallos, F. A.; Cava, R. J. PNAS **2015**, 112 (51), E7048–E7054.

[12] Miller, G. J. Eur. J. Inorg. Chem. **1998**, 1998 (5), 523–536.

[13] Matthias, B. T.; Geballe, T. H.; Compton, V. B. Rev. Mod. Phys. **1963**, 35 (1), 1–22.

[14] Matthias, B. T. Phys. Rev. **1955**, 97 (1), 74–76.

[15] Matthias, B. T.; Geballe, T. H.; Geller, S.; Corenzwit, E. Phys. Rev. **1954**, 95 (6), 1435–1435.

[16] Xie, W.; Luo, H.; Phelan, B. F.; Cava, R. J. J. Mater. Chem. C **2015**, 3 (31), 8235–8240.

[17] Xie, W.; Luo, H.; Seibel, E. M.; Nielsen, M. B.; Cava, R. J. Chem. Mater. **2015**, 27 (13), 4511–4514.

[18] Xie, W.; Fuccillo, M. K.; Phelan, B. F.; Luo, H.; Cava, R. J. J. Solid State Chem. **2015**, 227, 92–97.

[19] Snyder, G. J.; Toberer, E. S. Nat. Mater. **2008**, 7 (2), 105–114.

[20] Cava, R. J. Phys. C Superconductivity **1997**, 282, 27–33.

[21] Cava, R. J.; Batlogg, B.; van Dover, R. B.; Krajewski, J. J.; Waszczak, J. V. Nature **1990**, 345, 602–604.

[22] Takagi, H.; Cava, R. J.; Marezio, M.; Batlogg, B.; Krajewski, J. J.; Peck, W. F.; Bordet, P.; Cox, D. E. Phys. Rev. Lett. **1992**, 68 (25), 3777–3780.

[23] Cava, R. J. Chemistry of Bismuth and Lead based Superconducting Perovskites. United States: American Ceramic Society Inc.,1990.

[24] Kauzlarich, S. M.; Brown, S. R.; Snyder, G. J. Dalton Trans. **2007**, 21, 2099–2107.

[25] Schäfer, H.; Eisenmann, B.; Müller, W. Angew. Chem. Int. Ed. Engl. **1973**, 12 (9), 694–712.

[26] Cava, R. J.; Batlogg, B.; Krajewski, J. J.; Farrow, R.; Rupp, L. W.; White, A. E.; Short, K.; Peck, W. F.; Kometani, T. Nature **1988**, 332 (6167), 814–816.

[27] Cava, R. J.; Siegrist, T.; Peck, W. F.; Krajewski, J. J.; Batlogg, B.; Rosamilia, J. Phys. Rev. B **1991**, 44 (17), 9746–9748.

[28] Retoux, R.; Studer, F.; Michel, C.; Raveau, B.; Fontaine, A.; Dartyge, E. Phys. Rev. B **1990**, 41 (1), 193–199.

[29] Goodenough, J. B. Rep. Prog. Phys. **2004**, 67 (11), 1915.

[30] Timusk, T.; Statt, B. Rep. Prog. Phys. **1999**, 62 (1), 61.

[31] Mattheiss, L. F.; Hamann, D. R. Phys. Rev. B **1983**, 28 (8), 4227–4241.

[32] Hosono, H.; Tanabe, K.; Takayama-Muromachi, E.; Kageyama, H.; Yamanaka, S.; Kumakura, H.; Nohara, M.; Hiramatsu, H.; Fujitsu, S. Sci. Technol. Adv. Mater. **2015**, 16 (3), 33503.

[33] Cava, R. J.; Takagi, H.; Zandbergen, H. W.; Krajewski, J. J.; Peck, W. F.; Siegrist, T.; Batlogg, B.; van Dover, R. B.; Felder, R. J.; Mizuhashi, K.; Lee, J. O.; Eisaki, H.; Uchida, S. Nature **1994**, 367 (6460), 252–253.

[34] Miller, G. J. Am. Chem. Soc. 1994, **116**, 63324336.

[35] Montorsi, A. The Hubbard Model: A Reprint Volume; World Scientific, Singapore 1992.

[36] Compton, V. B.; Matthias, B. T. Acta Crystallogr. **1959**, 12 (9), 651–654.

[37] Blöchl, P. E.; Jepsen, O.; Andersen, O. K. Phys. Rev. B **1994**, 49 (23), 16223.

[38] Andersen, O. K.; Jepsen, O. Phys. Rev. Lett. **1984**, 53 (27), 2571–2574.

[39] Glötzel, D.; Segall, B.; Andersen, O. K. Solid State Commun. **1980**, 36 (5), 403–406.

[40] Becke, A. D. J.Chem. Phys. **1993**, 98 (2), 1372–1377.

[41] Methfessel, M.; Rodriguez, C. O.; Andersen, O. K. Phys. Rev. B **1989**, 40 (3), 2009–2012.

[42] Monkhorst, H. J.; Pack, J. D. Phys. Rev. B **1976**, 13 (12), 5188–5192.

[43] Dronskowski, R.; Bloechl, P. E. J. Phys. Chem. **1993**, 97 (33), 8617–8624.

[44] Schwarz, K.; Blaha, P. Comput. Mater. Sci. **2003**, 28 (2), 259–273.

[45] Schwarz, K. J. Solid State Chem. **2003**, 176 (2), 319–328.

[46] Perdew, J. P.; Burke, K.; Ernzerhof, M. Phys. Rev. Lett. **1996**, 77 (18), 3865–3868.

[47] Blaha, P.; Schwarz, K.; Sorantin, P.; Trickey, S. B. Comp. Phys. Commun. **1990**, 59 (2), 399–415.

[48] Jain, A.; Ong, S. P.; Hautier, G.; Chen, W.; Richards, W. D.; Dacek, S.; Cholia, S.; Gunter, D.; Skinner, D.; Ceder, G.; Persson, K. A. APL Mater. **2013**, 1 (1), 11002.

[49] Cava, R. J.; Murphy, D. W.; Zahurak, S.; Santoro, A.; Roth, R. S. J. Solid State Chem. **1984**, 53 (1), 64–75.

[50] Johnston, D. C.; Prakash, H.; Zachariasen, W. H.; Viswanathan, R. Mater. Res. Bull. **1973**, 8 (7), 777–784.

[51] Morosan, E.; Zandbergen, H. W.; Dennis, B. S.; Bos, J. W. G.; Onose, Y.; Klimczuk, T.; Ramirez, A. P.; Ong, N. P.; Cava, R. J. Nat. Phys. **2006**, 2 (8), 544–550.

[52] Burdett, J. K. Acta Cryst. Sect. B Struct. Sci. **1995**, 51 (4), 547–558.

[53] Hoffmann, R. Acc. Chem. Res. **1971**, 4 (1), 1–9.

[54] Murphy, D. W.; Sunshine, S.; van Dover, R. B.; Cava, R. J.; Batlogg, B.; Zahurak, S. M.; Schneemeyer, L. F. Phys. Rev. Lett. **1987**, 58 (18), 1888–1890.

[55] Nagata, S.; Hagino, T.; Seki, Y.; Bitoh, T. Phys. B Condens. Matter **1994**, 194, 1077–1078.

[56] Radaelli, P. G.; Horibe, Y.; Gutmann, M. J.; Ishibashi, H.; Chen, C. H.; Ibberson, R. M.; Koyama, Y.; Hor, Y.-S.; Kiryukhin, V.; Cheong, S.-W. Nature **2002**, 416 (6877), 155–158.

[57] Robbins, M.; Willens, R. H.; Miller, R. C. Solid State Commun. **1967**, 5 (12), 933–934.

[58] Burdett, J. K. Chem. Soc. Rev. **1994**, 23 (5), 299–308.

[59] Tachibana, M. Solid State Commun. **2012**, 152 (10), 849–851.

[60] Okada, H.; Koyama, K.; Watanabe, K. the AIP Conference Proceedings **2006**, 850, 1317–1318.

[61] Fleming, R. M.; DiSalvo, F. J.; Cava, R. J.; Waszczak, J. V. Phys. Rev. B **1981**, 24 (5), 2850–2853.

[62] Luo, H.; Klimczuk, T.; Müchler, L.; Schoop, L.; Hirai, D.; Fuccillo, M. K.; Felser, C.; Cava, R. J. Phys. Rev. B **2013**, 87 (21), 214510.

Section 3

Spinel-Based Ferrofluids

Chapter 7

Manganese-Zinc Spinel Ferrite Nanoparticles and Ferrofluids

Rajender Singh and Gadipelly Thirupathi

Additional information is available at the end of the chapter

http://dx.doi.org/10.5772/66522

Abstract

The nanoparticles and ferrofluids of spinel ferrites are useful in bio-sensors, transducers, storage devices, optical devices, and so on. The Mn-Zn ferrite (MZF) is generalized soft spinel ferrite having high saturation magnetization at low applied magnetic field. This chapter covers the synthesis of nanoparticles of various sizes and compositions of $Mn_{1-x}Zn_x Fe_2O_4$ with x = 0–1 by co-precipitation method. The structural and magnetic properties of the nanoparticles are discussed. The ferrofluids of superparamagnetic and ferromagnetic MZF nanoparticles were synthesized. The magneto-viscosity of ferrofluids with the dispersion of nanoparticles in different colloidal was studied. The Herschel-Bulkley model is applied to analyse the data for low viscosity ferrofluids.

Keywords: spinel, Mn Zn ferrite, nanoparticles, ferrofluids

1. Introduction

Ferrites are ceramic materials with magnetic properties. According to the crystal structure, the ferrites are mainly classified as spinels, hexaferrites, garnets and perovskites. Spinel is the first class among these ferrites having mineral structure with significant magnetic behaviour. The nanoparticles and ferrofluids of spinel ferrites are useful in bio-sensors, transducers, storage devices, energy conversion devices, heat absorbers and generators, shock absorbers, lubricants, magneto-optical devices, and so on.

Among all ferrite types, the spinel ferrites are easy to form with control on the size of their nanoparticles. The Mn-Zn ferrite (MZF) is generalized soft spinel ferrite having high magnetization, and it saturates at low applied magnetic field. In this work, the optimization of the synthesis procedures was done to obtain stable nanoparticles and ferrofluids of Mn-Zn spinel ferrites. The structural and magnetic properties of the nanoparticles and magneto-viscosity of the MZF ferrofluids were studied to understand the correlation between physical properties of nanoparticles and flow behaviour of ferrofluid in applied magnetic field.

2. Synthesis of nanoparticles of spinel ferrite

There are several methods for the synthesis of ferrite nanoparticles, that is, photo synthesis, microemulsion, sol-gel, hydrothermal, ball-milling, co-precipitation, catalyst-based methods and so on discussed in the literature [1–4].

In this work, the nanoparticles of $Mn_{1-x}Zn_x Fe_2O_4$ (MZF) with $x = 0$–1 were synthesized by soft chemical approach of co-precipitation method [5–10]. The metal salts with high solubility around room temperature, that is, $MnCl_2$, $ZnSO_4$ and $FeNO_3$, were chosen for the synthesis. The metal salts in stoichiometric mole ratio were dissolved in distilled water, and the mixture was heated at 353 K. The size of nanoparticles was controlled by controlling moles of OH^- ions in the solution (n) according to the following chemical formula:

$$(1\text{-}x)\, Mn^{2+} + x\, Zn^{2+} + 2\, Fe^{3+} + n\, OH^- \rightarrow Mn_{1-x}\, Zn_x\, Fe_2\, O_4 + H_2\, O$$

The solution was washed with distilled water until the pH = 7 was reached and the slurry was heated at 373 K to get the MZF nanoparticles. The MZF nanoparticles of different compositions in $Mn_{1-x}Zn_x Fe_2O_4$ with $x = 0$, 0.25, 0.5, 0.75 and 1 were synthesized.

3. Synthesis of the ferrofluids of spinel ferrite

The properties of ferrofluid depend on the hydrodynamic distribution of magnetic nanoparticles. The following ferrofluids were synthesized using MZF nanoparticles:

(i) Ferrofluids with the dispersion of superparamagnetic (SPM) nanoparticles (no surface coating) in ethylene glycol.

(ii) Ferrofluids with the dispersion of ferromagnetic nanoparticles in different colloidal with suitable surface coating of the nanoparticles.

Finding suitable colloidal is necessary for the stability of ferrofluid in various technological applications. A number of colloidal, with their properties listed in **Table 1**, were used for synthesizing different MZF ferrofluids using methods given the literature [11–15].

The synthesized (MZF) ferrofluids are listed in **Table 2**. These ferrofluids were studied for their magneto-viscosity properties. The table has three categories of ferrofluids:

(i) Ethylene glycol-based ferrofluids synthesized from SPM $Mn_{0.75}Zn_{0.25}Fe_2O_4$ nanoparticles.

(ii) Water, kerosene and toluene-based ferrofluids synthesized with surfactant-coated Mn-ferrite (MF) nanoparticles.

(iii) The ferrofluids synthesized with surfactant-coated $Mn_{0.75}Zn_{0.25}Fe_2O_4$ nanoparticles in water, kerosene, toluene and paraffin and $Mn_{0.9}Zn_{0.1}Fe_2O_4$ nanoparticles in paraffin.

http://dx.doi.org/10.5772/66522

S. No.	Colloidal	Dynamic viscosity of colloidal at 300 K (mPa. s)	Density (g/ml)	Vapour pressure (k Pa)
1	Ethylene glycol	16.2	1.1	0.5
2	Toluene	0.55	0.86	8.7
3	Kerosene	1.64	0.78	0.5
4	Water	0.8	1	4.3
5	Paraffin oil	25 to 80	1.1	0.5

Table 1. Properties of colloidal used for making ferrofluids.

Name of ferrofluid	Composition and magnetic behaviour of the nanoparticles	Colloidal	Surfactant
MZFE1	$Mn_{0.75}Zn_{0.25}Fe_2O_4$ (SPM)	Ethylene glycol	Not used
MFW	$MnFe_2O_4$ (FM)	Water	Tetra-methyl ammonia (TMA)
MFK		Kerosene	Oleic acid
MFT		Toluene	Oleic acid
MZFW	$Mn_{0.75}Zn_{0.25}Fe_2O_4$ (FM)	Water	TMA
MZFK		Kerosene	Oleic acid
MZFT		Toluene	Oleic acid
MZFP1		Paraffin oil	Oleic acid
MZFP2	$Mn_{0.9}Zn_{0.1}Fe_2O_4$ (FM)	Paraffin oil	Oleic acid

Table 2. Different MZF-based ferrofluids investigated in this work.

4. Structural and magnetic properties of spinel ferrites nanoparticles

The chemical formula of spinel ferrite is generally expressed, where 'Me' represents a divalent metal ion (e.g. Fe^{2+}, Ni^{2+}, Mn^{2+}, Mg^{2+}, Co^{2+}, Cu^{2+}, etc.) and '*i*' is inversion parameter which is 0 for normal spinel and 1 for inverse spinel. The inverse parameter varies from 0 to 1 for mixed spinel. The ionic radii of cation, crystal field effect, ionic charge, and so on are the main factors for the cation distribution in the nanoparticles of spinel ferrites. In the cubic close-packed arrangement of the spinel structure, the unit cell has 32 octahedral sites (B-site) and 64 tetrahedral sites (B-site), out of which only 16 octahedral sites and eight tetrahedral sites are occupied.

The Mn-Zn ferrite nanoparticles of various sizes and compositions were synthesized. Their structural properties were studied by X-ray diffraction (XRD) analysis. The morphological and microstructure of these nanoparticles was analysed using transmission electron microscopy (TEM). The magnetic properties of the nanoparticles were studied by measuring magnetization as a function of temperature and applied magnetic field, $M(T,H)$ and temperature-dependent ferromagnetic resonance (FMR) spectra.

The MZF nanoparticles of various sizes were synthesized by varying metal ions to hydroxide ratio (r = Me/OH^-) by co-precipitation method [16]. The value of r was varied from 0.375 to 0.17. The XRD patterns of all the samples (**Figure 1**, left) are analysed using Rietveld refinement method. The structural refinements are fitted with a single-phase spinel structure

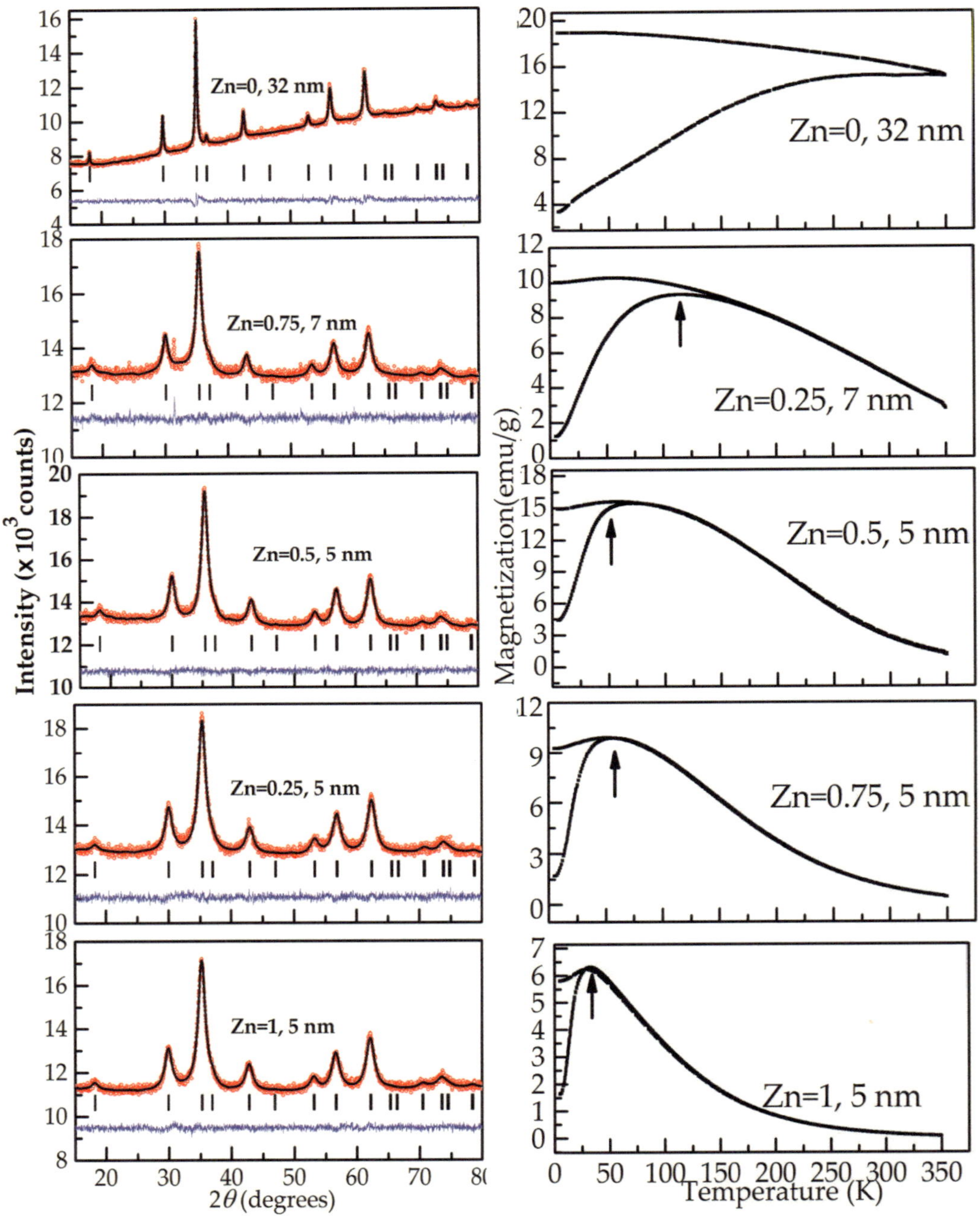

Figure 1. XRD pattern (left) and magnetization as a function of temperature at 100 Oe (right) of MZF nanoparticles for different Zn content.

Zn content	0		0.25		0.5		0.75		1	
S. No	D (nm)	a (Å)	D (nm)	a (Å)	D (nm)	a (Å)	D (nm)	a (Å)	D (nm)	a (Å)
1	32	8.483	7	8.411	4	8.418	3	8.439	2	8.461
2	60	8.478	10	8.427	5	8.410	4	8.431	5	8.451
3	104	8.477	11	8.412	7	8.397	5	8.419	9	8.447

Table 3. Listed here are lattice parameter (a) and crystallite size D determined from XRD for various compositions of $Mn_{1-x}Zn_x Fe_2O_4$ with $x = 0–1$ (MZF) nanoparticles.

having space group Fd$\bar{3}$m by Fullprof suite program for the XRD pattern [17, 18]. The structural parameters (lattice parameter (a), crystallite size, strain, etc.) are extracted from the fits. The atomic position coordinates of A-, B- and O-sites are (1/8, 1/8, 1/8), (1/2, 1/2, 1/2) and (1/4+u, 1/4+u, 1/4+u), respectively. Here, u is the oxygen position coordinate shift parameter. The crystallite size and strain are calculated from Williamson-Hall plot for XRD pattern (see **Table 3**). Since the ionic radii of Mn^{2+} is comparable in coordination of octahedral and tetrahedral sites, so Mn-ferrite (MF) shows more inverse spinel ferrite.

The lattice parameter of MZF nanoparticles with $x = 0$, 0.5, 0.75 and 1 decreases with an increase in the size of the nanoparticle. This indicates the lattice contraction due to reduction of lattice disorder with an increase in nanoparticle size. The nanoparticles have several types of disorders such as oxygen deficiency, lattice disorders, dangling bonds, and so on. It is pointed out that the lattice expansion takes place with a decrease in nanoparticle size [19]. However, in the present study the lattice parameter of MZF nanoparticles with $x = 0.25$ increases initially when the particle size increases from 7 to 10 nm and then decreases with further increase in particle size. The nanoparticles of 7-nm size have impurity phase (Fe_3O_4 phase due to orthorhombic structure). The secondary impurity phase shares the lattice. So, the lattice parameter is lower for these nanoparticles. But in general, the resultant unit cell volume from both the phases will be higher.

The variation of crystallite size of MZF nanoparticles with Zn content is shown in **Figure 2** (left). The drastic decrease in crystallite size occurs at $x = 0.25$. This is attributed to the fact that when Zn^{2+} ions are substituted in Mn-ferrite, it causes a change from mixed spinel to normal spinel structure. The Zn^{2+} ions prefer to occupy tetrahedral site due to its stable valence

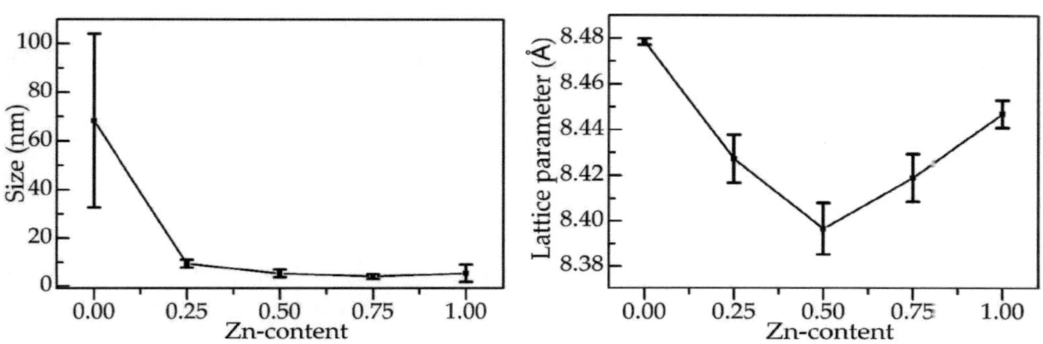

Figure 2. The variation of crystallite size (left) and lattice parameter (right) with Zn content of MZF nanoparticles.

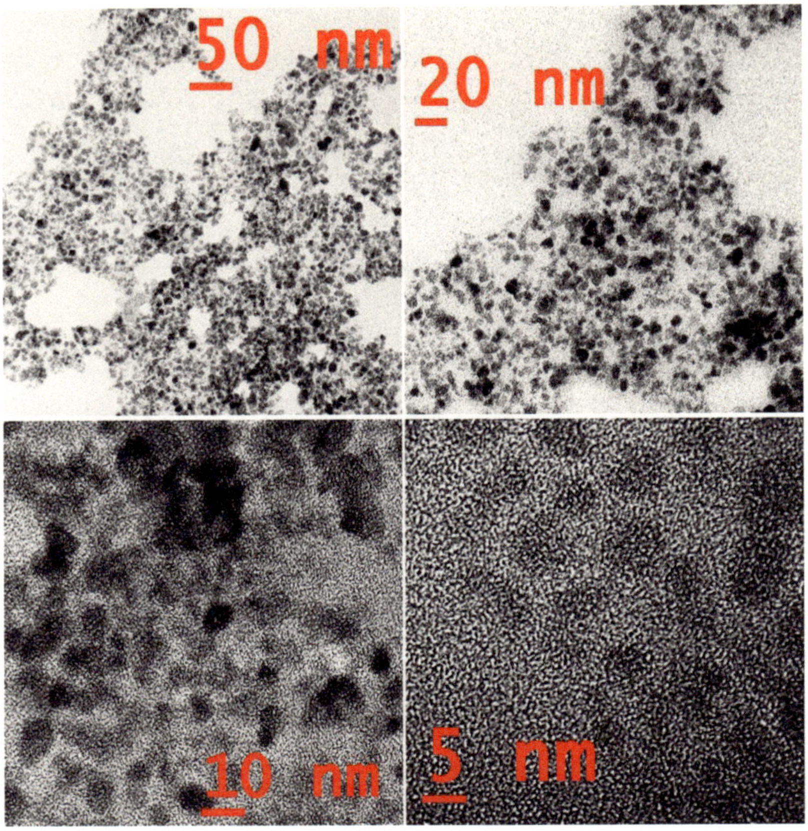

Figure 3. The TEM micrographs of $ZnFe_2O_4$ nanoparticles of crystallite size 5 nm.

and ion size. The decrease in crystallite size is marginal with further increase in Zn content. **Figure 3** (right) shows the variation of lattice parameter with Zn content in MZF nanoparticle. The lattice parameter decreases when x increases from 0 to 0.5 and then increases when Zn content increases from 0.5 to 1. This is due to the transition from mixed spinel to normal spinel phase as x increases from 0 to 0.5. The MZF nanoparticles again return to mixed spinel phase when x increases from 0.5 to 1. This is attributed to the nanophase formation due to the existence of lattice disorders in the system of nanoparticles.

The temperature dependence of the magnetization, $M(T)$, of Mn-ferrite and $Mn_{0.75}Zn_{0.25}Fe_2O_4$ nanoparticles measured at H = 100 Oe (see **Figure 1**, right) shows the bifurcation (irreversible) temperature (T_{irr}). The T_{irr}-value is high (>350 K) for Mn-ferrite nanoparticles. The irreversibility also depends on the crystalline size. The T_{irr} deceases when Zn content x = 0.25 is added in Mn-ferrite nanoparticles. This indicates the decrease in ferromagnetic interactions due to drastic decrease in size with Zn addition and the change in cation distribution which is influenced by the structural changes. With further increase in Zn content, the crystallite size becomes smaller than the critical superparamagnetic size. A single superparamagnetic particle or a system of mono-dispersed nanoparticles shows overlapped-blocking temperature (T_B) and irreversible temperature in $M(T)$ plots. If the size distribution of nanoparticles is taken

http://dx.doi.org/10.5772/66522

into account, each particle size will show a T_B-value. The resultant will be a distribution in T_B-value. The distribution of particle size and magnetic anisotropy is explained by thermomagnetic plots in earlier reports [20–22]. The T_B-value also depends on the magnetic field used for FC-ZFC mode and the measurement time. This is due to the slow relaxation of spins near the blocked state. The size distribution can be found from Neel's model. This model explains the relaxation of non-interacting single- domain nanoparticles experiencing uniaxial anisotropy. The uniaxial anisotropy gives a double well potential in two directions of spin alignment. The wells are separated by an energy barrier E_B to overcome thermal activation with the relaxation time [23]. The relaxation time is: $\tau = \tau_0 \exp(E_B/k_B T)$ whereas the $1/\tau_0$ is attempt frequency. In the small field limit: $E_B = K_{eff} V$ where the K_{eff} is the effective anisotropy constant and V is the particle volume. The volume distribution $f(V)$ can be estimated from M_{ZFC} plot (the thermomagnetic plot in zero-field-cooled mode). Then, the M_{ZFC} can be written in terms of distribution of $T_{B,}$ that is, $f(T_B)$ as follows:

$$m_{ZFC}(T) = \frac{H m_S^2(T)}{3 k_B T} \int_0^T f(T_B)\, d\, T_B$$

The equation implies to $f(T_B)\ \alpha\ \frac{\mathrm{d}}{\mathrm{dT}} (T\, m_{ZFC}(T))$

This equation is modified into the following by considering the log-normal size distribution $f(D)$ [21]:

$$f(D) \ = f_0(1/T_B^2) \frac{\mathrm{d}}{\mathrm{dT}} (T\, m_{ZFC}(T))$$

The decrease of T_B is observed in $M(T)$ plots of superparamagnetic MZF nanoparticles (**Figure 1**, right) with increase in Zn content from 0.5 to 1. This is due to the decrease in size and creation of short-range magnetic order in the nanoparticles. The T_B-value shift is attributed to the magnetic cluster formation and its size distribution [24–28]. The T_B-value and T_{irr} will overlap for the small-sized nanoparticles, whereas they are distinctly different for the large-sized particles. This is attributed to the existence of narrow-size distribution in the smaller nanoparticles compared to large nanoparticles [29–31]. The $M(T)$ results are in good agreement with the size distributions obtained from TEM data analysis. The uniform particle size distribution is observed in $ZnFe_2O_4$ nanoparticles of crystallite size 5 nm from the transmission electron microscope (TEM) micrograph analysis as shown in **Figure 3**.

Figure 4 (left) shows the magnetic hysteresis loops $M(H)$ plots of MZF nanoparticles at 5 and 325 K. These plots show soft ferrimagnetic behaviour for Mn-ferrite and $Mn_{0.75}Zn_{0.25}Fe_2O_4$ nanoparticles [32]. The coercivity is more for the $Mn_{0.75}Zn_{0.25}Fe_2O_4$ nanoparticles compared to Mn-ferrite at 5 and 325 K. This indicates that the domain size is more effective to slow domain wall motion in $Mn_{0.75}Zn_{0.25}Fe_2O_4$ nanoparticles due to non-magnetic Zn inclusion in Mn-ferrite. But there is not much change in saturation magnetization with $x = 0.25$ doping. Here, the surface-to-volume ratio plays more prominent role as the crystallite size decreases from 104 to 11 nm when Zn content varies from 0 to 0.25. Apart from this, Mn-ferrite is cubic spinel ferrite with a partial inverse spinel structure. Inter-sublattice super-exchange interactions of the cations on the (A–B) are much stronger than the (A–A) and (B–B) intra-sublattice exchange interactions. So, the cation distribution plays a major role on the magnetic properties of Mn-ferrite nanoparticles. The Zn

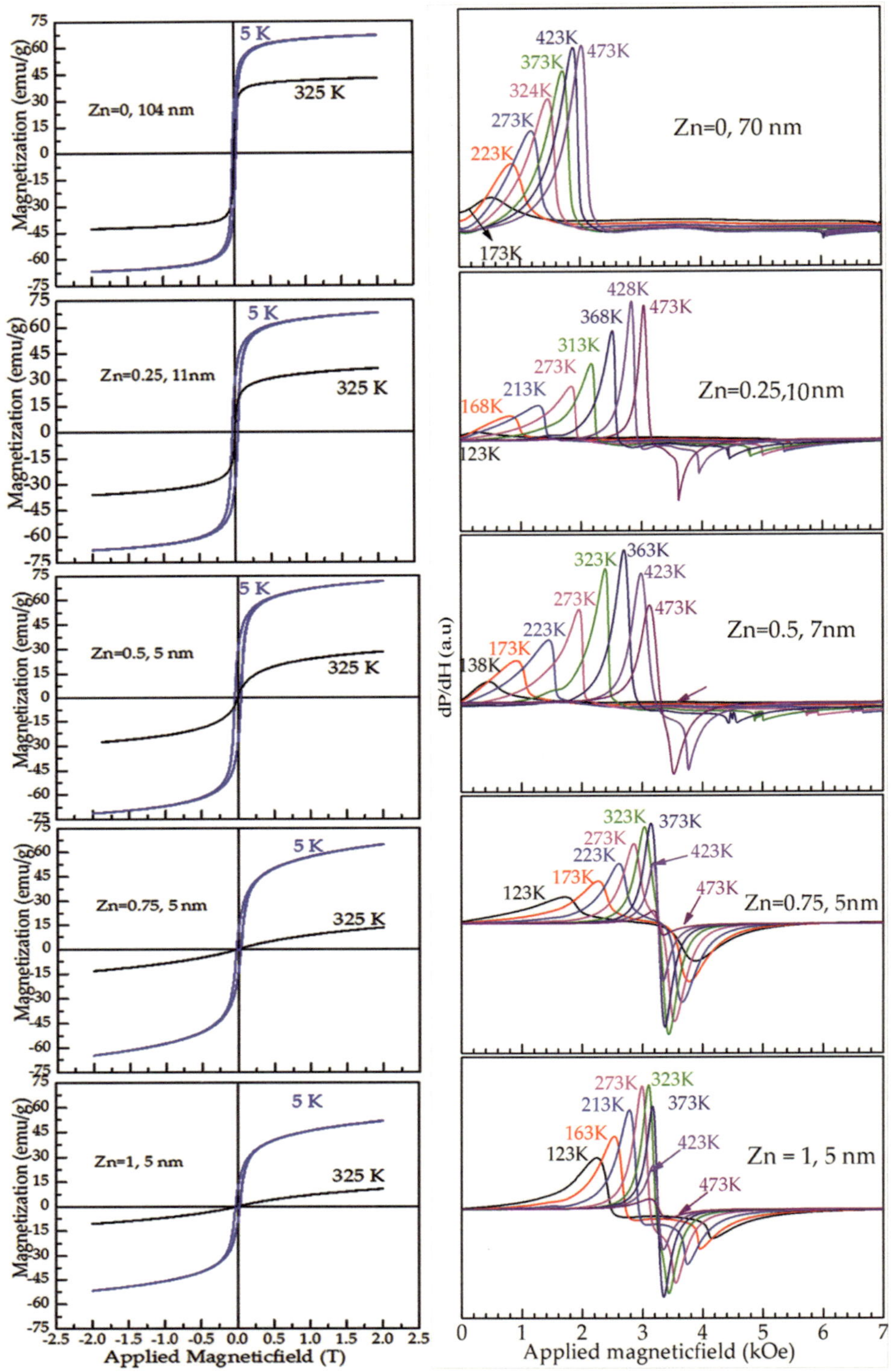

Figure 4. The M(H) plots at 5 and 325 K (left) and the FMR spectra (right) of MZF nanoparticles with different Zn content.

doping in Mn-ferrite decreases the crystallite size due to structural changes towards normal spinel structure. But the net magnetization is not reduced due to a decrease in the A-sublattice magnetization. So, the ferrimagnetism does not show much change in saturation magnetization.

The *M*(*H*) plots of MZF nanoparticles with Zn content of 0.5, 0.75 and 1 having an average crystallite size of 5 nm show zero coercivity at 325 K and small coercivity at 5 K. This is because spins below blocking temperature do not relax. This can be attributed to spin canting and surface spin disorder in the nanoparticles [33, 34]. The coercivity at 5 K decreases with an increase in Zn content from 0.5 to 1. So, the decrease in blocking temperature can be expected with increasing Zn content. The shape of magnetization curve also depends on the measurement time and Neel relaxation of the nanoparticles. The *M*(*H*) plots show the change in M_S which is due to the influence of the cationic stoichiometry and occupancy of cations in specific sites. In addition, random canting of particle surface spins and non-saturation effects due to a random distribution of particles are also responsible for the shape of *M*(*H*) plots. The ferromagnetic resonance spectra shown in **Figure 4** (right) confirm the increase in the ferromagnetic component as Zn content is decreased in the MZF. This is indicated by progressive shifting of shoulder peak towards lower field value with a decrease in Zn content in MZF.

5. Investigation of magnetic properties of spinel ferrites based on Mössbauer studies

Mössbauer measurements on MZF nanoparticles were carried out at room temperature in transmission geometry using ^{57}Fe nuclei. Mössbauer spectrum of Mn-ferrite nanoparticles

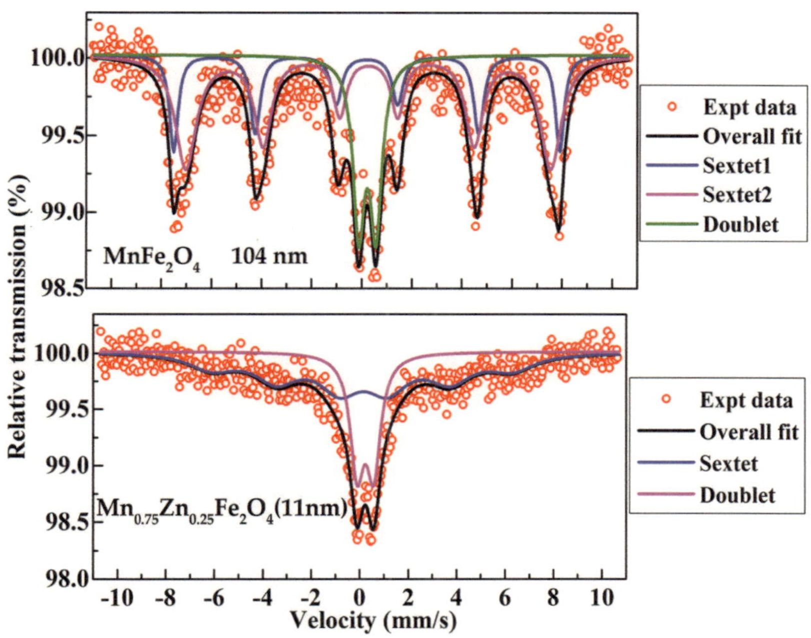

Figure 5. The Mössbauer spectra of Mn-ferrite and $Mn_{0.75}Zn_{0.25}Fe_2O_4$ nanoparticles.

with crystallite size of 104 nm is resolved into two sextets and one doublet (**Figure 5**). This system is soft ferromagnetic at room temperature. Two sextets come from two different Fe co-ordinations.

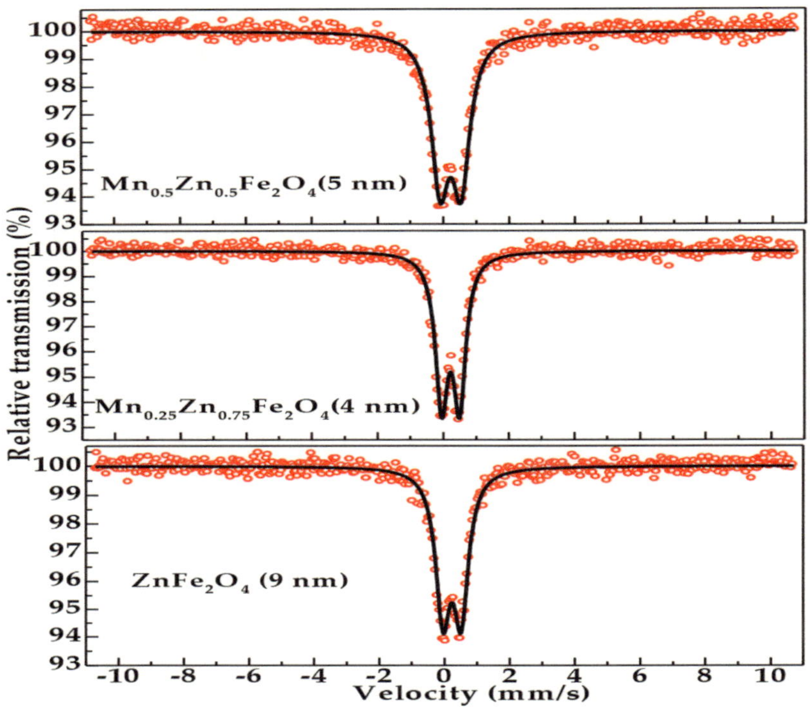

Figure 6. The Mössbauer spectra of MZF nanoparticles.

Sample	Sub-spectrum	Mössbauer parameters				
		H_{hf} (kOe)	Quadrupole shift (ΔE_q) (mm/s)	Isomer shift δ (mm/s)	Line width WV (mm/s)	Relative intensity (%)
$MnFe_2O_4$	Sextet-1 (A-site)	478	0.002	0.195	0.40	20
	Sextet-2 (B-site)	452	0.009	0.262	0.86	58
	Doublet (SPM)	–	0.691	0.223	0.55	22
$Mn_{0.75}Zn_{0.25}Fe_2O_4$	Sextet	385	0.055	0.129	1.86	67
	Doublet (SPM)	–	0.684	0.234	0.65	33
$Mn_{0.5}Zn_{0.5}Fe_2O_4$	Doublet (SPM)	–	0.633	0.222	0.65	100
$Mn_{0.25}Zn_{0.75}Fe_2O_4$	Doublet (SPM)	–	0.541	0.225	0.46	100
$ZnFe_2O_4$	Doublet (SPM)	–	0.553	0.231	0.53	100

Table 4. Parameters determined from the analysis of the Mössbauer spectra are listed here.

http://dx.doi.org/10.5772/66522

Sextet 1 has the hyperfine field value of 478 kOe and the absence of quadrupole splitting indicates that the Fe site is octahedral coordinated in cubic symmetry with 3+ valence state. Sextet 2 has a hyperfine field value of 452 kOe with quadrupole splitting of 0.009 mm/s due to 3+ valence state of Fe in tetrahedral coordination [35–37]. In addition, the doublet of relative area of around 22% is observed and this is due to the existence of small-sized magnetic particles showing superparamagnetic behaviour.

When 25 mole % of Zn (x = 0.25) is introduced in Mn-ferrite nanoparticles (with the crystallite size of 11 nm), the Mössbauer spectra are fitted with resolved one sextet and one doublet with a relative area of 67 and 33%, respectively. The sextet has a hyperfine field value of 385 Oe with quadrupole splitting of 0.055 mm/s. This indicates the possibility of tetrahedral coordination of Fe^{3+} ions. Mössbauer spectra of MZF nanoparticles (x = 0.5, 0.75 and 1) showed single-doublet pattern shown in **Figure 6** for the crystallite size of 5, 4 and 9 nm, respectively. This indicates that the nanoparticles are small-sized magnetic particles showing superparamagnetic behaviour at room temperature. Mössbauer parameters of all the samples investigated here are listed in **Table 4**.

6. Magneto-viscosity of ferrofluids of spinel ferrites

When the magnetic nanoparticles are suspended in colloidal, there are three main forces acting on the particles, that is, internal magnetic force, surface tension of the fluid and gravitational force. The ferrofluids show spikes when the fluid is subjected to the magnetic field. The chain-like or spot-like structure formation depends on these forces. The magneto-viscosity of the ferrofluid mainly depends on the magnetic properties of the nanoparticles used in the colloidal. The ferrofluids of superparamagnetic and ferromagnetic nanoparticles were chosen

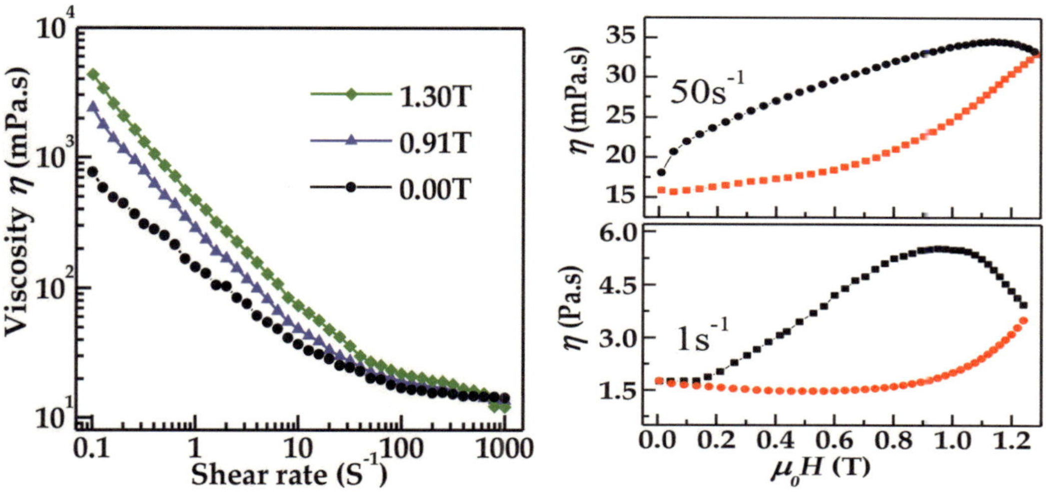

Figure 7. Flow curves at different applied fields (left) and magneto-viscosity curves at different shear rates (right) of MZFE ferrofluid.

for the study of magneto-viscosity. The surfactant coating to the nanoparticles increases the stability of the ferrofluid.

6.1. Magneto-viscosity of ferromagnetic nanoparticles-dispersed ferrofluid without surfactant coating

The $Mn_{0.75}Zn_{0.25}Fe_2O_4$ nanoparticles of 3-nm size were dispersed in ethylene glycol to prepare MZFE ferrofluid. The volume ratio of nanoparticles to the colloidal is 1:4. The superparamagnetic nature of the nanoparticles is confirmed by magnetization studies. **Figure 8** shows shear viscosity versus shear flow plot at various magnetic field values. The viscosity decreases with an increase in shear rate at zero magnetic field. It shows a shear-thinning effect at low shear rate and Newtonian behaviour over a wide range of shear rate. This can be explained by considering that at low shear rate the aggregates or clusters of nanoparticles offer high resistance to fluid flow leading to high viscosity. With increases in shear rate, the aggregates break into smaller units leading to fluid flow with low viscosity [38]. The interaction between aggregates or clusters increases when magnetic field is applied and aligns them in the direction of magnetic field forming chain-like structures which offer high resistance to the fluid flow leading to an increase in viscosity. Magneto-viscosity of other ferrofluids also shows similar behaviour [39, 40].

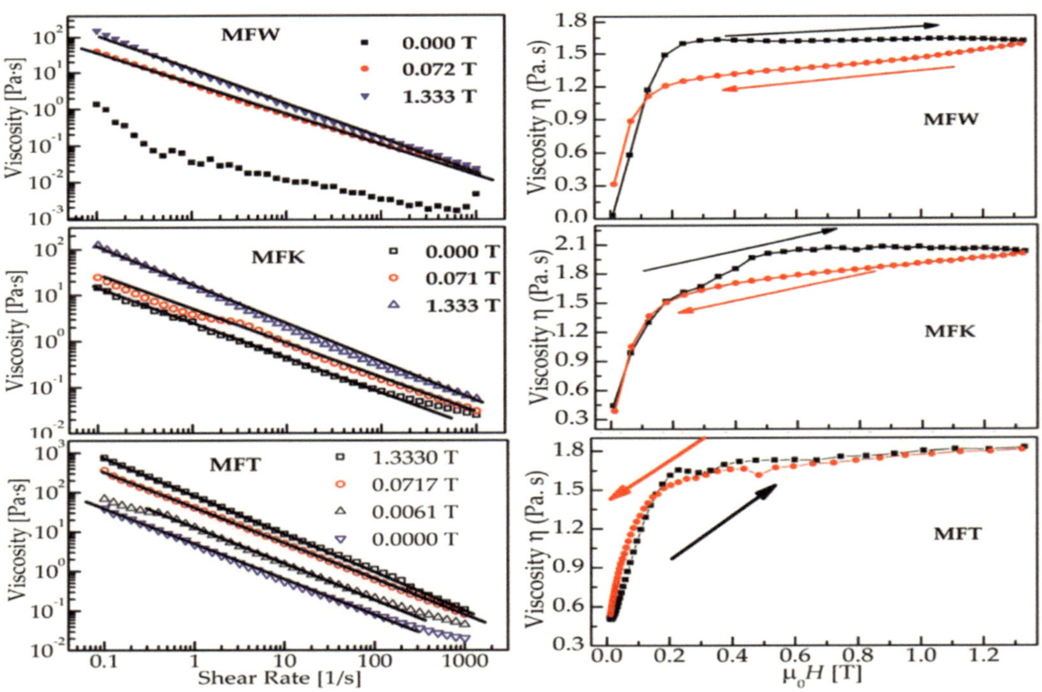

Figure 8. Viscosity as a function of shear rate at different applied magnetic fields (left) and viscosity as a function of applied magnetic field at a shear rate of 10 S^{-1} (right) of water-based (MFW), kerosene-based (MFK) and toluene-based (MFT) Mn-ferrite ferrofluids.

Odenbach [41] studied the magneto-viscous effect at various shear rates in a commercial magnetite-based ferrofluid and explained it in terms of interactions between the particles in the agglomerations aligned in straight chains and not due to the interaction between the chains. **Figure 7** shows irreversible nonlinear behaviour in η versus applied magnetic field plots at the steady shear rates of 1 and 50 s^{-1}. The irreversibility of magneto-viscosity curve is larger at low shear rate similar to the earlier reports [42]. The field dependence of viscosity indicates the particle size distribution and formation of chains with an increase in magnetic field. With a decrease in field, the viscosity increases due to the magnetic interactions and mechanical alignment of particles and slow loss of aggregation of particles. Then, the system tries to retain its initial state. The open hysteresis loop is due to unbroken chains and aggregates present in the system [43]. Odenbach [41] and Rhodesa [43] explained magnetic influence on the viscous properties of ferrofluid in terms of four forces acting on the magnetic particles in the ferrofluid. These forces arise from liquid media, internal magnetic field between magnetic particles, gravitational field and the applied magnetic field. The four regions in the plots of **Figure 7** are according to the limitations of these forces. Apart from this, the irreversibility is due to the relaxation of the nanoparticles in ferrofluids. The Brownian and Neel relaxations play a role for the observed behaviour of SPM nanoparticles in liquid media [13, 44].

6.2. Magneto-viscosity of surfactant-coated ferromagnetic nanoparticles-dispersed ferrofluids

It is quite interesting and more specific to study the magneto-viscosity of magnetic nanoparticles in less viscous colloidal such as water, kerosene and toluene. The ferromagnetic MF and MZF nanoparticles are used in different colloidal media with suitable surfactant coating for the preparation of MF and MZF ferrofluids. The surfactants have hydrophobic and hydrophilic ends. The hydrophilic end will be on the nanoparticle surface and hydrophobic end will overcome the formation of agglomerates of the nanoparticles. This property increases the stability of the ferrofluids.

6.2.1. Magneto-viscosity of Mn-ferrite ferrofluids

The magneto-viscosity of Mn-ferrite ferrofluid is investigated in flow field (flow curves) and in external magnetic field (magneto-viscosity plots). Mn-ferrite ferrofluids (i.e. MFW, MFK and MFT) were synthesized using Mn-ferrite nanoparticles suspension in different colloidal, that is, water, kerosene and toluene, respectively. The flow behaviour is studied for these ferrofluids in various applied magnetic fields. The respective volume ratio of Mn-ferrite nanoparticles, surfactant and colloidal (water, kerosene and toluene for respective ferrofluids) is taken as 1:0.5:1.5 for the preparation of Mn-ferrite ferrofluids. The MF nanoparticles are coated with tetramethyl ammonia (TMA) for the water-based MF ferrofluid (MFW). The kerosene-based ferrofluid (MFK) and toluene-based ferrofluid (MFT) are synthesized by using oleic acid-coated nanoparticles.

6.2.2. Flow curves of Mn-ferrite ferrofluids

Figure 8 (left) shows the flow curves of MF ferrofluid at various applied magnetic fields in the range from 0 to 1.33 T. The MFW ferrofluid shows non-Newtonian behaviour at lower shear rates for zero magnetic field. The behaviour changes to Newtonian behaviour with increasing shear rate. The TMA coating of the nanoparticles exhibits more hydrophobic nature of the nanoparticles, leading to decrease in friction between the layers of the ferrofluid. So the power law behaviour cannot be observed in zero field magneto-viscosity plot for the water-based ferrofluid. For all the range of shear rates, it follows single behaviour, that is, power law behaviour without any discrepancy.

$$f(D) = f_0(1/T_B^2)\frac{\mathrm{d}}{\mathrm{dT}}(T\,m_{ZFC}(T))$$

Here, K is the consistency coefficient and the exponent n is the power law index.

Similar behaviour is observed in the MFK and MFT ferrofluids but the magnetic response is more. This is because the MFK and MFT are synthesized using FM nanoparticles in less viscous colloidal. So the hydrodynamic force is less compared to MFW ferrofluid. The very small n-value (0.1–0.05) indicates the high shear-thinning behaviour as reported in an earlier work [45]. With increase in applied field, n-value decreases indicating that the viscosity is affected more by magnetic field at low shear rates than at high shear rates [46, 47]. This is due to the competition between the flow field and the applied magnetic field. In this process, at higher shear rates (shear rate greater than 200 s^{-1}), the flow field dominates, whereas applied field dominates at lower shear rates. These ferrofluids showed completely non-Newtonian behaviour whereas the SPM nanoparticles-based ferrofluids (MZFE1) show the change in behaviour from non-Newtonian to Newtonian with an increase in shear rate similar to that reported in our earlier work [48, 49]. Above the applied field of 0.3 T, the magnetic nanoparticles (magnetic nano-dipoles) orient completely in the field direction with long-chain formation in the fluid.

6.2.3. Magneto-viscosity of Mn-ferrite ferrofluids

The magneto-viscosity plots at a shear rate of 10 s^{-1} (**Figure 8**, right) show a rapid increase with an increase in magnetic field initially, followed by its saturation at higher fields, with low hysteresis when the applied field is decreased to zero. Since the ferromagnetic particles are dispersed in less viscous fluid, as the field increases the nanoparticles try to rotate in the field direction [50–52]. Above the applied field of 0.3 T, the viscosity saturates. Because of long-chain formation along the field direction, the viscosity reaches the maximum possible value at the given shear rate. Magneto-viscosity at different shear rates is described in Refs. [53–56]. This behaviour is similar to magnetization plots of nanoparticles as a function of magnetic field.

6.2.4. Magneto-viscosity of Mn-Zn ferrite ferrofluids

Mn-Zn ferrite ferrofluids were synthesized using the MZF nanoparticles suspension in different colloidal, that is, water, kerosene, toluene and paraffin, respectively. The respective volume ratio of Mn-Zn ferrite nanoparticles, oleic acid and colloidal (water, kerosene, toluene or paraffin for respective ferrofluids) is taken as 1:0.5:1.5 for the preparation of Mn-Zn ferrite ferrofluids.

6.2.5. Flow curves of Mn-Zn ferrite ferrofluids

The flow behaviour is studied using flow curves shown in **Figure 9** (left)) for $Mn_{0.75}Zn_{0.25}Fe_2O_4$ ferrofluids in different applied magnetic fields. The fluid shows power law behaviour with

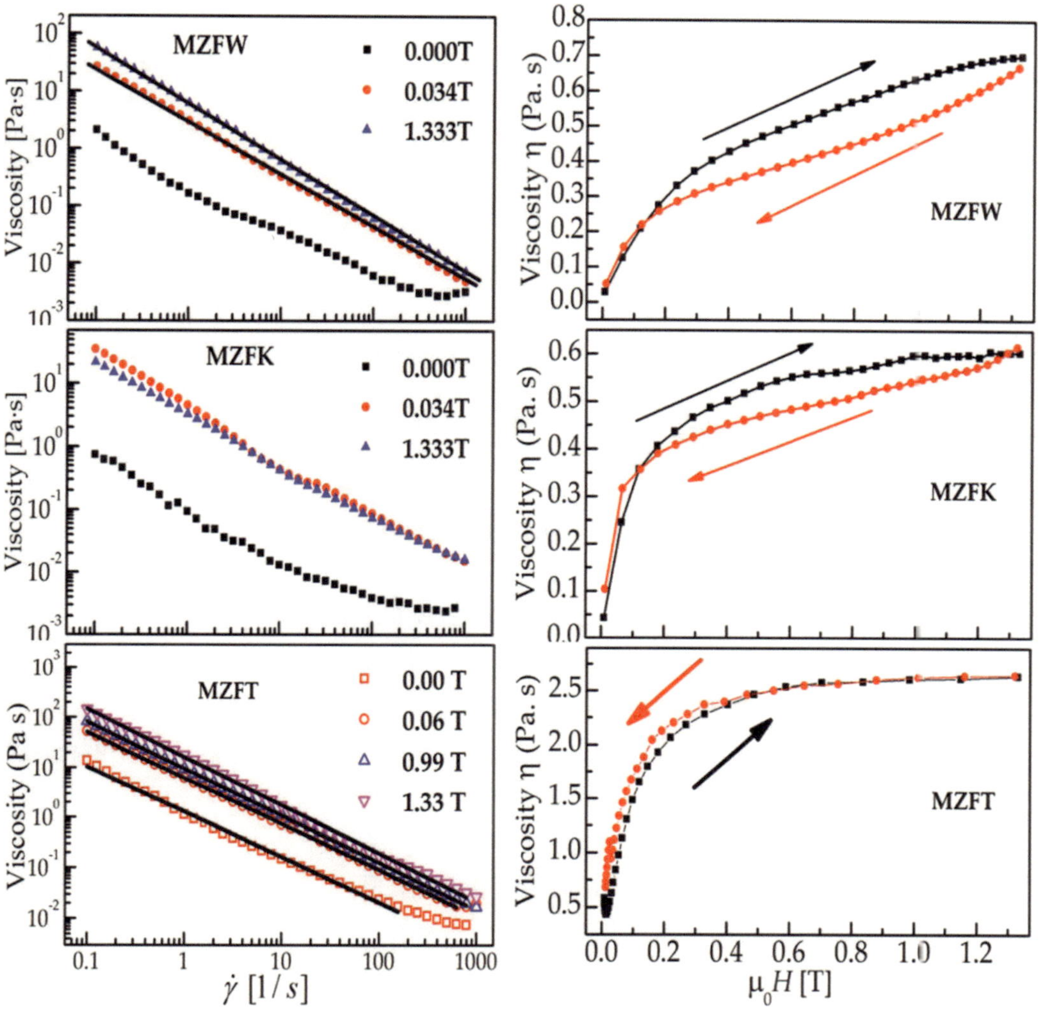

Figure 9. Viscosity as a function of shear rate at different applied magnetic fields (left) and viscosity as a function of applied magnetic field at a shear rate of 10 S^{-1} (right) of MZF ferrofluids synthesized with various colloidal.

small n-value between 0.1 and 0.04 for various magnetic fields in the range of 0–1.33 T indicating non-Newtonian behaviour with higher shear thinning [45]. The viscosity versus shear rate plot show the change from non-Newtonian to Newtonian behaviour at higher shear rates (shear rate greater than 300 s^{-1}), as we described in our earlier work [48]. The fluid viscosity behaviour is accounted to only the applied magnetic field. So for all the range of shear rate, it follows single power law behaviour without any discrepancy. The non-dimensional parameter, Mason number (*Ma*), explains the flow behaviour. Similar to electro-rheological fluids, the 'Ma' is the ratio of shear forces or hydrodynamic forces (F_H) to the magnetic forces (F_M), that is, $Ma = F_H/F_M$. The viscosity as a function of Mason number also follows the power law behaviour. The critical *Ma*-value determines the transition from magnetization to hydrodynamic control of the suspension structure [57].

6.2.6. Magneto-viscosity plots of Mn-Zn ferrite ferrofluids

The viscosity versus applied magnetic field at a shear rate of 10 S^{-1} is shown in **Figure 9** (right). The magneto-viscosity plots show a rapid increase with an increase in magnetic field initially. The viscosity is saturated at higher fields (around 0.2 T) and show low hysteresis when the applied field is decreased to zero. This behaviour is similar to the magnetization plots of the nanoparticles as a function of magnetic field.

6.2.7. Magneto-viscosity of paraffin-based ferrofluids

Paraffin is more stable colloidal compared with other colloidals used for the synthesis of ferrofluids. The dispersion of FM nanoparticles in paraffin gives fine control of viscosity in magnetic field. This study is quite interesting and useful for ferrofluid applications. Similar to the synthesis of other ferrofluids, the paraffin-based $Mn_{0.75}Zn_{0.25}Fe_2O_4$ (MZFP1) and $Mn_{0.9}Zn_{0.1}$ Fe_2O_4 (MZFP2) ferrofluids are synthesized using oleic acid-coated nanoparticles. The magneto-viscosity plots of MZFP1 and MZFP2 ferrofluids are shown in **Figure 10** at the shear rate of 1 and 10 s^{-1}.

The comparison between the magneto-viscosity of these ferrofluids is as follows:

(i) The gradual increment of viscosity is observed in MZFP2 ferrofluid with an increase in magnetic field whereas sharp increment is observed in MZFP1 ferrofluid.

(ii) When the magnetic field is decreased, the magneto-viscosity plots of MZFP1 ferrofluid show less hysteresis and the plot is not relaxing to initial position at zero magnetic field.

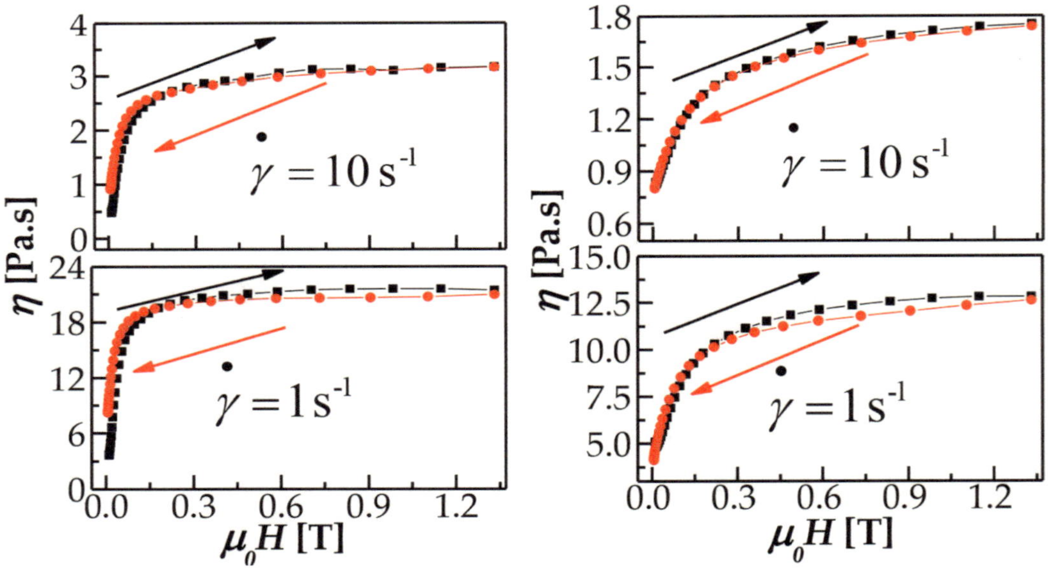

Figure 10. Viscosity as a function of applied magnetic field at different shear rates of paraffin-based ferrofluids of composition $Mn_{0.75}Zn_{0.25}Fe_2O_4$ (MZFP1) (left) and $Mn_{0.9}Zn_{0.1}$ Fe_2O_4 (MZFP2) (right).

http://dx.doi.org/10.5772/66522

(iii) The magnetic nanoparticles are aligned in field direction and the nanoparticles are rotated along the easy axis of the system.

(iv) Resultant unbroken large chains are expected even when the magnetic field is removed. The MZFP2 ferrofluid shows different behaviour due to smaller-size particles and lower initial susceptibility.

6.2.8. The Herschel-Bulkley behaviour in ferrofluids

The MF nanoparticles dispersed in less viscous colloidal (toluene) are free to move in the fluid and so it is easy to study the structural changes in spot-like and chain-like structures as a function of applied magnetic field. MFT ferrofluid is chosen to explain the Herschel-Bulkley (H-B) behaviour in ferrofluids. Shear stress (τ) as a function of shear rate ($\ddot{\gamma}$) plots is shown in **Figure 11**.

The H-B model combines the power law model with the yield stress as per the equation, $\tau = \tau_0 + K\dot{\gamma}^n$

Here, τ_0 is the yield stress. The yield stress can be determined from the plots using H-B model.

Apart from the magnetic behaviour of magnetic particles and the colloidal behaviour, the dosage of surfactant and the concentration of the magnetic nanoparticles in ferrofluids play a major role for the change of rheological behaviour at lower shear rates. This influences the flow field interaction in the ferrofluid. With the application of field, the shear stress versus shear rate plots show the H-B fluid behaviour. It is also observed as the magnetic field increases, the shear stress increases. The yield stress was determined by the extrapolation of the plots using the B-H model. The plots shifted upward due to the drastic change in yield stress. It increases from 4 to 27 Pa as magnetic field increases from 0 to 0.0717 T. The deviation from H-B model fit can be observed in shear stress (τ) versus shear rate plot at 1.33 T applied magnetic field [58]. As the applied field increases, the formation of chain-like and spot-like structure takes

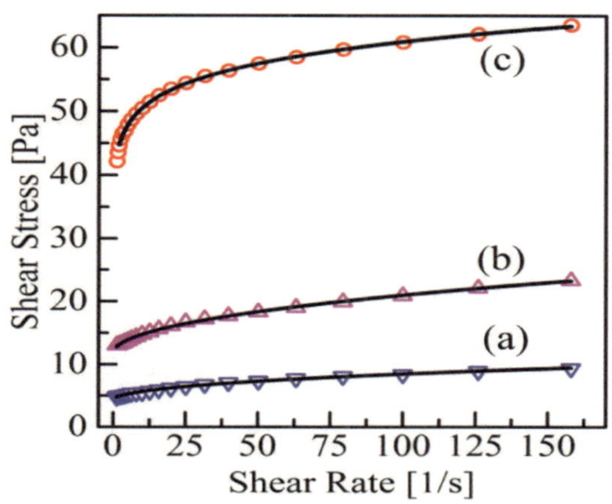

Figure 11. Shear stress as a function of shear rate, whereas solid line is H-B model fit for applied fields 0.000 (a), 0.0061 (b) and 0.0717 T (c) of MFT ferrofluid.

place [59]. The fluid remains as H-B fluid in the presence of small number of chains formed in the low field region with the increase in magnetic field. The strong- and long-chain formation occurs with the increase in magnetic field further. The magnetic field force dominates the other forces present in the fluid, as the magnetic field is increased leading to deviation from H-B fluid behaviour.

7. Conclusions

The nanoparticles of various sizes and compositions $Mn_{1-x}Zn_x Fe_2O_4$ with $x = 0–1$ were synthesized by co-precipitation method. The detailed structural and magnetic properties of MZF nanoparticles were carried out. The lattice parameter decreases with an increase in Zn content reaching a minimum value at $x = 0.5$ followed by an increasing trend with further increase in Zn content. The saturation magnetization and blocking temperature increase with a decrease in Zn content. The ferrofluids based on superparamagnetic and ferromagnetic MZF nanoparticles were synthesized. The magneto-viscosity of ferrofluids with the dispersion of nanoparticles in different colloidal was studied. The viscosity versus shear rate plot in applied magnetic field follows single power law behaviour without any discrepancy. The data on ferrofluids based on MF nanoparticles dispersed in less viscous colloidal (toluene) are analysed in view of Herschel-Bulkley model.

Author details

Rajender Singh* and Gadipelly Thirupathi

*Address all correspondence to: rsinghsp@gmail.com

School of Physics, University of Hyderabad, Hyderabad, India

References

[1] L. Yu, G. Hao, J. Gu, S. Zhou, N. Zhang, and W. Jiang, *Journal of Magnetism and Magnetic Materials*, vol. 394, pp. 14–21, Nov. 2015.

[2] J.-H. Yu, D.-W. Lee, B.-K. Kim, and T. Jang, *Journal of Magnetism and Magnetic Materials*, vol. 304, no. 1, pp. e16–e18, Sep. 2006.

[3] L. B. Tahar, H. Basti, F. Herbst, L. S. Smiri, J. P. Quisefit, N. Yaacoub, J. M. Grenèche, and S. Ammar, *Materials Research Bulletin*, vol. 47, no. 9, pp. 2590–2598, Sep. 2012.

[4] M. Staruch, D. Hires, D. Violette, D. Navarathne, G. A. Sotzing, and M. Jain, *Integrated Ferroelectrics*, vol. 131, no. 1, pp. 102–109, 2011.

[5] Y. Xuan, Q. Li, and G. Yang, *Journal of Magnetism and Magnetic Materials*, vol. 312, no. 2, pp. 464–469, May 2007.

[6] M. Sertkol, Y. Köseoğlu, A. Baykal, H. Kavas, and A. C. Başaran, *Journal of Magnetism and Magnetic Materials*, vol. 321, no. 3, pp. 157–162, Feb. 2009.

[7] R. Justin Joseyphus, A. Narayanasamy, K. Shinoda, B. Jeyadevan, and K. Tohji, *Journal of Physics and Chemistry of Solids*, vol. 67, no. 7, pp. 1510–1517, Jul. 2006.

[8] G. Thirupathi and R. Singh, *IEEE Transactions on Magnetics*, vol. 48, no. 11, pp. 3630–3633, Nov. 2012.

[9] R. M. Freire, T. S. Ribeiro, I. F. Vasconcelos, J. C. Denardin, E. B. Barros, G. Mele, L. Carbone, S. E. Mazzetto, and P. B. A. Fechine, *Journal of Nanoparticle Research*, vol. 15, p 2041, May 2013.

[10] K. Ali, A. K. Sarfraz, I. M. Mirza, A. Bahadur, S. Iqbal, and A. ul Haq, *Current Applied Physics*, vol. 15, no. 8, pp. 925–929, Aug. 2015.

[11] L. Vékás, M. V. Avdeev, and D. Bica, in *NanoScience in Biomedicine*, D. Shi, Ed. Springer: Berlin, Heidelberg, 2009, pp. 650–728.

[12] F. A. Tourinho, R. Franck, and R. Massart, *Journal of Material Science*, vol. 25, no. 7, pp. 3249–3254, Jul. 1990.

[13] I. Sharifi, H. Shokrollahi, and S. Amiri, *Journal of Magnetism and Magnetic Materials*, vol. 324, no. 6, pp. 903–915, Mar. 2012.

[14] N. Maleki-Jirsaraei, B. Ghane-Motlagh, F. Ghane-Golmohamadi, R. Ghane-Motlagh, and S. Rouhani, in *2010 International Conference on Nanoscience and Nanotechnology (ICONN)*, 2010, pp. 91–93.

[15] D. Maity and D. C. Agrawal, *Journal of Magnetism and Magnetic Materials*, vol. 308, no. 1, pp. 46–55, Jan. 2007.

[16] Z. X. Tang, C. M. Sorensen, K. J. Klabunde, and G. C. Hadjipanayis, *Journal of Applied Physics*, vol. 69, no. 8, pp. 5279–5281, Apr. 1991.

[17] E. Solano, C. Frontera, I. Puente Orench, T. Puig, X. Obradors, S. Ricart, and J. Ros, *Journal of Applied Crystallography*, vol. 47, no. 4, pp. 1478–1478, Aug. 2014.

[18] P. K. Manna, S. M. Yusuf, M. Basu, and T. Pal, *Journal of Physics: Condensed Matter*, vol. 23, no. 50, p. 506004, Dec. 2011.

[19] K. C. George, S. Kurien, and J. Mathew, *Journal of Nanoscience and Nanotechnology*, vol. 7, no. 6, pp. 2016–2019, Jun. 2007.

[20] B. D. Plouffe, D. K. Nagesha, R. S. DiPietro, S. Sridhar, D. Heiman, S. K. Murthy, and L. H. Lewis, *Journal of Magnetism and Magnetic Materials*, vol. 323, no. 17, pp. 2310–2317, Sep. 2011.

[21] R. S. DiPietro, H. G. Johnson, S. P. Bennett, T. J. Nummy, L. H. Lewis, and D. Heiman, *Applied Physics Letters*, vol. 98, no. 21, p. 216103, 2011.

[22] R. S. DiPietro, H. G. Johnson, S. P. Bennett, T. J. Nummy, L. H. Lewis, and D. Heiman, *Applied Physics Letters*, vol. 96, no. 22, p. 222506, 2010.

[23] A. Aharoni, *Introduction to the Theory of Ferromagnetism*. Oxford University Press, Great Clarendon Street, Oxford OX2 6DP ,2000.

[24] Q. Song and Z. J. Zhang, *Journal of Physical Chemistry B*, vol. 110, no. 23, pp. 11205–11209, Jun. 2006.

[25] P. Jeppson, R. Sailer, E. Jarabek, J. Sandstrom, B. Anderson, M. Bremer, D. G. Grier, D. L. Schulz, A. N. Caruso, S. A. Payne, P. Eames, M. Tondra, H. He, and D. B. Chrisey, *Journal of Applied Physics*, vol. 100, no. 11, p. 114324, Dec. 2006.

[26] K. Nadeem, H. Krenn, W. Sarwar, and M. Mumtaz, *Applied Surface Science*, vol. 288, pp. 677–681, Jan. 2014.

[27] K. Maaz, W. Khalid, A. Mumtaz, S. K. Hasanain, J. Liu, and J. L. Duan, *Physica E: Low-Dimensional Systems and Nanostructures*, vol. 41, no. 4, pp. 593–599, Feb. 2009.

[28] N. T. Lan, T. D. Hien, N. P. Duong, and D. V. Truong, *Journal of the Korean Physical Society*, vol. 52, no. 5, p. 1522, May 2008.

[29] Y. Jun, J. Seo, and J. Cheon, *Accounts of Chemical Research*, vol. 41, no. 2, pp. 179–189, Feb. 2008.

[30] M. Han, C. R. Vestal, and Z. J. Zhang, *Journal of Physical Chemistry B*, vol. 108, no. 2, pp. 583–587, Jan. 2004.

[31] S. Dey, S. K. Dey, B. Ghosh, P. Dasgupta, A. Poddar, V. R. Reddy, and S. Kumar, *Journal of Applied Physics*, vol. 114, no. 9, p. 093901, Sep. 2013.

[32] Q. Song and Z. J. Zhang, *Journal of the American Chemical Society*, vol. 134, no. 24, pp. 10182–10190, Jun. 2012.

[33] Rajshree B. Jotania and Hardev S. Virk, Ferrites and Ceramic Composites Trans Tech Publications,Inc, Reinhardstrasse 18, CH-8008 Zurich, Switzerland, 2013.

[34] S. Bedanta and W. Kleemann, *Journal of Physics D: Applied Physics*, vol. 42, no. 1, p. 013001, 2009.

[35] R. Malik, S. Annapoorni, S. Lamba, V. Raghavendra Reddy, A. Gupta, P. Sharma, and A. Inoue, *Journal of Magnetism and Magnetic Materials*, vol. 322, no. 23, pp. 3742–3747, Dec. 2010.

[36] L. B. Tahar, M. Artus, S. Ammar, L. S. Smiri, F. Herbst, M.-J. Vaulay, V. Richard, J.-M. Grenèche, F. Villain, and F. Fiévet, *Journal of Magnetism and Magnetic Materials*, vol. 320, no. 23, pp. 3242–3250, Dec. 2008.

[37] S. Kubickova, J. Vejpravova, P. Holec, and D. Niznansky, *Journal of Magnetism and Magnetic Materials*, vol. 334, pp. 102–106, May 2013.

[38] S. Odenbach and H. Störk, *Journal of Magnetism and Magnetic Materials*, vol. 183, no. 1–2, pp. 188–194, Mar. 1998.

[39] M. Chand, S. Kumar, A. Shankar, R. Porwal, and R. P. Pant, *Journal of Non-Crystalline Solids*, vol. 361, pp. 38–42, Feb. 2013.

[40] D. Soto-Aquino and C. Rinaldi, *Physical Review E*, vol. 82, no. 4, p 046310, Oct. 2010.

[41] S. Odenbach, *Colloids and Surfaces A: Physicochemical and Engineering Aspects*, vol. 217, no. 1–3, pp. 171–178, Apr. 2003.

[42] S. Odenbach, *Journal of Physics: Condensed Matter*, vol. 16, no. 32, pp. R1135–R1150, Aug. 2004.

[43] S. Rhodes, X. He, S. Elborai, S.-H. Lee, and M. Zahn, *Journal of Electrostatics*, vol. 64, no. 7–9, pp. 513–519, Jul. 2006.

[44] P. C. Fannin, L. Kinsella, and S. W. Charles, *Journal of Physics D: Applied Physics*, vol. 30, no. 4, p. 533, 1997.

[45] M. T. López-López, A. Gómez-Ramírez, L. Rodríguez-Arco, J. D. G. Durán, L. Iskakova, and A. Zubarev, *Langmuir*, vol. 28, no. 15, pp. 6232–6245, Apr. 2012.

[46] S. Odenbach, *Magnetoviscous Effects in Ferrofluids*. Springer: Berlin, 2002.

[47] K. Shah, R. V. Upadhyay, and V. K. Aswal, *Smart Materials and Structures*, vol. 21, no. 7, p. 075005, Jul. 2012.

[48] G. Thirupathi and R. Singh, *Physica B: Condensed Matter*, vol. 448, pp. 346–348, Sep. 2014.

[49] A. Y. Zubarev and L. Y. Iskakova, *Journal of Physics: Condensed Matter*, vol. 18, no. 38, p. S2771, Sep. 2006.

[50] L. J. Felicia and J. Philip, *Physical Review E*, vol. 89, no. 2, p. 022310, Feb. 2014.

[51] J. Nowak, E. Dohmen, and S. Odenbach, *IEEE Transactions on Magnetics*, vol. 50, no. 11, pp. 1–4, Nov. 2014.

[52] H. Shahnazian, D. Gräf, D. Y. Borin, and S. Odenbach, *Journal of Physics D: Applied Physics*, vol. 42, no. 20, p. 205004, Oct. 2009.

[53] A. Y. Zubarev and L. Y. Iskakova, *Physica A: Statistical Mechanics and Its Applications*, vol. 376, pp. 38–50, Mar. 2007.

[54] L. M. Pop and S. Odenbach, *Journal of Physics: Condensed Matter*, vol. 18, no. 38, p. S2785, Sep. 2006.

[55] D. Susan-Resiga and L. Vékás, *Rheologica Acta*, vol. 53, no. 8, pp. 645–653, Aug. 2014.

[56] K. Shah, R. V. Upadhyay, and V. K. Aswal, *Smart Materials and Structures*, vol. 21, no. 7, p. 075005, Jul. 2012.

[57] N. Gautam, G. Thirupathi, and R. Singh, *IEEE Transactions on Magnetics*, Article #: 4600204,Volume: 52, Issue: 7, July 2016

[58] R. Y. Hong, Z. Q. Ren, Y. P. Han, H. Z. Li, Y. Zheng, and J. Ding, *Chemical Engineering Science*, vol. 62, no. 21, pp. 5912–5924, Nov. 2007.

[59] M. Klokkenburg, B. H. Erné, A. Wiedenmann, A. V. Petukhov, and A. P. Philipse, *Physical Review E*, vol. 75, no. 5, p. 051408, May 2007.

Chapter 8

Mechanochemical Synthesis of Water-Based Magnetite Magnetic Fluids

Tomohiro Iwasaki

Additional information is available at the end of the chapter

http://dx.doi.org/10.5772/66075

Abstract

Magnetite, Fe_3O_4 ($Fe^{II}O-Fe^{III}{}_2O_3$), is a member of the spinel group as well as a common ferrite with a cubic inverse spinel structure. Aqueous colloidal solutions of Fe_3O_4, i.e. water-based Fe_3O_4 magnetic fluids, have attracted substantial attention in biomedical applications, such as drug delivery, magnetic resonance imaging, and magnetic hyperthermia. In this study, to readily prepare water-based magnetic fluids with biocompatible dispersant-coated Fe_3O_4 nanoparticles that are stably dispersed in water medium, a mechanochemical synthesis method was developed. In this method, an iron-free citric acid solution is milled in a tumbling ball mill with steel balls at room temperature, reducing the production costs and environmental impacts. The initial gas phase in the milling vessel is air, and pressure is varied to control the formation of Fe_3O_4 nanoparticles. Although no iron species are contained in the starting solution, Fe_3O_4 nanoparticles form in the solution according to the reaction mechanism based on the oxidation-reduction processes of the corrosion of steel. At the same time, the Fe_3O_4 nanoparticle surface is modified with citrate ions, resulting in a stable dispersion. The magnetic fluids prepared using this mechanochemical method possess good induction heating properties in an alternating current magnetic field.

Keywords: magnetite nanoparticles, inverse spinel, magnetic fluid, mechanochemical reaction, oxidation-reduction, hyperthermia

1. Introduction

Magnetic fluids are colloidal solutions containing magnetic nanoparticles stably dispersed in a liquid medium so that the entire fluid behaves like a ferromagnet. Moreover, no solid-liquid separation occurs, even under centrifugal force fields [1]. Therefore, magnetic fluids are widely used in industrial products, such as rotating-axis seal material, lubricants, and liquid damper. In magnetic fluids, iron-based oxide nanoparticles are often employed as the magnetic material. In particular, magnetic fluids that consist of superparamagnetic magnetite (Fe_3O_4) nanoparticles are used in biomedical and environmental fields, such as magnetic resonance imaging (MRI) contrast agents [2], magnetic hyperthermia in cancer therapy [3], drug delivery systems (DDS) [4], and so on, due to their high performance, low toxicity, and low environmental impact. The appropriate size of the Fe_3O_4 nanoparticles depends on their application. For example, when used for MRI, hyperthermia, and DDS, the particle size is required to be below 30 nm. Appropriate surface modification of the Fe_3O_4 nanoparticles can bind them to biomolecules as a result of the large surface area, leading to good bio-circulation [5]. In addition, the enhanced permeability and retention (EPR) effect can selectively accumulate Fe_3O_4 nanoparticles into cancer tissue [6]. Accordingly, the sizes of primary particles and aggregates must be controlled, and good stable dispersion is needed. Therefore, many studies on controlling the particle size as well as dispersion and aggregation using additives and organic solvents have been conducted. Wen et al. [7] prepared Fe_3O_4 nanoparticles with a size of 4–5 nm in an organic solvent by adding sodium oleate as a surfactant. Zhang et al. [8] synthesized Fe_3O_4 nanoparticles with good dispersibility in an organic solvent using surface modification with polystyrene. Frimpong et al. [9] controlled the sizes of the primary particle and aggregate using citric acid. Elisa de Sousa et al. [10] obtained citrate-adsorbed Fe_3O_4 nanoparticles that were dispersed under neutral conditions. In Fe_3O_4 nanoparticles that are used in vivo, biocompatible molecules, such as dextran, polyethylene glycol, polyvinyl alcohol, citric acid, polyacrylic acid, and phospholipid, are frequently employed as additives. In particular, citric acid is non-toxic and can form ultrafine primary particles. Furthermore, citrate ions can adsorb onto Fe_3O_4 nanoparticles, leading to aggregation inhibition due tothe steric hindrance and electrostatic repulsion. Therefore, citric acid is widely used as an anti-aggregation agent.

Magnetic fluids were originally developed by Papell of the United States National Aeronautics and Space Administration (NASA) in 1965 for position control of liquid fuel under zero gravity [11]. The initially developed magnetic fluid was prepared by ball-mill grinding of Fe_3O_4 grains in kerosene containing oleic acid for several hundred hours, followed by the removal of coarse particles by centrifugal separation. At present, various methods for synthesizing Fe_3O_4 nanoparticles using chemical reactions in gas, liquid, and solid phases have been developed. Among them, liquid phase synthesis has been actively studied because the component and concentration of the reactants can be controlled fairly easily and the formation reaction can progress, even under moderate conditions. In particular, coprecipitation methods that produce Fe_3O_4 by adding a base as a precipitant to a solution containing Fe^{2+} ions and Fe^{3+} ions are industrially employed because they can easily provide homogeneous Fe_3O_4 nanoparticles with smaller primary sizes [12]. The thermal decomposition method is also frequently used. Jeyadevan et al. [13] synthesized Fe_3O_4 nanoparticles that are suitable for magnetic hyperthermia via thermal decomposition of iron pentacarbonyl in oleic acid-containing dioctyl ether. Sun and Zeng [14] prepared mono-dispersed superparamagneticFe_3O_4 nanoparticles using iron acetylacetonate, oleic acid, and amine oleate. As a result, conventional methods can provide magnetic fluids that are desirable for various applications. For the previously

http://dx.doi.org/10.5772/66075

mentioned biomedical applications, water-based Fe_3O_4 magnetic fluids are suitable. However, in many cases, the conventional methods require iron salt, base, and organic solvents, requiring the removal of unnecessary components from the product to clean water-based magnetic fluids. This may increase the environmental impact and production cost. As a result, an innovative process is required to more readily prepare water-based Fe_3O_4 magnetic fluids without any additional operations.

To meet the demand, we developed a new mechanochemical process with an iron-free aqueous solution containing citric acid (CA) as a reaction accelerator and anti-aggregation agent milled at room temperature with a ball mill using steel balls, resulting in the production of water-based Fe_3O_4 magnetic fluid [15]. The formation of crystalline Fe_3O_4 nanoparticles in this process may consist of several steps, as follows: (1) corrosion of steel balls, (2) oxidation of released ferrous ions, (3) reduction of ferric ions, and (4) formation and crystal growth of Fe_3O_4. Therefore, the mechanochemical effect can enhance the formation and crystallization of Fe_3O_4. Furthermore, this method does not use iron salts as iron sources or a base as the precipitant because iron ions and hydroxide ions form during the corrosion, which can provide a water-based Fe_3O_4 magnetic fluid without any post-operations, such as the removal of unnecessary ions and solvent displacement.

This paper presents the properties of Fe_3O_4 magnetic fluids prepared by this mechanochemical process and a detailed analysis of the reaction mechanism based on changes in the composition of gas, liquid, and solid phases in the formation of Fe_3O_4 as well as ferrous, ferric, and hydroxide ions. Furthermore, the formation reaction of Fe_3O_4 is kinetically analysed under different gas phase conditions and mechanical energy fields.

2. Experimental

All chemicals used in this work were purchased from Wako Pure Chemical Industries and were used without further purification. Some preliminary experiments determined that the CA concentration in the starting solution was desirable to be 5 mmol/L, and a single Fe_3O_4 phase was finally obtained. Ninety millilitres of the CA solution (pH = 2.7) was placed in a tumbling ball mill consisting of a Teflon-lined gas-tight vessel (capacity 500 mL, diameter 90 mm) and carbon steel balls (Fe > 99 mass%, diameter 3 mm). The charged volume of the balls (includes the voids among balls) was 40% of the vessel capacity. The initial gas phase in the vessel was air at atmospheric pressure. The solution was milled at room temperature for anappropriate time period. The rotational speed of the vessel was 140 rpm, corresponding to the theoretically determined critical rotational speed. After milling, the fluid was removed from the vessel and characterized. The weight loss of balls during milling was also measured.

The particle size distribution and zeta potential of samples were determined using a particle size analyser (Malvern Zetasizer Nano ZS), and the phase evolution was evaluated using a powder X-ray diffractometer (Rigaku RINT-1500) after drying the sample. The average crystallite size was determined using Scherrer's equation for the diffraction peak from the Fe_3O_4 (311) plane at $2\theta = 35.4°$. The FT-IR spectrum was measured using a Fourier transform infrared spectrophotometer (Shimadzu IRAffinity-1). The magnetic properties (i.e. magnetization-magnetic field hysteretic cycle) were analysed with a superconducting quantum interference device (SQUID) magnetometer (Quantum Design MPMS XL) at room temperature over a magnetic field range of −10 kOe to +10 kOe. The magnetic hyperthermia-related induction heating properties of the fluid were evaluated by an alternating magnetic field generator consisting of a radio frequency power source (Thamway T162-5723A), an impedance matching box (Thamway T020-5723F), and a solenoid coil (inner diameter 70 mm) with 21 copper tube (outer diameter of 4 mm and inner diameter of 3 mm) turns [13]. Cooling water flowed inside the copper tube. A proper amount (approximately 1.0 g) of fluid was charged in a glass tube with a diameter of 16 mm, and the test tube was placed in the coil centre. The temperature increase in the fluid from 37°C (corresponding to normal body temperature) in an alternating magnetic field was measured with an optical fibre thermometer (FISO Technologies FTI-10 equipped with FOT-L-NS-967). The frequency and amplitude of the magnetic field were 600 kHz and 3 kA/m, respectively.

In the analysis of the reaction mechanism, 0.5 mL of the gas in the vessel was sampled with a syringe after milling, and the gas component was analysed by a gas chromatograph (Shimadzu GC-8A). The pH of the resultant fluid was measured with a pH meter (Horiba D-21 equipped with 9625-10D electrode), and the iron(II) and iron(III) concentrations were determined by a colorimetric method with 1,10-phenanthroline and a spectrophotometer (JASCO Ubest V-530).

The influence of gas phase conditions in the vessel on the Fe_3O_4 formation was studied. The 5 mmol/L CA solution was placed in the vessel; then, compressed air was charged into the vessel before milling. The total pressure in the gas phase was varied from 1 atm (atmospheric pressure) to 6 atm, corresponding to initial oxygen partial pressures of 0.21–1.26 atm. The milling time was 24 h at total pressures lower than 2.5 atm but 48 h at higher than 3 atm. The pressure in the vessel before and after milling was measured, and the oxygen consumption was determined from the pressure change. Furthermore, the rotational speed of the vessel was varied from 0 to 140 rpm at a total pressure of 1 atm, which can alter the intensity of the mechanical energy field. Based on the obtained results, the kinetics of the Fe_3O_4 formation reaction were investigated.

Lastly, scaleup of the process was examined. In this investigation, a Teflon-lined milling vessel with a capacity of 2.6 L (diameter 150 mm) was used. The charged volume of the steel balls (diameter 3 mm) was 40% of the vessel capacity, and 476 mL of the 5 mmol/L CA solution was placed in the vessel and milled for 24 h. The gas phase was air at atmospheric pressure. The rotational speed of the large vessel was adjusted so that either the Froude number or the peripheral velocity in both vessels agreed with each other.

3. Results and discussion

3.1. Formation of Fe_3O_4 magnetic fluids

Figures 1 and **2** show the X-ray diffraction (XRD) patterns and average crystallite sizes of solid products obtained after milling the CA solution, respectively. At milling times of less

http://dx.doi.org/10.5772/66075

than 1.5 h, broad peaks that were attributed to ferrihydrite ($Fe_5O_8H{\cdot}4H_2O$), which is an amorphous or low crystalline oxyhydroxide [16–18], were observed in the XRD patterns. After 2 h, formation and crystal growth of Fe_3O_4 occurred, and milling for more than 18 h provided relatively high crystalline Fe_3O_4. However, most of the Fe_3O_4 particles settled in the fluid due to aggregation. To improve the dispersion, the total citrate concentration of fluids was increased by adding anhydrous CA and trisodium citrate dehydrate to the fluid obtained by milling for 18 h; and then the zeta potential was measured. The pH of the fluids was kept constant at approximately 8 in all fluids by adding proper amounts of CA and trisodium citrate. **Figure 3** illustrates the change in the zeta potential with the changing citrate concentration. The absolute value of the zeta potential increased with the increasing citrate concentration. When the citrate concentration was 14 mmol/L, corresponding to an iron/citrate molar ratio of approximately 3 in the fluid, the zeta potential was less than −40 mV, leading to good dispersion. It was inferred that an increase in the citrate concentration can improve the dispersion.

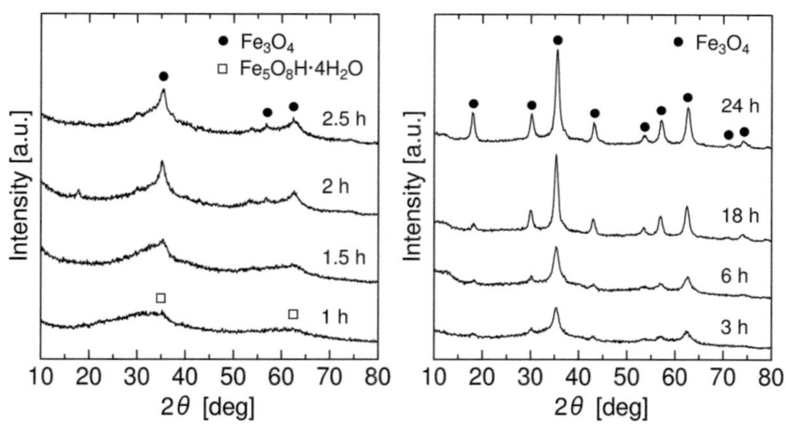

Figure 1. Effect of the milling time on the Fe_3O_4 phase evolution.

Using the fluid obtained at a milling time of 24 h, fluid with a total citrate concentration of 27 mmol/L was prepared and characterized. As shown in **Figure 4**, this fluid had good dispersion even after 2 weeks. Furthermore, when a permanent magnet was placed beside the glass bottle, the fluid was attracted to magnet, suggesting that the Fe_3O_4 nanoparticles have superparamagnetic properties. **Figure 5** shows the particle size distribution, magnetization-magnetic field curve, and FT-IR spectrum of the solid product obtained by drying the fluid after magnetic separation. The median diameter was 7.3 nm, while the crystallite size was 8.8 nm. The primary particle size was almost the same as the crystallite size, suggesting that the obtained Fe_3O_4 nanoparticles were monocrystalline. The saturation magnetization was 27 emu/g, which was much lower than that of bulk Fe_3O_4 (92 emu/g) due to the smaller particle size. The residual magnetization was approximately zero, and the coercivity was very low, indicating superparamagnetism. Additionally, some strong absorption bands in the FT-IR spectrum were observed, which were attributed to citrate ions, indicating that the Fe_3O_4

nanoparticle surface was modified by citrate ions. **Figure 6** illustrates a temperature adjustment of the fluid within 43 ± 0.5°C, corresponding to a typical temperature range in hyperthermia treatments, in an on-off-controlled alternating magnetic field. This fluid was found to exhibit good magnetic hyperthermia properties. In addition, the temperature was successfully controlled within the temperature range, suggesting that the fluid can be used in hyperthermia therapies.

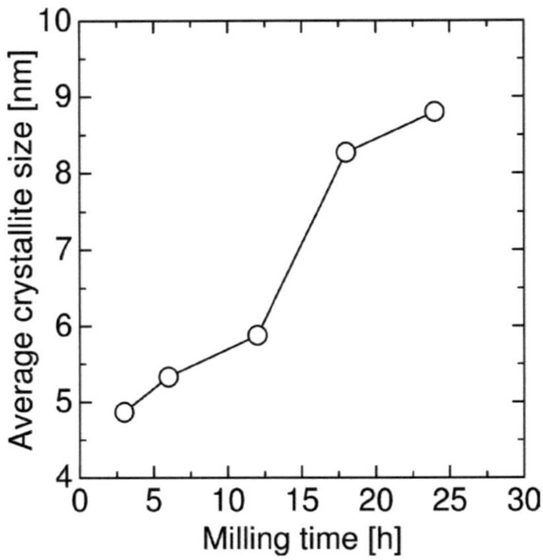

Figure 2. Change in the average crystallite size of Fe_3O_4 with milling time.

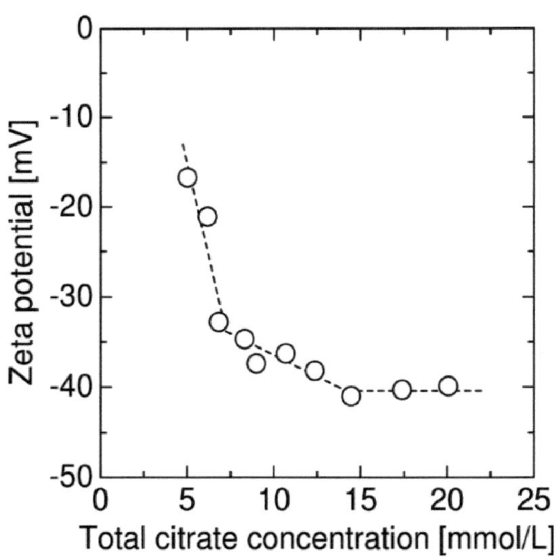

Figure 3. Effect of citrate concentration on the zeta potential of the fluid obtained with milling time of 18 h.

http://dx.doi.org/10.5772/66075

Figure 4. A fluid with a total citrate concentration of 27 mmol/L after 2 weeks.

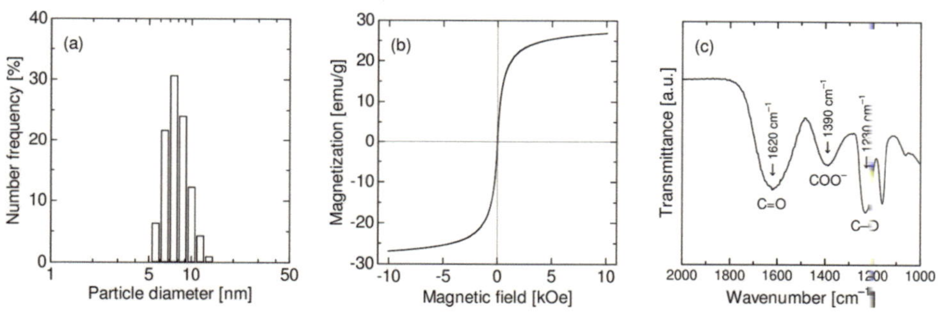

Figure 5. (a) Particle size distribution, (b) magnetization-magnetic field hysteresis cycle, and (c) FT-IR spectrum of the solid product.

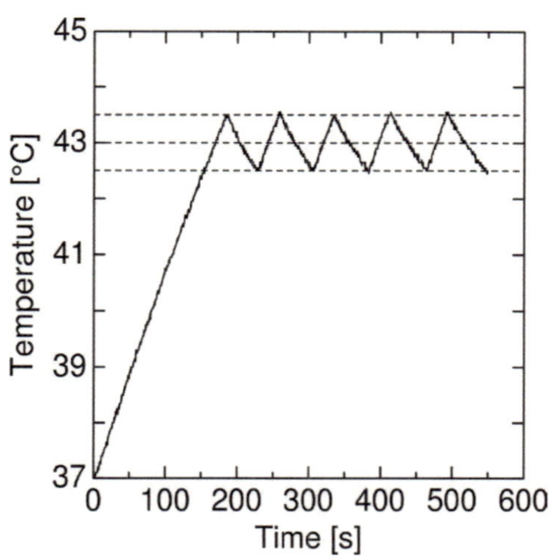

Figure 6. Temperature adjustment of the fluid in an on-off-controlled alternating magnetic field.

3.2. Reaction mechanism

Figure 7 shows the changes in the weight loss of balls and the iron concentration of the fluid with the milling time. With increasing milling time, the mass of the balls decreased and the iron concentrations increased, implying that ferrous (Fe^{2+}) ions were released from steel balls during milling due to corrosion, as expressed by Eq. (1).

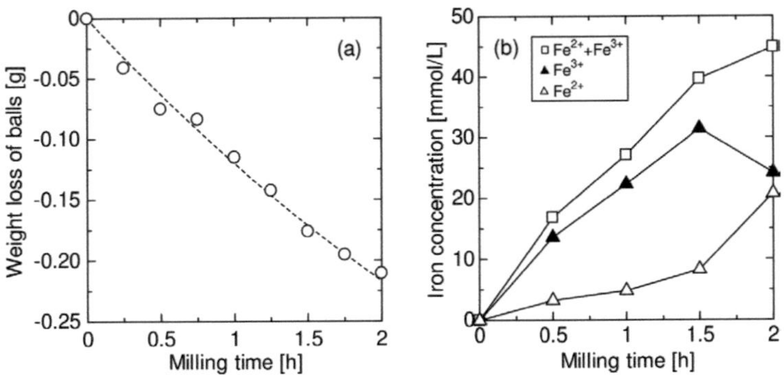

Figure 7. Change in the (a) weight loss of balls and (b) iron concentration with milling time at the initial stages.

$$Fe \rightarrow Fe^{2+} + 2e^{-} \tag{1}$$

Figure 8 shows the variation in the pH of fluids with the milling time. The fluid was acidic before milling and then immediately neutral. In general, free electrons (e^-) generated simultaneously with Fe^{2+} ion release are received by oxygen molecules (O_2) in a solution under acidic conditions, according to a cathode reaction expressed by Eq. (2).

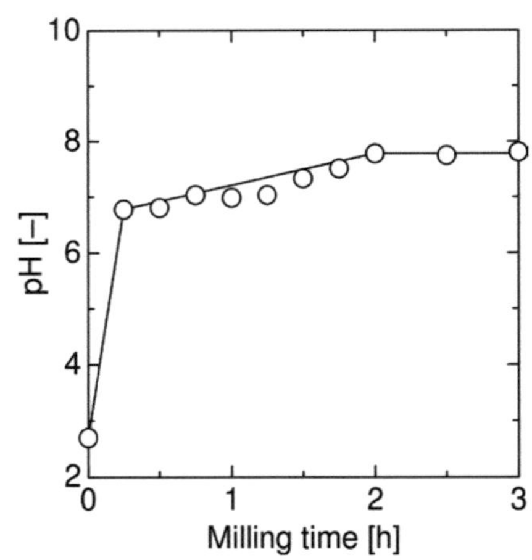

Figure 8. Change in the pH with milling time at the initial stages.

http://dx.doi.org/10.5772/66075

$$O_2 + 4H^+ + 4e^- \rightarrow 2H_2O \tag{2}$$

As hydrogen (H^+) ions in the solution are consumed, the solution pH increases and the solution becomes neutral. Under neutral conditions, the following cathode reaction occurs and hydroxide (OH^-) ions form.

$$O_2 + 2H_2O + 4e^- \rightarrow 4OH^- \tag{3}$$

On the other hand, released Fe^{2+} ions react with the OH^- ions formed according to Eq. (3), producing ferrous hydroxide ($Fe(OH)_2$) by Eq. (4).

$$Fe^{2+} + 2OH^- \rightarrow Fe(OH)_2 \tag{4}$$

As seen in **Figure 7**, most $Fe(OH)_2$ is immediately oxidized by dissolved O_2; then, ferric hydroxide ($Fe(OH)_3$) forms according to Eq. (5).

$$4Fe(OH)_2 + 2H_2O + O_2 \rightarrow 4Fe(OH)_3 \tag{5}$$

$Fe(OH)_3$ can be transformed to ferrihydrite ($Fe_5O_8H \cdot 4H_2O$) as below:

$$5Fe(OH)_3 \rightarrow Fe_5O_8H \cdot 4H_2O + 3H_2O \tag{6}$$

From Eqs. (1)–(6), Eq. (7) and Eq. (8) are obtained.

$$2Fe + O_2 + 2H_2O \rightarrow 2Fe(OH)_2 \tag{7}$$

$$20Fe(OH)_2 + 5O_2 \rightarrow 4Fe_5O_8H \cdot 4H_2O + 2H_2O \tag{8}$$

Eqs. (7) and (8) include O_2 as a reactant. Therefore, by analysing the dissolved O_2 concentration, the validity of the proposed reaction mechanism was confirmed. The dissolved O_2 concentration was estimated from the O_2 partial pressure in the gas phase using Henry's law:

$$p = Hx \tag{9}$$

where x is the molar fraction of O_2 in a liquid phase, p is the O_2 partial pressure in a gas phase in equilibrium with the liquid phase, and H is the Henry constant (= 4.38×10^4 atm at 25°C [19]).

In this process, the dissolved O_2 concentration can always vary during milling. However, the liquid phase is well mixed with the gas phase by milling, resulting in a relatively large gas-liquid interfacial area. Therefore, when the dissolved O_2 is consumed, O_2 can immediately be supplied from the gas phase. Accordingly, it can be assumed that the liquid and gas phases are always in equilibrium with each other. **Figure 9** shows the gas phase composition during milling. The open circles in this figure indicate the estimated values that were calculated using Eqs. (7) and (8) from the iron concentrations shown in **Figure 6**. It was confirmed that the volume of nitrogen (N_2) gas was almost constant and that hydrogen (H_2) gas evolution hardly occurred. In contrast, the O_2 gas was completely consumed within 2 h. Furthermore, the experimental data of the O_2 gas volume mostly agreed with the calculated values. The results demonstrate the validity of the reaction mechanism according to Eqs. (1)–(8).

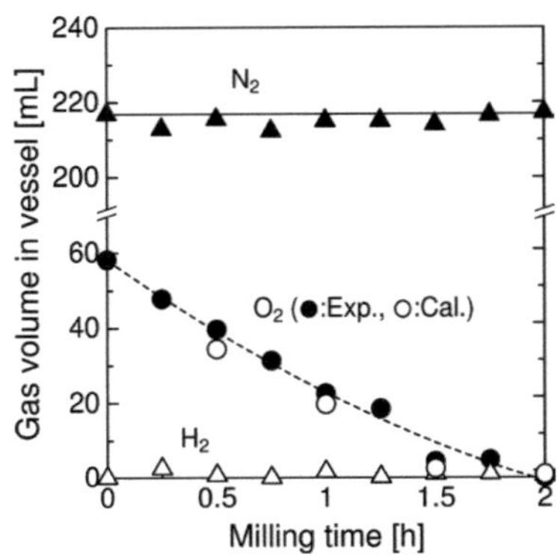

Figure 9. Change of the gas volumes in the vessel with milling time at the initial stages.

As seen in **Figure 8**, after 2 h, O_2 was absent from the vessel and the pH was almost constant at 7.8. **Figure 10** shows the change in the Fe^{2+} and Fe^{3+} ion concentrations with a milling time of up to 24 h. Fe^{3+} ions were reduced after 2 h. These results indicate that the free electrons formed at the steel corrosion are incorporated into $Fe_5O_8H{\cdot}4H_2O$, namely Fe^{3+} ions are reduced, according to Eq. (10) instead of Eq. (3).

$$Fe_5O_8H \cdot 4H_2O + 3H_2O + 5e^- \rightarrow 5Fe\left(OH\right)_2 + 5OH^- \tag{10}$$

Although this chemical reaction produces OH^- ions, the OH^- ions can be completely consumed to form $Fe(OH)_2$ according to Eq. (4), which keeps the pH constant. In the absence of dissolved O_2, the formed $Fe(OH)_2$ can react with $Fe_5O_8H{\cdot}4H_2O$ without being oxidized, resulting in the formation of Fe_3O_4 according to Eq. (11).

$$2Fe_5O_8H \cdot 4H_2O + 5Fe\left(OH\right)_2 \rightarrow 5Fe_3O_4 + 14H_2O \tag{11}$$

Consequently, the overall reaction under low O_2 conditions can be described by Eq. (12).

http://dx.doi.org/10.5772/66075

$$5Fe + 8Fe_5O_8H \cdot 4H_2O \rightarrow 15Fe_3O_4 + 36H_2O \quad (12)$$

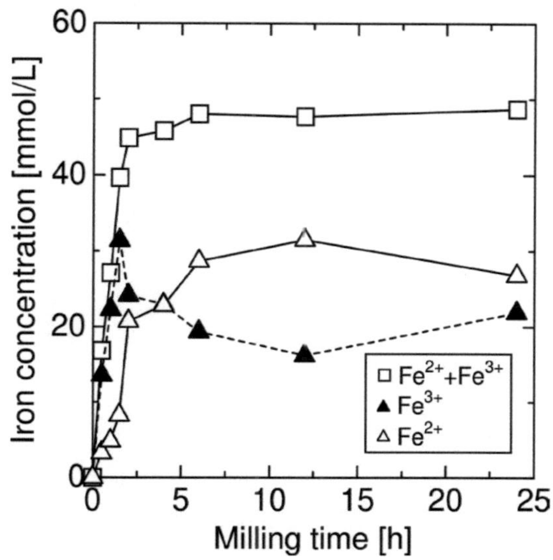

Figure 10. Change in iron concentrations with a milling time of up to 24 h.

Figure 11 illustrates the evolution of H_2 during milling. The formation of H_2 was noticed for long milling times. As seen in **Figure 10**, the iron concentrations slightly increased even after 6 h. Thus, the following reduction reaction may occur, resulting in H_2 evolution.

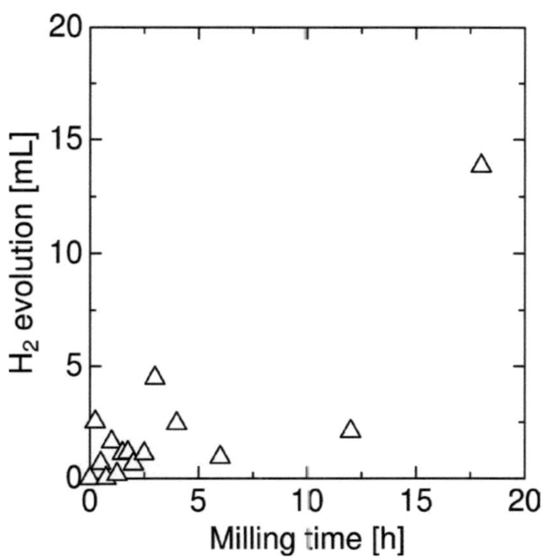

Figure 11. Evolution of H_2 during milling.

$$2H_2O + 2e^- \rightarrow 2OH^- + H_2 \quad (13)$$

OH^- ions thus formed react with Fe^{2+} ions, resulting in the formation of $Fe(OH)_2$. According to Eq. (4), $Fe(OH)_2$ can transform to Fe_3O_4 according to Eq. (14).

$$3Fe(OH)_2 \rightarrow Fe_3O_4 + 2H_2O + H_2 \quad (14)$$

From Eqs. (1), (4), (13), and (14), Eq. (15) is derived as the overall reaction for the formation of Fe_3O_4:

$$3Fe + 4H_2O \rightarrow Fe_3O_4 + 4H_2 \quad (15)$$

In summary, it has been established that a magnetic fluid consisting of well-dispersed superparamagnetic Fe_3O_4 nanoparticles can be formed by milling a CA solution with steel balls, according to the following reaction mechanism.

1. Under high O_2 conditions, milling the solution can result in an immediate mass transfer of O_2 from the gas phase to the liquid phase. Fe^{2+} ions are released from the steel balls. At the same time, using formed free electrons, OH^- ions are produced from O_2 and H_2O and then react with Fe^{2+} ions, resulting in the formation of $Fe(OH)_2$. $Fe(OH)_2$ is oxidized by dissolved O_2; then, $Fe_5O_8H{\cdot}4H_2O$ forms.

2. Under low O_2 conditions, $Fe_5O_8H{\cdot}4H_2O$ is reduced to $Fe(OH)_2$, and the reaction between $Fe(OH)_2$ and $Fe_5O_8H{\cdot}4H_2O$ produces Fe_3O_4.

3. After the consumption of $Fe_5O_8H{\cdot}4H_2O$, OH^- ions and H_2 are produced via reduction of H_2O. $Fe(OH)_2$ forms from Fe^{2+} and OH^- ions, and Fe_3O_4 and H_2 are produced from $Fe(OH)_2$.

3.3. Effect of the oxygen partial pressure

Figures 12–14 illustrate the XRD pattern of the obtained solid products, pH after milling, and internal pressure change of the vessel, respectively, under various initial total pressures in the vessel. Regardless of the internal total pressure, crystalline Fe_3O_4 was obtained, and the pH after milling was approximately 9. As seen in **Figure 14**, the internal pressure change was in agreement with the theoretical values (shown by a broken line) that had been calculated by assuming that O_2 in the vessel was completely consumed during milling, suggesting that under an initial oxygen partial pressure of less than 1.26 atm, O_2 was used in the Fe_3O_4 formation process.

From Eqs. (1), (3)–(6), and (11), the overall reaction for Fe_3O_4 formation under neutral and high O_2 conditions is expressed by Eq. (16).

http://dx.doi.org/10.5772/66075

$$3Fe + 2O_2 \rightarrow Fe_3O_4 \quad (16)$$

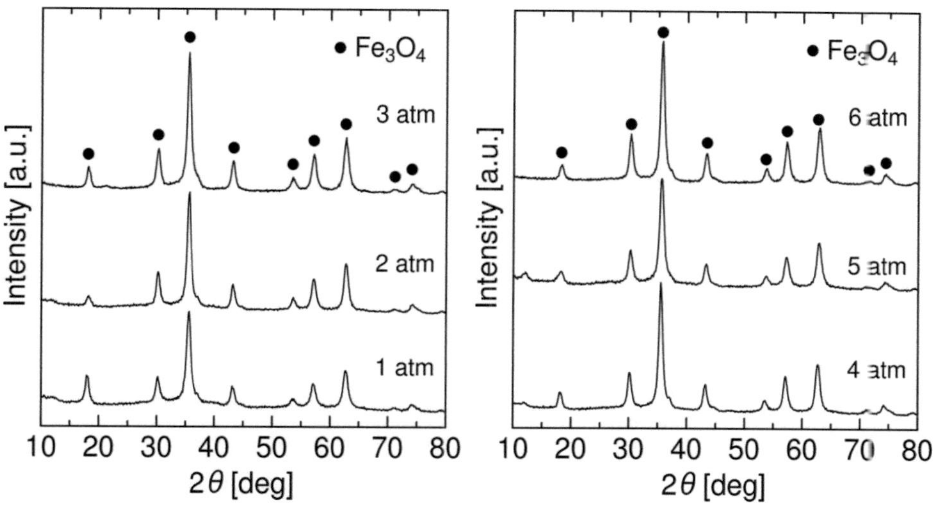

Figure 12. Effect of the initial total pressure on Fe_3O_4 phase evolution.

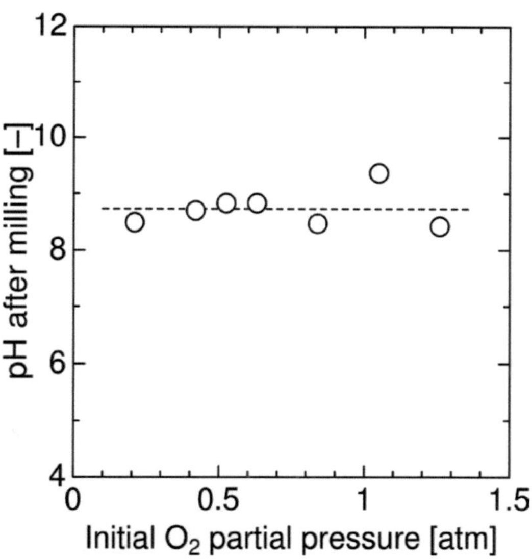

Figure 13. Change in the pH after milling with the initial O_2 partial pressure.

Figure 15 shows the iron concentration of fluids estimated based on the internal pressure change, as shown in **Figure 14** with Eq. (16). The iron concentration increased with increasing initial O_2 partial pressure, indicating that the iron concentration can be controlled by the initial O_2 partial pressure.

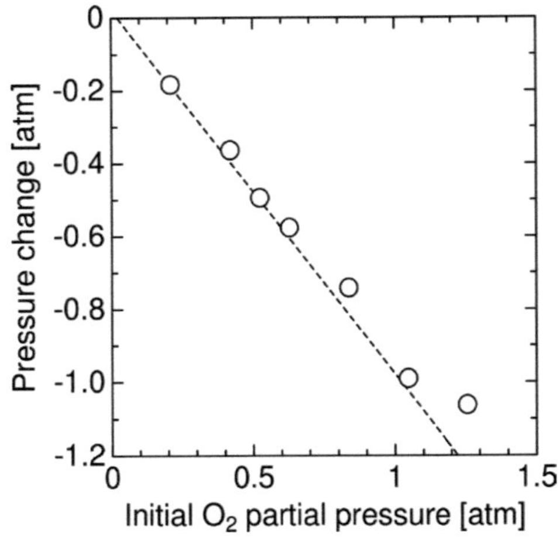

Figure 14. Change in the pressure after milling with the initial O_2 partial pressure.

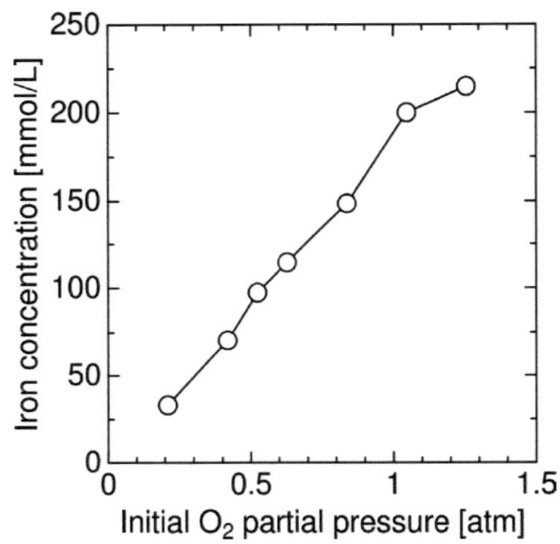

Figure 15. Change in the estimated iron concentration with the initial O_2 partial pressure.

3.4. Kinetics of the reaction for the formation of Fe_3O_4

Based on the concentrations of Fe^{2+} and Fe^{3+} ions during milling, the Fe_3O_4 formation reaction was kinetically analysed. **Figure 16** shows the change in the O_2 partial pressure and iron concentrations with the milling time in the initial stages at an initial total pressure of 1 atm. A monotonous decrease in the O_2 partial pressure with increased milling time was observed. Because the dissolved O_2 concentration is proportional to the O_2 partial pressure, the rate of reactions related to the dissolved O_2 concentration, as shown by Eqs. (7) and (8), can also vary depending on the milling time. However, at less than 1.5 h, O_2 was consumed at a constant rate, and the rates of Fe^{2+} and Fe^{3+} ion formation were almost constant regardless of the milling

http://dx.doi.org/10.5772/66075

time, as seen in **Figure** 7b. The results imply that the reactions are independent of the dissolved O_2 concentration. Therefore, the reaction rates of Eqs. (7) and (8) can be described by a zero-order model. This model is effective because O_2 quickly dissolves into the solution due to vigorous gas-liquid mixing by milling and milling accelerates the corrosion of steel due to the improvement in the diffusion rate of O_2 to the steel surface. Accordingly, the Fe_3O_4 formation process may be the oxidation-reduction reaction control, and both the dissolution of O_2 from the gas phase to the liquid phase and diffusion rate of O_2 in the liquid phase can be much faster than the rate of the oxidation-reduction reaction.

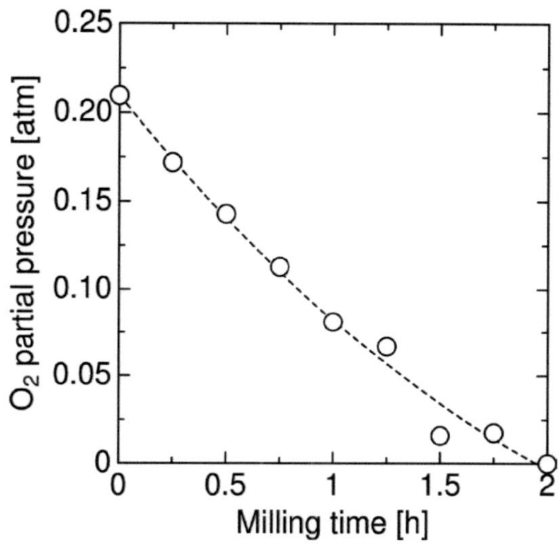

Figure 16. Changes in the O_2 partial pressure with milling time at an initial total pressure of 1 atm.

Using the data shown in **Figure 16**, the rates of Fe^{2+} and Fe^{3+} ion formation were calculated. Based on Eqs. (7) and (8), the O_2 consumption rate was determined to be approximately 0.40 μmol/s, which nearly agreed with the experimental data (0.41 μmol/s) shown in **Figure 16**. This result suggests that the consumed O_2 was spent on the release and oxidation of Fe^{2+} ions.

Eqs. (7) and (8) are rewritten as follows:

$$\mathrm{Fe} + \frac{1}{2}\mathrm{O_2} + \mathrm{H_2O} \rightarrow \mathrm{Fe(OH)_2} \tag{17}$$

$$\mathrm{Fe(OH)_2} + \frac{1}{4}\mathrm{O_2} \rightarrow \frac{1}{5}\mathrm{Fe_5O_8H \cdot 4H_2O} + \frac{1}{10}\mathrm{H_2O} \tag{18}$$

Assuming that the reaction rates of Eqs. (17) and (18), r_1 and r_2, can be described by a zero-order model and expressed as follows:

$$r_1 = k_1 \tag{19}$$

$$r_2 = k_2 \tag{20}$$

Here k_1 and k_2 are the rate constants of Eqs. (17) and (18), respectively. As a result, the rate of $Fe(OH)_2$ formation reaction, r, can be expressed by Eq. (21).

$$r = r_1 - r_2 = k_1 - k_2 = K \tag{21}$$

K indicates the rate constant of the overall $Fe(OH)_2$ formation reaction.

Figure 17 shows the change in the O_2 partial pressure with a milling time of up to 24 h. Regardless of the milling time, the rate of O_2 consumption was constant under all of the initial total pressures. **Figure 18** illustrates the rate constant K calculated using Eqs. (19)–(21) as a function of the initial O_2 partial pressure. K was found to decrease with the increasing initial O_2 partial pressure. As shown in **Figure 19**, when the initial O_2 partial pressure was relatively high, goethite (α-FeOOH) and $Fe_5O_8H \cdot 4H_2O$ formed as intermediates. In general, the rate of steel corrosion is affected by the mass fraction of iron oxyhydroxides and iron oxides in a corrosion product on the steel. In particular, when a high level of α-FeOOH is contained in the corrosion product, the corrosion rate can decrease because α-FeOOH prevents O_2 from penetrating into the steel surface [20]. At high initial O_2 partial pressures, $Fe_5O_8H \cdot 4H_2O$ can make a phase transition to α-FeOOH, and a dense corrosion product layer may form on the steel surface. This inhibits the mass transfer of O_2, resulting in a decrease in the rate of $Fe(OH)_2$ formation reaction.

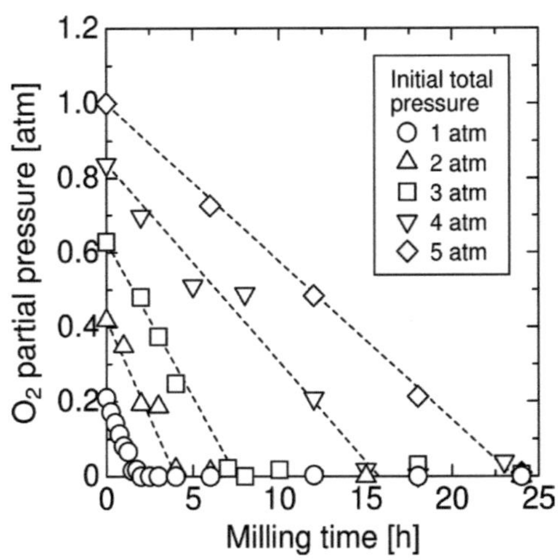

Figure 17. Change in the O_2 partial pressure with a milling time of up to 24 h.

http://dx.doi.org/10.5772/66075

Next, the effect of the vessel rotational speed on the O_2 consumption was studied in atmospheric pressure. **Figure 20** depicts the change in the O_2 partial pressure with the milling time. The O_2 consumption rate was almost the same, regardless of the rotational speed. **Figure 21** illustrates the relationship between the rotational speed and K calculated assuming that the O_2 consumption can be expressed by a zero-order equation. At rotational speeds higher than 10 rpm, K was almost constant, suggesting that the milling generated a sufficient level of energy required for the mass transfer and consumption of O_2, even at low rotational speeds.

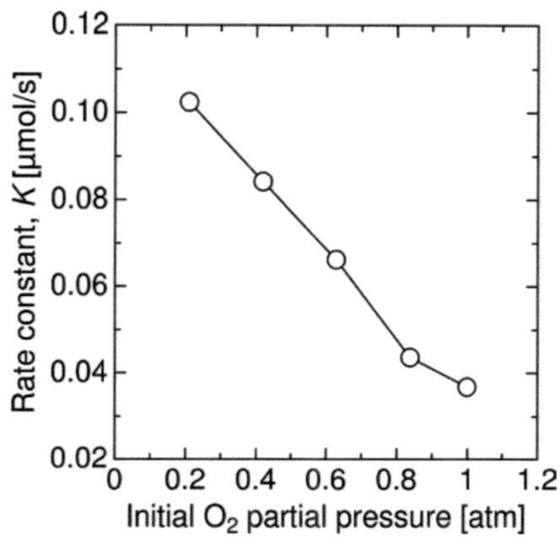

Figure 18. Change in the rate constant K with an initial O_2 partial pressure.

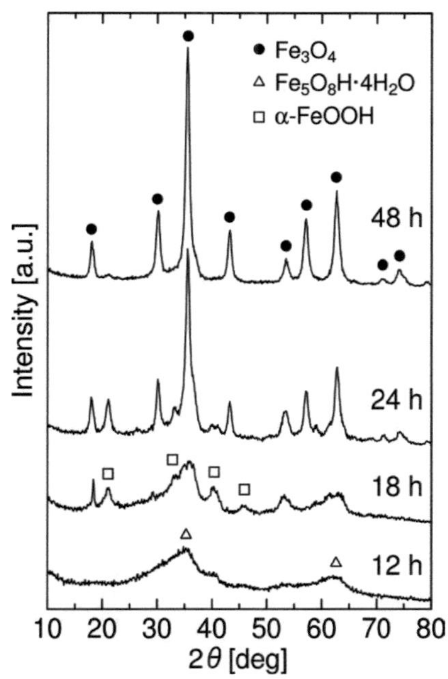

Figure 19. XRD patterns of a solid product obtained at an initial total pressure of 3 atm.

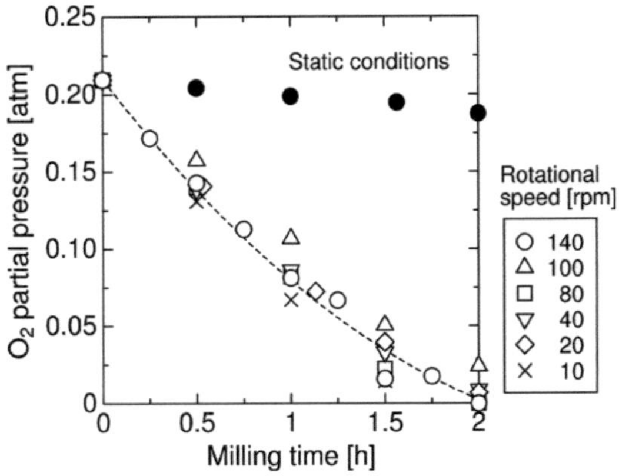

Figure 20. Effect of the rotational speed on the O_2 partial pressure after milling.

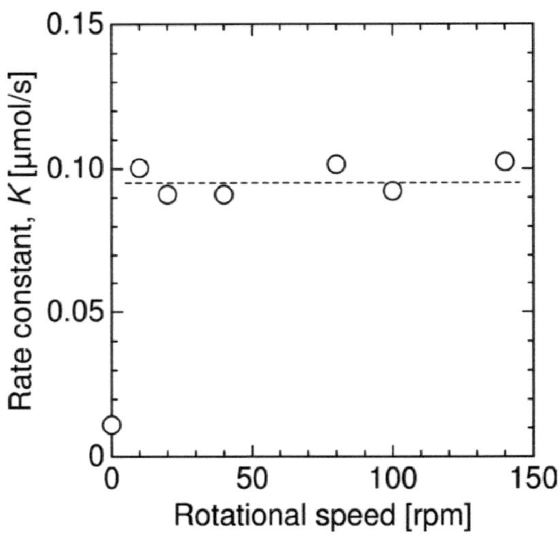

Figure 21. Effect of the vessel rotational speed on the rate constant *K*.

3.5. Scaleup of the process

Using a milling vessel with a capacity of 2.6 L, which was approximately 5 times as large as the small vessel used in the above investigations, the scaleup of the process was studied based on two cases, constant Froude number and constant peripheral velocity of the vessels. The rotational speed of 140 rpm for the small vessel corresponds to 108.4 rpm for the large vessel at a constant Froude number (= 0.0500) and 87.7 rpm at a constant peripheral velocity (= 0.735 m/s). **Figures 22–24** show the XRD pattern of solid products, average crystallite size, and pH after milling, respectively. The results demonstrate that the large scale process can fabricate a similar fluid, suggesting that scale-up of the process can be successful based on either the Froude number or peripheral velocity of the vessel.

http://dx.doi.org/10.5772/66075

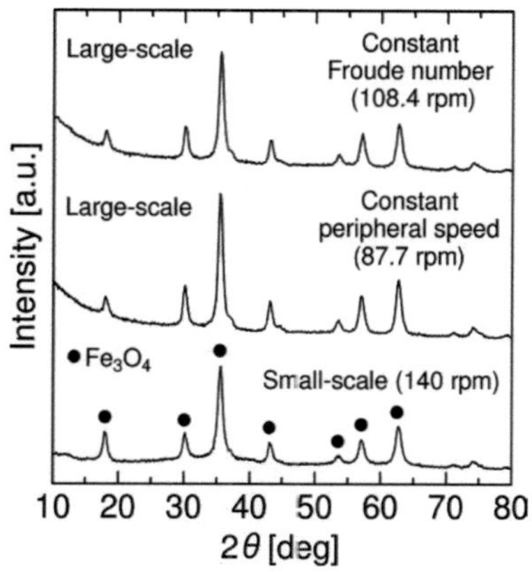

Figure 22. XRD patterns of solid products obtained in both processes.

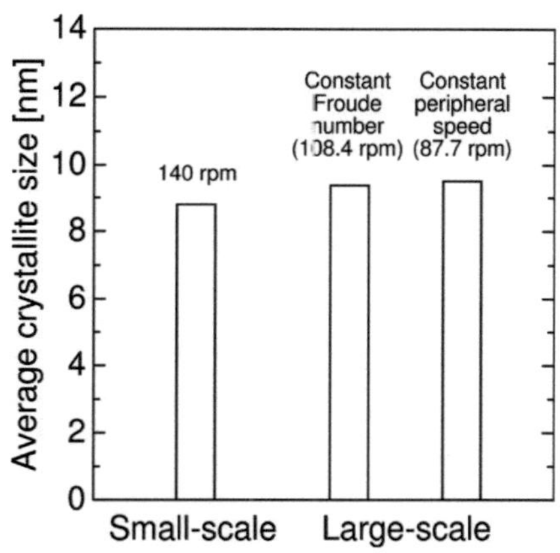

Figure 23. Comparison of the average crystallite size of solid products obtained in both processes.

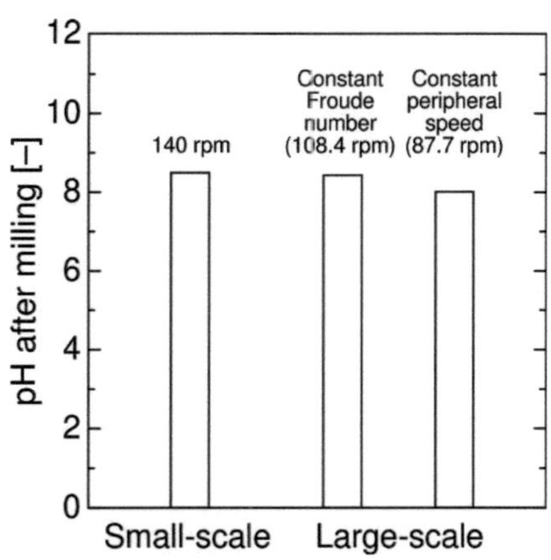

Figure 24. Comparison of the final pH of fluids obtained in both processes.

4. Conclusions

This study analysed in detail a new mechanochemical process for readily synthesizing water-based magnetic Fe_3O_4 fluids. Major conclusions are summarized as follows:

a. The Fe_3O_4 formation mechanism in this process has been clarified, which can be described by several oxidation-reduction reactions, such as the corrosion of steel, oxidation of $Fe(OH)_2$, and reduction of $Fe_5O_8H \cdot 4H_2O$. Additionally, O_2 plays an important role in the reaction mechanism.

b. The Fe_3O_4 concentration of fluids can be controlled by the initial O_2 partial pressure in the vessel.

c. Kinetic analysis of the process clarified the effects of the dissolved O_2 and the vessel rotational speed on the rate of the $Fe(OH)_2$ formation reaction. In the process, the reaction rate of O_2 consumption can be expressed by a zero-order equation.

d. Scaleup of the process can be successful by considering either the Froude number or peripheral velocity of the vessel.

Acknowledgements

This work was financially supported by JSPS KAKENHI Grant Number JP24686090. The author thanks Mr. Tsukasa Kagawa for his supports during the experiments.

Author details

Tomohiro Iwasaki

Address all correspondence to: iwasaki@chemeng.osakafu-u.ac.jp

Osaka Prefecture University, Sakai, Japan

References

[1] Mitamura Y. Medical applications of magnetic fluids. J Jpn Soc Prec Eng. 2006;72:834–837.

[2] Tiefenauer LX, Tschirky A, Kühne G, Andres RY. In vivo evaluation of magnetite nanoparticles for use as a tumor contrast agent in MRI. Magn Reson Imaging. 1996;14:391–402.

[3] Lima-Tenório MK, Gómez Pineda EA, Ahmad NM, Fessi H, Elaissari E. Magnetic nanoparticles: In vivo cancer diagnosis and therapy. Int J Pharm. 205;493:313–327.

[4] Mürbe J, Rechtenbach A, Töpfer J. Synthesis and physical characterization of magnetite nanoparticles for biomedical applications. Mater Chem Phys. 2008;110:426–433.

[5] Moghimi SM, Hunter AC, Murray JC. Long-circulating and target-specific nanoparticles: Theory to practice. Pharmacol Rev. 2001;53:283–318.

[6] Matsumura Y, Maeda H. A new concept for macromolecular therapeutics in cancer chemotherapy: Mechanism of tumoritropic accumulation of proteins and the antitumor agent smancs. Cancer Res. 1986;46:6387–6392.

[7] Wen X, Yang J, He B, Gu Z. Preparation of monodisperse magnetite nanoparticles under mild conditions. Curr Appl Phys. 2008;8:535–541.

[8] Zhang F, Su Z, Wen F, Li F. Synthesis and characterization of polystyrene-grafted magnetite nanoparticles. Colloid Polym Sci. 2008;286:837–841.

[9] Frimpong RA, Dou J, Pechan M, Hilt JZ. Enhancing remote controlled heating characteristics in hydrophilic magnetite nanoparticles via facile co-precipitation. J Magn Magn Mater. 2010;322:326–331.

[10] Elisa de Sousa M, Fernández van Raap MB, Rivas PC, Mendoza Zélis P, Girardin P, Pasquevich GA, Alessandrini JL, Muraca D, Sánchez FH. Stability and relaxation mechanisms of citric acid coated magnetite nanoparticles for magnetic hyperthermia. J Phys Chem C. 2013;117:5436–5445.

[11] Papell SS. Low viscosity magnetic fluid obtained by the colloidal suspension of magnetic particles. U.S. Patent 3215572. 1965.

[12] Kim DK, Zhang Y, Voit W, Rao KV, Muhammed M. Synthesis and characterization of surfactant-coated superparamagnetic monodispersed iron oxide nanoparticles. J Magn Magn Mater. 2001;225:30–36.

[13] Jeyadevan B. Present status and prospects of magnetite nanoparticles-based hyperthermia. J Ceram Soc Jpn. 2010;118:391–401.

[14] Sun S, Zeng H. Size-controlled synthesis of magnetite nanoparticles. J Am Chem Soc. 2002;124:8204–8205.

[15] Iwasaki T, Kagawa T, Nakamura H, Watano S. Control of aggregation and dispersion of mechanochemically synthesized magnetite nanoparticles in aqueous magnetic fluid. In: Proceedings of the 10th International Symposium on Agglomeration (Agglos 10); 2–4 September 2013; Kobe.

[16] Zhao J, Huggins FE, Feng Z, Lu FL, Shah N, Huffman GP. Structure of a nanophase iron oxide catalyst. J Catal. 1993;143:499–509.

[17] Feng Z, Zhao J, Huggins FE, Huffman GP. Agglomeration and phase transition of a nanophase iron oxide catalyst. J Catal. 1993;143:510–519.

[18] Zhao J, Huggins FE, Feng Z, Huffman GP. Ferrihydrite; surface structure and its effects on phase transformation. Clay Clay Miner. 1994;42:737–746.

[19] Han KN. Fundamentals of Aqueous Metallurgy. Littleton: Society for Mining, Metallurgy, and Exploration; 2002. 197 p.

[20] Yamashita M, Maeda A, Uchida H, Kamimura T, Miyuki H. Crystalline rust compositions and weathering properties of steels exposed in nation-wide atmospheres for 17 years. J Jpn Inst Metals. 2001;65:967–971.

Section 4

Synthesis and Applications

Chapter 9

Nanostructured Spinel Ferrites: Synthesis, Functionalization, Nanomagnetism and Environmental Applications

Oscar F. Odio and Edilso Reguera

Additional information is available at the end of the chapter

http://dx.doi.org/10.5772/67513

Abstract

Nanostructured spinel ferrites have gained a great deal of attention. It comes from the possibility of tuning their magnetic properties by careful manipulation of the synthetic conditions. At the same time, since the nanoparticle (NP) surface is reactive toward many chemical groups, it provides great versatility for further functionalization of the nanosystems. Such characteristics make ferrite nanoparticles excellent candidates for environmental applications. First, the chapter deals with the basics of the synthetic methodologies, functionalization strategies and magnetic properties of nanoparticles, with emphasis on how surface manipulation is reflected in the properties of the materials. Next, we review some of the applications of ferrites as magnetic sorbents for several hazardous substances in aqueous medium and try to systematize the adsorption mechanism as a function of the coating material. Finally, a short summary concerning the main uses of ferrites as magnetic catalysts in oxidation technologies is included.

Keywords: spinel ferrites, superparamagnetism, surface complexes, heavy metals, dyes

1. Introduction

Magnetic nanoparticles (NPs) have been the focus of intense studies, both at the fundamental and at the technological level. Among many promising materials, nanostructured spinel ferrites occupy a special place. These iron oxide-based materials are easy and cheap to synthesize, are stable under a wide range of conditions and some family members present low toxicity for living organisms. Besides, due to their high reactivity toward several organic groups, ferrite surface offers a great versatility for ligand functionalization, which in many cases defines the ultimate application. In addition, one of the most prominent properties of spinel ferrite NPs is the onset of superparamagnetism. This phenomenon is a crucial feature for several biomedical applications [1], catalytic processes [2, 3] and environmental remediation strategies [1, 4–7]. Currently, there are available in the literature several extensive reviews covering these issues in detail [8, 9]. In this chapter, we focus primarily on adsorption and oxidation technologies for water decontamination using nanostructured spinel ferrites where particle functionalization plays a major role. In particular, we focus on basic topics concerning spinel ferrite NPs with an emphasis on the surface manipulation by chemical methods and how it is reflected in the properties and performances of the ultimate nanomaterial. Also, attention is paid to the machinery that governs the adsorption process in order to try to systematize the available data. Every step in this direction is aimed to improve and design newer and better solutions for the great challenge of water remediation.

2. Structural and magnetic properties

Spinel ferrites are mixed valence oxides where oxygen anions form a close-packed cubic array, while metallic cations occupy randomly one-eighth of the tetrahedral (**A**) sites and one-half of the octahedral (**B**) interstices. This family is classified into the *Fd3m* space group with the general formula $[M_{(1-i)}Fe_i]^A[M_iFe_{(2-i)}]^BO_4$, where M (Fe, Ni, Mg, Mn, Zn, Co, etc.) is a divalent cation that shares with Fe(III) cations the occupancies of **A** and **B** sites, while i is defined as the inversion parameter. There are three possibilities [10]: (i) $i = 0$ yields a *normal* spinel such as $ZnFe_2O_4$; (ii) $i = 1$ yields an *inverse* spinel such as Fe_3O_4 and $CoFe_2O_4$ and (iii) for $0 < i < 1$ cations are distributed on both sites yielding a partially inverted spinel such as $MnFe_2O_4$ in which $i = 0.2$. In addition, it is possible to synthesize mixed ferrites in the sense that different divalent cations could coexist in the same compound [11, 12]. All these possibilities open an amazing range for tailoring different properties [11, 13].

In the spinel structure (left panel of **Figure 1**), magnetic moments of sublattice **A** are coupled with magnetic moments of sublattice **B** in an antiferromagnetic fashion by superexchange interactions mediated by the oxygen anions [14]. Since spins in both lattices are generally uncompensated, the resulting net magnetic moment causes the material to display ferrimagnetic behavior [10]. As can be inferred, either the type of divalent ion or the average ion distribution plays a critical role in the magnetic properties of the material. This effect can be illustrated by varying the Zn content in the mixed $Zn_xFe_{3-x}O_4$ ferrite [11] in which Zn^{2+} presents a zero magnetic moment ($\mu = 0$) and a high tendency for tetrahedral **A** sites. In the interval $0 < x < 0.4$, the saturation magnetization (M_S) increases as x increases, since the antiferromagnetic **A-B** interaction is progressively weakened, thus enhancing the net magnetic moment in sublattice **B** (see the right panel of **Figure 1**). However, for values of x near 1, **A-B** interactions are no longer dominant, but the magnetism now depends on the very weak **B-B** interactions, thus leading to a marked decrease of M_S, which becomes zero at $x = 1$.

2.1. Nanomagnetism

Magnetic NPs differ from bulk magnetic materials mainly due to the finite size and surface effects. The reduction of size leads to a single magnetic domain at a particular size and the onset

http://dx.doi.org/10.5772/67513

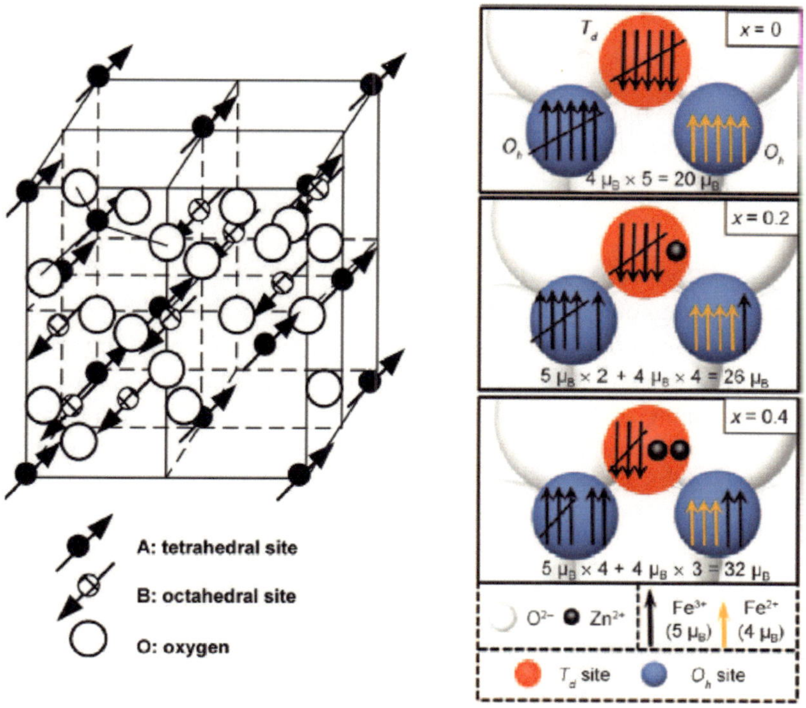

Figure 1. *Left:* Representation of a partial spinel ferrite unit cell and the ferrimagnetic order of the structure (Reproduced from Ref. [10] with permission of the American Society of Chemistry). *Right:* Schematic view of the spin organization in sublattice **A** and **B** as a function of the content of Zn in $Zn_xFe_{3-x}O_4$ (Adapted from Ref. [11] with permission of Wiley).

of superparamagnetism, while surface effects result in symmetry breaking of the crystal structure, which could also alter the magnetic properties. These new features are treated briefly below.

2.1.1. Single-domain limit

Large magnetic particles usually have a multidomain structure, each domain separated from its neighbors by domain walls. As the particle diameter D or volume V is decreased, domain wall creation is no longer energy favorable for a specific size, leading to single-domain NPs with all atomic spins aligned in the same direction. This critical particle diameter is characteristic for each material and is of the order of tens of nanometers (128 nm for Fe_3O_4) [9]. Since the spins are parallel within the NPs, magnetic reversal does not depend on wall displacement but is only possible by the coherent rotation of spins, which depends entirely on the effective anisotropy (K_{eff}). Given that coercitivity (H_C) is proportional to K_{eff}, it is higher for nanomaterials with respect to their bulk counterparts. For single-domain particles with uniaxial anisotropy, the anisotropy energy is defined as:

$$E(\theta) = K_{eff} V \sin^2\theta \tag{1}$$

Here, θ is the angle between the net magnetization and the easy axis of magnetization. For spherical particles, K_{eff} can be expressed as:

$$K_{eff} = K_V + \frac{6}{D}K_S \tag{2}$$

Here, K_V and K_S are the volume and surface anisotropies, respectively. As can be seen, for ultra-small NPs, the surface term may dominate the total anisotropy of the material.

2.1.2. Superparamagnetism

The product $K_{eff}V$ is the energy barrier for the coherent rotation of all atomic spins between the two equivalent easy axes of magnetization. As D is decreased, the thermal energy k_BT eventually overcomes the energy barrier for a particular size, thus leading to thermal equilibrium of the total magnetic moment of the system. In this state, the resulting H_C is zero and the system behaves like a paramagnet but with a huge magnetic moment. For an assembly of non-interacting single-domain NPs, each magnetic moment fluctuates with a relaxation time τ given by the Arrhenius-Néel law [15]:

$$\tau = \tau_0 e^{(K_{eff}V/k_BT)} \tag{3}$$

Here, τ_0 is a characteristic time of the system and the actual magnetic state at a given T depends on the measuring time τ_m. If $\tau < \tau_m$, the systems are in the superparamagnetic state, and for $\tau > \tau_m$, the spins appear in a blocked state. The temperature for which $\tau = \tau_m$ is the so-called blocking temperature T_B and is given by:

$$T_B = \frac{K_{eff}V}{k_B}\ln \quad (\tau_0/\tau_m) \tag{4}$$

One of the main advantages of ferrites is the possibility for tuning the magnetic properties by varying simply either the divalent cation or the arrangement of the metals into the spinel structure. For instance, in a series of nanoparticle ferrites MFe_2O_4 (M = Mn^{2+}, Fe^{2+}, Co^{2+}, Ni^{2+}, Zn^{2+}), Mohapatra *et al.* [16] could vary M_S and T_B by taking advantage of the differing magnetic moments and spin-orbit coupling strengths of M^{2+} cation. Similar conclusions can be extracted from other reports [17]. Likewise, for ultra-small $CoFe_2O_4$ NPs, the progressive variation in the inversion index (from a total to a partially inverted spinel) as D decreases is reflected in a decrease in the magnetocrystalline anisotropy and hence, in the H_C of the material [18]. In other works, Tahar *et al.* [13] found that as Zn^{2+} substitutes Co^{2+} cations in 5-nm-mixed $Zn_xCo_{1-x}Fe_2O_4$ ferrite NPs both H_C and T_B decrease, in accordance with progressive reduction of magnetocrystalline anisotropy, which is higher for octahedral coordinated Co^{2+} cations.

2.1.3. Surface effects

Progressive decrease in NPs' size makes the number of atoms on the surface comparable with the number of atoms in the bulk. For magnetic NPs, this trend lowers M_S and increases K_{eff} as D decreases [19, 20]. The decrease in M_S is associated with the presence of a magnetic dead layer, occurrence of spin canting or spins glass-like behavior at the surface level. The increase in K_{eff} is associated with the onset of surface anisotropy following Eq. (2).

The effect of surface coating in order to tune the magnetic properties of NPs is another area of active investigations. The adsorption of organic ligands could alter the particle size distribution, the interparticle interactions and the spin canting at the surface [21]. The overall effect seems to be the result of a complex interplay between the coordination mode, the capping density and the surface disorder in the synthesized sample [22]. In a careful study of adsorption of stilbene carboxylates and phosphonates as capping agents of 39 nm Fe_3O_4 NPs, Daou *et al.* [23] reported that carboxyl ligands tend to enhance spin canting at the surface of the oxide, leading to M_S reduction; on the other hand, phosphonate ligands seem to mimic the iron coordination in the bulk, hence, M_S of the NPs is unaltered as for uncovered Fe_3O_4. However, a direct correlation between magnetic measurements and the nature of the coordination bonding at the organic-inorganic interface is still needed. Regarding the important case of carboxylates, Odio *et al.* [24] found that spin disorder is larger in chelating than in bridging complexes; it suggests that occurrence of this last geometry makes possible the partial reconstruction of the crystal field of iron ions in the bulk phase. In another report, Aslibeiki *et al.* [25] noted that tetraethylene glycol ligands attached to Fe_3O_4 surface also contribute to decrease the surface spin disorder. Similarly, Jia *et al.* [26] noted that in Co- and Ni-mixed ferrite with SO_4^{2-} attached to the surface, M_S values increase with the content of superficial anions. In contrast, adsorption of carbonyl groups belonging to poly(vinyl pyrrolidone) (PVP) capping chains of Mn ferrites seems to decrease M_S values [27].

In an interesting paper, Vestal and Zhang [28] performed a systematic study of the correlation between the nature of the capping ligand (substituted benzenes) and the magnetic properties of $MnFe_2O_4$ ferrites with different particles' diameter. They found that for very small NPs (4 nm), H_C is reduced by ligand interaction with respect to the uncovered ferrite, while M_S is enhanced. Such behavior is consistent with the fact that metal-ligand interactions at the surface reduce K_S, leading to a reduction of H_C, and at the same time, they induce spin order at the surface. Furthermore, the authors showed that the higher the crystal field splitting energy of the surface complex, the smaller the magnitude of the spin-orbit interaction parameter and hence, the smaller the K_S value. The fact that such trends are less pronounced for larger particles (12 nm and 25 nm) reveals the importance of surface effects in H_C and M_S of single-domain NPs. This point has been verified in other reports [16, 29]. Besides, the contribution of K_S to K_{eff} is revealed when comparing the H_C variation of $MnFe_2O_4$ and $CoFe_2O_4$ particles with the same diameter and organic ligands: in the case of Co ferrite, the larger magnetocrystalline anisotropy masks K_S variation, leading to nearly unchanged values of H_C [30].

2.1.4. *Magnetic interparticle interactions*

The presence of magnetic interactions between particles has a great influence on superparamagnetism [31–33]. This effect alters the energy barrier for coherent rotation, which is no longer governed by only anisotropic contributions. The system becomes very complex and results in the difficulty to separate the contributions of different factors [34]. For ferrite NPs, the ordinary kinds of magnetic interactions are dipolar-dipolar and direct exchange interactions between spins at the interface of particles in close contact [35]. The first contribution is almost ubiquitous in any system, given its anisotropic and long-range nature, which could favor either ferro or antiferromagnetic alignment of the spins. The minimization of such an effect can only be achieved in samples where individual particles are well separated from each other, either by steric or by coulombic repulsions [25, 36]. The dipolar magnetic field generated by a single spherical particle is proportional to its volume; hence, the effect is more pronounced for large particles. In that case, the magnetic energy between two spheres decays with d^{-3}, but smaller systems in the superparamagnetic state (where moments fluctuate) partially destroy the order induced by dipolar interactions, and hence, the energy associated decays with d^{-6}. In the limit of strong interactions, particles do not relax according to their own energy barrier, but the magnetic evolution depends on the energy of the whole ensemble [22, 34, 36, 37].

In the presence of interparticle interactions (IPI), Eq. (4) for T_B is modified to [38–40]:

$$T_B = To + \frac{K_{eff}V}{k_B}\ln(\tau_0/\tau_m) \tag{5}$$

Here, $T_o < T_B$ is a measure of the strength of the IPI in the system. To determine T_o, T_B needs to be measured at several different measuring frequencies $f_m = 1/\tau_m$. The presence of IPI is also determined by the quantity $\Phi = \Delta T_B/[T_B \Delta \log_{10} f_m]$ with $\Delta T_B = T_B(2) - T_B(1)$ being the difference in T_B determined at two sufficiently different measuring frequencies $f_m(2) > f_m(1)$ [41]. The magnitude of $\Phi < 0.13$ signifies the presence of IPI with its strength increasing with decreasing magnitude of Φ. In a recent paper [42], Φ and T_o have been shown to be related by the equation:

$$\Phi = \Phi_o\left[1 - \left(T_o/T_B(1)\right)\right] \tag{6}$$

In Eq. (6), $\Phi_0 = 2.3026/\{\ln[f_0/f_m(2)]\}$. As noted above, measurements of T_B at several different frequencies are essential in order to determine K_{eff} since f_o and T_o must also be determined simultaneously [40].

3. Synthesis

Synthesis of spinel ferrite NPs is a challenging task owing to their colloidal nature. A good methodology must yield well-dispersed particles with uniform size and good crystallinity; besides, it is desirable that the synthetic setup allows for the tuning of structure and properties of the materials by simple modification of the conditions. Other important features entail the use of nontoxic reagents, low-temperature processes and the requirement of simple scalable operations. The procedures for the synthesis of ferrite NPs are given in several reviews [2, 8]. Here, we outline some of the most common examples.

3.1. Co-precipitation

Co-precipitation is straightforward and efficient and can be extended for a wide variety of simple and mixed ferrites [12, 43, 44]. This method, developed by Massart, consists of the joint precipitation of an aqueous solution containing inorganic salts in the proper stoichiometry by increasing the solution pH. Ageing of the resulting particles can be assessed at room or higher temperatures. By changing experimental conditions (e.g., concentration of metal precursors, pH of the final solution, anion of initial salts, reaction time, reaction temperature and ionic

http://dx.doi.org/10.5772/67513

strength), it is possible to obtain a wide variety of particle sizes and shapes [45, 46]. The main drawback relies on the difficulties for a proper separation of nucleation and growth stages, which leads to relative broad size distribution [45]. Besides, in some instances, the resulting powder is subjected to thermal annealing to enhance crystallinity [12, 43].

Nucleation and growth of NPs can be affected by the addition of surfactant molecules like sodium dodecyl sulfate [47], poly(acrylic) acid (PAA) chains [48] and hexadecyl trimethylammonium bromide (CTAB) [49]. Variations in the surfactant content give rise to different particle sizes and morphologies of the as-synthesized material. Other employed additives aiming to decrease the particle size dispersion are polymeric matrices like cellulose [50] and chitosan [51, 52]. A similar approach entailed the use of graphene oxide (GO) during the co-precipitation step. After the formation of the ferrite/GO composite, GO is reduced to yield porous nanocomposites containing superparamagnetic ferrite NPs and reduced GO (rGO) is used as a functional material. The resulting material possesses high surface area since rGO avoids ferrite particle agglomeration. Ni, Co and Mn ferrites/rGO nanocomposites have been synthesized with this strategy [53–55]. Alternatively, some investigations have reported the use of organic amines, which can act as precipitating and stabilizing agents [17, 56, 57]. Alkanolamines limit and control the particle growth by forming surface complexes with M^{2+} cations resulting in a marked reduction in Co ferrite size as compared when using NaOH [56].

3.2. Thermal decomposition

This method utilizes thermal decomposition of organic metal complexes in a high boiling point solvent and in the presence of a surfactant. This approach yields monodisperse highly crystalline NPs and allows for the fine-tuning of NP size and morphology by controlling several parameters like the solvent nature, kind and concentration of surfactant, aging temperature and reaction time. The typical setup with oleic acid (OA)/oleylamine (OAm) as surfactants can be used to obtain $(Zn_xM_{1-x})Fe_2O_4$ (M = Fe^{2+}, Mn^{2+}) mixed ferrites with different Zn contents as the doping cation [11]. OAm is believed to assist OA deprotonation, which promotes the formation of iron carboxylates at the NP surface [58]. In a systematic study, Mohapatra *et al.* [16] have reported the synthesis of MFe_2O_4 (M = Mn^{2+}, Co^{2+}, Fe^{2+}, Ni^{2+}, Zn^{2+}) NPs using chloride salts as precursors and OAm acting as a solvent, reducing agent and stabilizing surface capping agent. By decreasing the amine content, it was possible to obtain uniform NPs with *D* values between 2 and 9 nm. OAm chains seem to control the growth process: high concentration enables an extended coverage of the initial nuclei, which hampers a fast growth and leads to uniform small NPs.

In an important report on the synthesis of Fe_3O_4 and other ferrite NPs, Hyeon *et al.* [59] used iron(III) oleate as an organometallic precursor with the purpose to avoid environmental harmful reagents like $Fe(CO)_5$ [60] and $Fe[acac]_3$ [61]. Also, there is no need for external reducing agents and extra surfactant stabilizers [36, 61–63]. The authors obtained high yields of well-crystallized monodispersed NPs with *D* values ranging from 9–22 nm by varying the solvent boiling point. Moriya *et al.* [64] introduced an interesting approach in which a pre-synthesized trimetallic complex containing two Fe^{3+} cations and one divalent cation (Fe, Co, Mn) is decomposed in dibenzyl ether in the presence of benzylic acid (BA). The resulting nanocrystals showed uniform size with variable shapes (from truncated octahedrons to cubes) that can be tuned as BA concentration is increased; the nature of the weak intermolecular interactions between adsorbed BA molecules seems to play a key role to bring about the final morphology.

One drawback of the thermal decomposition method is that the as-synthesized ferrite NPs do not disperse in water due to the hydrophobic surfactant adsorbed onto the surface, which leads to further phase transfer steps. To obtain directly water-soluble NPs, the groups of Li [29] and Verma [58] introduced a variation in which 4–5 nm Fe and Co ferrites are obtained through the decomposition of metal acetylacetonates in the presence of pyrrolidones that act either as solvents or as hydrophilic stabilizing agents.

3.3. Polyol method

This is a variation of the thermal decomposition in which a given polyol acts as a high-boiling solvent, reducing and stabilizer agent. Metal precursors are generally organic complexes like acetylacetonates and other carboxylate complexes [65]. Given that the reaction mixture is refluxed at the boiling point of the polyol, changing either the kind or concentration of the polyol leads to different particle *D* values [66–68], generally between 4 and 15 nm. The obtained NPs have a narrow size distribution and high crystallinity, although particle agglomeration could occur. There is question about the nature of the molecules adsorbed at the particle surface; some authors have stressed that polyol anchoring to the surface occurs through $R\text{-}O^{-}$ interactions [69, 70], but others have claimed that at high temperatures, hydroxyl groups are oxidized to carboxylic acids, which are further adsorbed onto the oxide surface by forming carboxylate complexes [71]. In many applications, there is no need for further treatment, since the NPs are stable in polar solvents [71]; however, in order to increase water stability and avoid particle aggregation, several polymers can be added to the reaction mixture like PVP and poly(ethylene imine) [67]. Many approaches also include functional materials like carbon nanotubes (CNT) in order to yield magnetic composites with enhanced properties [72]. This method has been used for Fe_3O_4 as well as for other nanospinels such as Co and Zn ferrites [13, 18, 70, 73].

3.4. Hydrothermal and solvothermal synthesis

Hydrothermal methodology consists of the formation of an aqueous (or aqueous-alcoholic) solution of the metal salts followed by the addition of a base until basic pH is reached. The resulting mixture is then transferred to a pressurized autoclave and subjected to $T > 180\,°C$ for many hours. The mechanism involves the initial formation of metal hydroxides, which are oxidized and converted into the crystalline spinel ferrite due to the thermal treatment at high pressures. The resulting NPs have high crystallinity and an acceptable narrow size distribution. Particle size and shape can be effectively tuned by varying the metal concentration, solvent composition, temperature and reaction time. The addition of surfactants like CTAB [74] and poly(ethylene glycol) (PEG) [75] can change the shape of NPs and aid to control the growth and avoid agglomeration. This method can be adapted for the in situ synthesis of ferrite composites with functional materials like rGO [76, 77]; after base addition, metal hydroxides are adsorbed onto GO, and both spinel crystallization and GO reduction (by the action of supercritical water) occur during hydrothermal treatment. As a result, the size of near-spherical Zn ferrite NPs decreases as the GO content increases. On the contrary, the absence of GO sheets yielded NPs with a rod-like shape [78]. An alternative procedure by Komarneni *et al.* [79] involves hydrothermal treatment under microwave

http://dx.doi.org/10.5772/67513

radiation, leading to a drastic reduction in reaction times to just few minutes; this approach was useful for many ferrites.

Solvothermal synthesis can be understood as a modified hydrothermal process where water is replaced by an organic solvent. For instance, n-octanol along with sodium dodecylbenzene-sulfonate has been employed for the preparation of mixed ferrite NPs of Ni and Co with several compositions and varying sizes (7–16 nm), which was tuned as a function of the reaction time [26]. OA can also be used as a steric stabilizer in the reaction mixture using n-pentanol as a solvent [22]; increasing OA content decreases *D* from 19 to 5 nm and changes the particle morphology from nanoplatelets to well-dispersed nanospheres. Other approaches reported the use of diol molecules as solvents [25, 80, 81]. Bastami *et al.* [80] introduced PEG and PVP as polymeric surfactants, which bind preferentially at the surface of the near normal (i = 0.2) $MnFe_2O_4$ spinel ferrite compared to the inverse spinel Fe_3O_4 NPs synthesized similarly. As a consequence, an increase in *D* was noted for magnetite relative to Mn ferrite. Such behavior can be rationalized by taking into account the larger content of highly reactive Fe(III) species in octahedral coordination in the normal ferrite. The same procedure has been used for including GO in the reaction mixture [81]; since polyols can act as reducing agents under these conditions, the simultaneous reduction of GO and the formation of ferrite NPs are verified. Other methods used for synthesizing ferrite NPs include sol-gel [28, 82, 83], micro-emulsion [10, 36, 84], biogenic [85, 86], auto-ignited combustion [87–89], electrochemical [90] and mechanical [91] methods.

4. Functionalization

Surface functionalization of nanostructured ferrites is a crucial step in the design of nanodevices for many applications since proper functionalization determines the final use and allows control over the physico-chemical processes at the surface, thus tuning several magnetic, optical and electrical properties in the desired direction. Although several synthetic methods allow in situ functionalization of obtained ferrite NPs, this approach is not always enough, and postsynthetic surface functionalization becomes necessary. For example, biomedical and environmental applications require hydrophilic NPs with definite chemical groups. The crucial feature that allows for surface functionalization is the availability of superficial transition metal *d* orbitals acting as Lewis acids in the presence of donor ligands. Fortunately, spinel ferrite surface is reactive toward many chemical groups, which provide room for multiple combinations. Ligands comprise many low and high molecular weight compounds [92]; functional groups available for surface complexation include carboxylic acids [93–95], amides [27, 29], hydroxyl [70, 96], phosphonic [73, 97] and hydroxamic acids [73].

There are mainly three approaches to make hydrophilic functional NPs: (i) ligand exchange reaction, (ii) silica coating and (iii) polymer coating. Ligand exchange reactions effectively transfer hydrophobic particles to aqueous medium by the replacement of hydrophobic ligands with hydrophilic ones, without affecting the magnetic core considerably. However, for some applications, magnetic NPs can also be transferred from polar to nonpolar mediums [98, 99]. Small ligands stabilize the NPs mainly by coulombic repulsion of ionized groups, like quaternary ammonium cations and carboxylates [100]; charged groups not only stabilize the magnetic suspension but also reinforce water affinity by facile solvation. Conversely, macromolecular ligands stabilize NPs by interparticle steric repulsions due to extended conformations that they can adopt in contact with good solvents [101]. In cases when the polymer carries ionisable groups, as PAA [34, 93, 101], coulombic repulsions enhance their capabilities as a stabilizer.

Silica coating has the advantage that provides excellent chemical stability to the magnetic core while preventing magnetic interactions, which is traduced into colloidal stability. Following the hydrolysis-condensation method established by Stöber [102], it is possible to achieve silica shells with controlled thickness by careful addition of tetraethyl orthosilicate (TEOS) to the NP dispersion without the appearance of individual silica particles, which in turn allows for a fine-tuning of magnetic interactions [103]. Furthermore, silica coating can be functionalized with several organosilanes containing suitable groups like -SH [104–106] and -NH [105], as depicted in **Figure 2**.

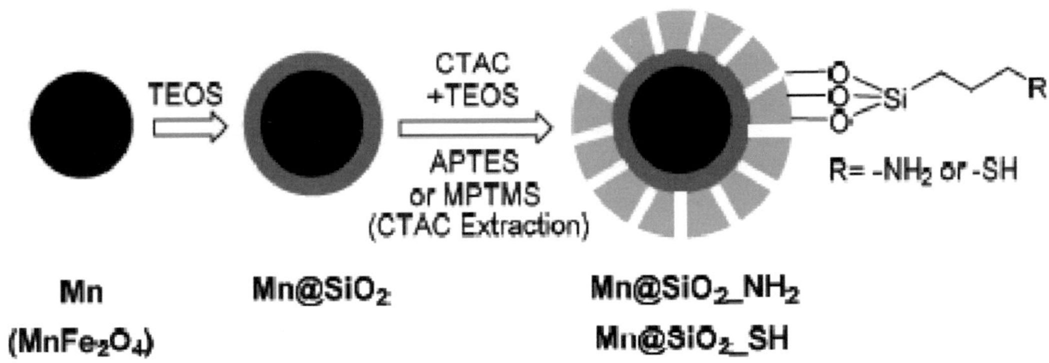

Figure 2. Scheme of the synthesis of amine and thiol functionalization of core magnetite NPs protected with a silica shell (Adapted from Ref. [105] with permission of the Royal Society of Chemistry).

Two main routes for polymer coating of NPs [107, 108] are: (i) functionalization of the NP surface with a molecule that acts as an initiator for further interfacial-controlled polymerization [109, 110] and (ii) synthesis of the polymer as the first step followed by surface anchoring [111–114]. The latter is simpler and allows for a wide variety of macromolecules, provided they bear suitable functional groups for surface binding. The former, although more laborious, has the advantage that it is possible to control the surface density of the grafted polymer and the length of the growing chains [115]. A shortcoming concerning macromolecular coating of magnetic NPs emerges when high mass magnetizations are required. Since polymers do not contribute to magnetization, mass magnetization of highly functionalized NPs drops noticeably, and so they might disable the whole system.

Conjugation after primary NP synthesis and water stabilization constitutes the final step prior to environmental and biomedical applications. It affords the ultimate precise chemical functions. Several strategies have been reported to achieve this goal entailing many known organic reactions [107]. For example, Zhao *et al.* [116] obtained hydroxamic acid-decorated Fe_3O_4/poly(acrylamide) (PAM) nanocomposites by treating the amide bonds with hydroxylamine solution and Zhao *et al.* [117] introduced amine groups by reacting ethylenediamine with a polymer containing epoxy moieties previously attached to Fe_3O_4 NPs. A very similar approach was employed in order to obtain thiol groups by adding NaSH to episulfide moieties [118]. Amide and ester formation are nice strategies to achieve NPs conjugation; in these reactions, carbodiimides are used as effective coupling agents. Following this strategy, Ren *et al.* [119] incorporated EDTA ligands to Fe_3O_4@SiO_2@chitosan particles through amide bonds between EDTA $-COO^-$ moieties and $-NH_2$ groups in chitosan shell (see **Figure 3**), while Ge *et al.* [120] incorporated a polycarboxylic acid into amine-decorated magnetite. Rare sulfur-containing functional groups have been incorporated from chitosan modification, as is the case of xanthate-decorated magnetite NPs [121]. Thiol-ene reactions have also employed to include phosphonic acid moieties in a thiol-decorated nanoplatform [122]; the kinetics and efficiency of this reaction prevent phosphonic acid depletion due to surface binding. The reaction is depicted in **Figure 4**.

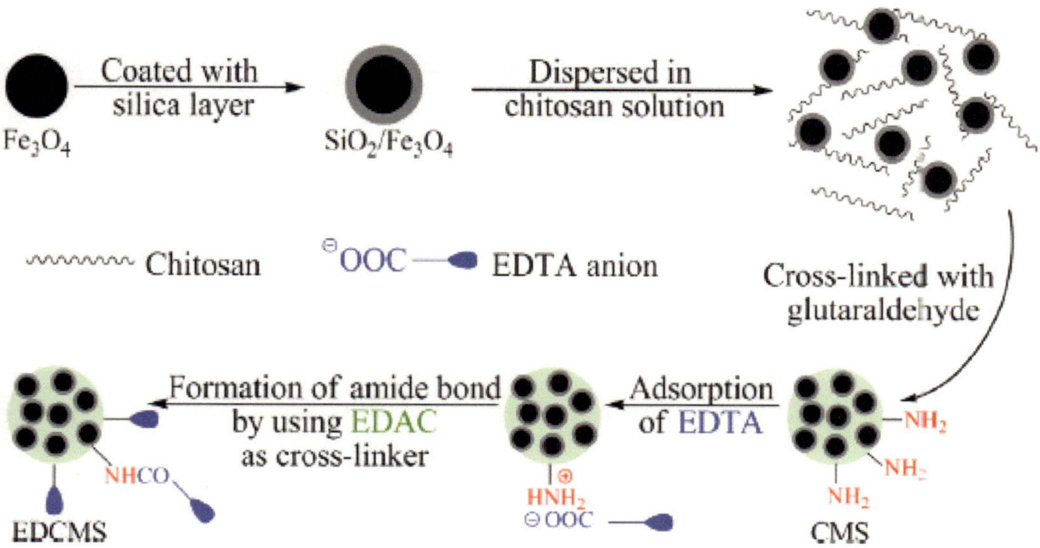

Figure 3. Synthesis of EDTA-containing magnetic NPs by amidation of chitosan (Adapted from Ref. [119] with permission of Elsevier).

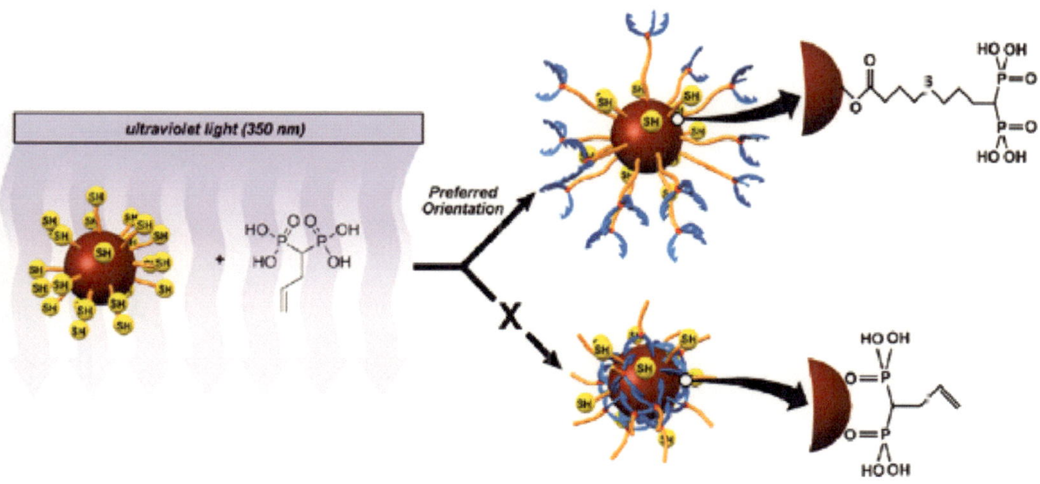

Figure 4. Synthesis of diphosphonic acid-containing magnetic NPs by a facile thiol-ene reaction (Adapted from Ref. [122] with permission of the American Chemical Society).

Coordination reactions like MOF construction have also been developed at ferrite surface. Fe_3O_4 NPs decorated with carboxyl groups were conjugated with a zeolitic imidazolate framework (ZIF-8) for the adsorption of contaminants [123, 124]. Such MOF was grown in a step-by-step assembly, initiated by the Zn^{2+} chelation to the oxide surface through carboxyl groups, resulting in a magnetic core surrounded by the ZIF shell. By varying the number of growth cycles, it is possible to tune the thickness of the MOF shell and hence, the interparticle distance between the magnetic cores. Another inorganic reaction at the interface of magnetic NPs reported recently [125] consists of the deposition of hydrous lanthanum oxide over $Fe_3O_4@SiO_2$

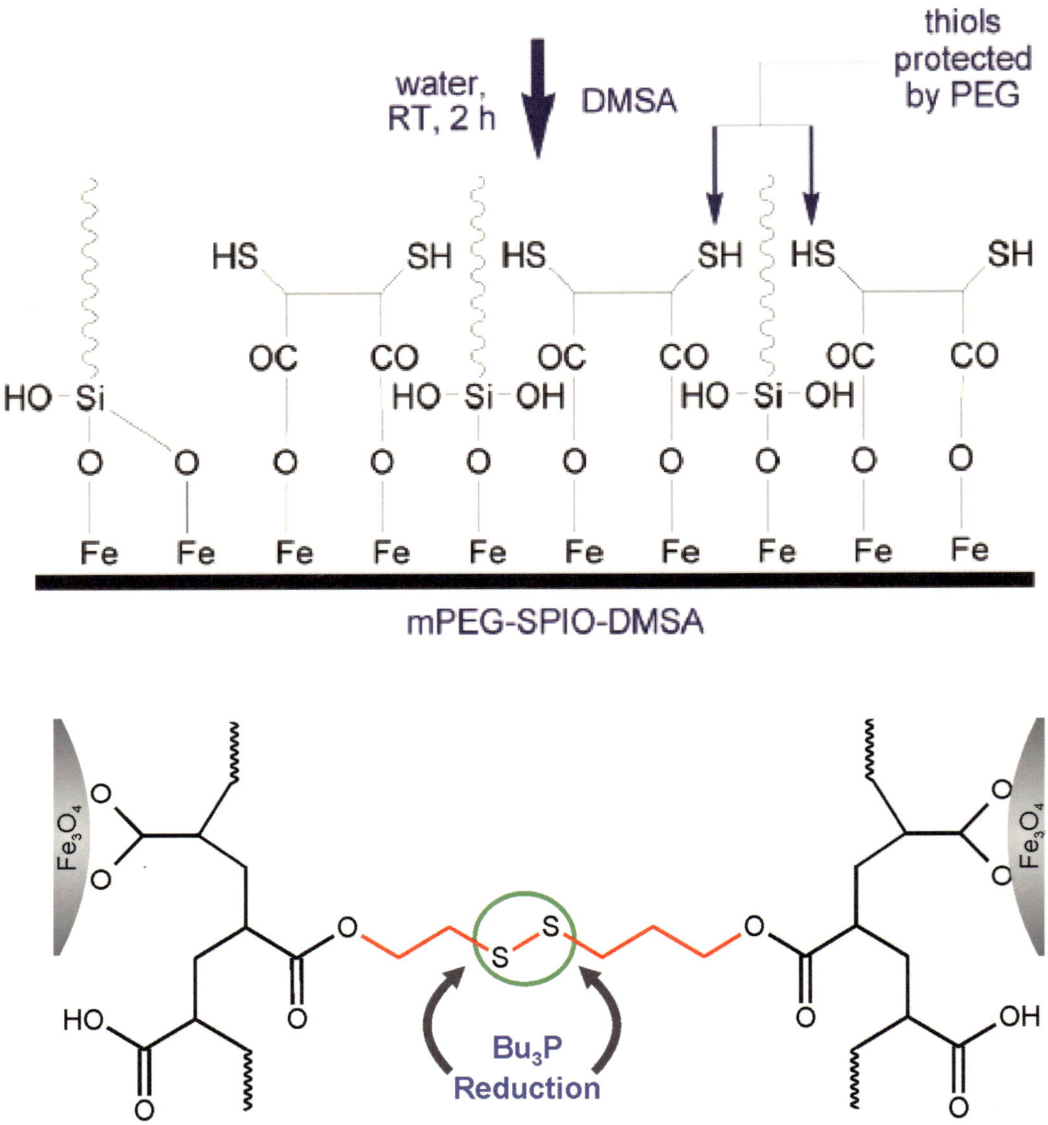

Figure 5. Two methods for thiol protection. *Top:* Steric hindrance by surface PEG grafting (Adapted from Ref. [130] with permission of the American Chemical Society). *Bottom:* Reversible thiol oxidation to disulfide bridges (Reproduced from Ref. [34] with permission of Elsevier).

core-shell NPs simply by adding $LaCl_3$ at basic pH in the presence of the magnetic material. Nanostructures composed of ferrites and noble metals have interesting and promising optical and magnetic properties; the synthesis of such materials can be easily performed by reduction of the corresponding metal salt in the presence of ferrite NPs [126].

In the case of surface thiol-decorated nanostructures, special care must be taken, since free thiol groups are prone to be oxidized during the synthetic procedures. For example, several papers have reported the oxidation of DMSA and cysteine to disulfide and sulfoxide compounds in the presence of Fe_3O_4 NPs [24, 127–129]; these undesirable processes not only reduce the effective amount of –SH groups but also could alter magnetite phase. To overcome this drawback, Maurizi *et al.* [130] grafted PEG chains onto the oxide surface as steric barriers in order to avoid the formation of intermolecular disulfide bridges between adjacent DMSA molecules. More recently, Odio *et al.* [34] designed a novel PAA copolymer containing disulfide bridges; after Fe_3O_4 functionalization with this polymer, resulting NPs were treated with tributyl phosphine in order to reduce disulfide bridges to free –SH groups. Both strategies are depicted in **Figure 5**.

5. Coordination chemistry at the surface

Since the nature of the metal-ligand interactions at the interface of the ferrites plays a key role in the properties of NPs, efforts have been devoted to unravel the structure and implications of the surface complexes occurring for different types of ligands. For this purpose, spectroscopic techniques like FTIR, XPS, EXAF and XANES are usually employed [22, 24, 58, 122, 131–133]. Mössbauer spectroscopy has also been used since iron spectra are sensitive to spin reorganization after ligand binding and to the kind of iron site that participates in the surface complexes [23, 58, 131].

Specifically, Daou *et al.* [131] showed that phosphate ligands bind to magnetite surface by forming bidentate binuclear complexes with octahedral Fe(III) cations. In contrast, Costo *et al.* [21] suggested that phosphonates bind to the surface through mono-dentate ligands. For carboxylates, it is reported that in solvothermal and thermal decomposition methods, OA is anchored to the surface by bridging bidentate complexes [22, 24, 134], a conclusion supported by a theoretical DFT study showing that the bidentate mode was the most stable configuration of iron-oleate complex no matter the surface plane exposed to the ligand [135]. However, after ligand exchange reactions with other carboxylic acids, carboxylate complexes could form chelates [24, 34, 136]. The use of amine functional groups showed that long-chain amine molecules adsorb at the surface of ferrites by N-metal interactions [16, 137]. In contrast, if both OA and OAm are used in the ferrite synthesis, the mode of coordination of each functional group could depend on the concentration and molar ratio of the surfactants. Thus, if the surfactants are diluted in N-methyl 2-pyrrolidone with a molar ratio 1:1, Verma and Pravarthana [58] suggested by means of IR analysis that OA complexes retain the bidentate mode while OAm molecules appear protonated and associated to the surface through coulombic interactions. However, XPS studies of Wilson and Langell [133] indicated that if the reaction is performed without a solvent and a higher OAm proportion is employed, OAm is anchored to the surface by Fe-N coordination, while OA binds to the surface though a monodentate complex. Using IR studies, alcohols are reported to be anchored to the oxide surface by metal-OH interactions [70]. Besides, ligands with thiol moieties anchor to magnetite surface by Fe(II)-S interactions, as suggested by IR and XPS studies [24, 138]. Finally, cyclic amides like PVP and 2-pyrrolidone seem to interact with ferrite surfaces through the carbonyl groups [27, 29]. Since this subject is not yet completely understood, computational methods are likely necessary to unravel the structural and electronic properties of surface complexes.

6. Environmental applications

In this section, we focus on two applications of nanostructured spinel ferrites for environmental remediation technologies in connection with water decontamination: adsorption and oxidation technologies.

6.1. Adsorption technologies for removal of inorganic and organic contaminants

Adsorption is often the most suitable choice for removal of toxic substances in drinking or waste waters, mainly due to its simplicity and high efficiency; the main disadvantage is the sorbent separation after the adsorption process, which can become tedious and energy consuming. However, the use of magnetic materials for adsorption makes the task of sorbent separation easier by allowing magnetic decantation with a permanent magnet. The high surface area of ferrite NPs along with their room temperature superparamagnetism and the great versatility for binding specific functional groups on their surfaces for specific contaminants makes them ideal candidates for the design and development of innovative adsorption strategies. Although several recent reviews covering this subject are available [6, 7, 139, 140], these have generally focused on the thermodynamics and kinetics of the adsorption process and relatively less attention has been paid to unravel the atomic and molecular nature of the interactions occurring at the interface. Although this is a difficult task, this information is crucial for the improvement and optimization of the nanoadsorbent.

6.1.1. Heavy metal cations

Heavy metal cations, found in natural and waste waters resulting from industrial activities, comprise a wide family of hazardous substances with a high impact on human health [141]. Here, we concentrate on those reported studies with a focus on two directions: (i) improving the adsorption capacity and/or selectivity toward a given contaminant by surface functionalization of ferrite NPs and (ii) shedding light on the adsorption mechanism at a molecular and atomic level.

6.1.1.1. Amine-functionalized nanosystems

Fe_3O_4 NPs functionalized with several amino-containing polymers were tested as Cr(VI) and Cu(II) sorbents in aqueous medium [142], showing the increase of adsorption capacity for both cations with the number of –NH moieties in the ligand incorporated to the magnetic nanoplatform. Adsorption and spectroscopic data suggested that metal removal involves coulombic interactions, ion exchange processes and formation of complexes between amine groups and metal ions, although the structure of such complexes was not revealed. Similar results were reported by Huang and Chen [113], in which Fe_3O_4@PAA NPs decorated with amine groups were proved as a good adsorbent for several heavy metals with positive and negative charges; based on pH studies, authors suggested that cations are adsorbed through chelate complexes while anions are incorporated after ion exchange mechanisms. New insights about Cr(VI) adsorption with an amino-decorated magnetic sorbent were reported by Zhao *et al.* [143]. This approach comprises nanocomposites of Fe_3O_4 and amino-functionalized GO sheets. Based on XPS measurements, authors suggested that after attractive coulombic interactions between chromate species and protonated amino groups, a fraction of Cr(VI) is reduced to Cr(III), which further forms amino-complexes; the source of electron for Cr(VI) reduction is provided by the GO sheets. Without extra evidence, these results should be taken with care because XPS deconvolution was not rigorous.

Amino-functionalized Fe_3O_4 NPs were tested as a sorbent for Cu(II), Cd(II) and Pb(II) [144]. Adsorption decreased at acid pH values and adsorption capacity for Cu(II) was higher than that for softer Lewis acids Cd(II) and Pb(II). Both results, along with thermodynamic and kinetic data, could indicate the prevalence of coulombic and complexing reactions between surface –NH moieties and the cations. Similar results were presented in other reports [52, 145]. Co-ferrite NPs coated with a polystyrene shell modified with amino and thioether groups were tested for Hg(II) adsorption [146]. Authors proposed the Hg(II) complexation by these functional groups, followed by partial reduction of Hg(II) to Hg(I), although no proof for this mechanism was presented.

6.1.1.2. Carboxyl-functionalized nanosystems

Several reports have focused on the adsorption of heavy metal cations by EDTA-modified magnetic nanosystems in order to take advantage of the high chelating ability of this multifunctional ligand. The key role of EDTA has been confirmed since the adsorption capacity decreases when no EDTA was used in the preparation of the sorbents. Indeed, Ren *et al.* [119] noted that the adsorption capacity follows the same order of the metal-EDTA complex stability: Cu(II) > Pb(II) > Cd(II). Similar conclusions are drawn from recent reports [147, 148] in which the formation of metal-carboxylates was verified by means of FTIR measurements. Carboxyl-decorated NPs from acrylic and crotonic acid copolymer were tested for Cu(II), Pb(II), Zn(II) and Cd(II) [120]. Although authors did not show spectroscopic evidences of metal-carboxylates interactions, the maximum adsorption capacity increases with the increase in Lewis acid hardness of the ion tested (Cu(II) > Zn(II) ≈ Pb(II) > Cd(II)). Likewise, Mahdavian *et al.* [149] studied the adsorption behavior of Pb(II), Cu(II), Ni(II) and Cd(II) with a nanoplatform consisting of PAA chains grown at the surface of magnetite NPs and found that the metal uptake increases with pH, suggesting chelate formation. Other carboxyl-based magnetite NPs for Pb(II) removal can be found elsewhere [150].

6.1.1.3. Thiol and other sulfur-containing compounds functionalized nanosystems

Fe_3O_4 NPs functionalized with a polythiolated ligand was probed as Hg(II) adsorbent [118]. XPS studies supported the occurrence of Hg(II)-S interactions and the simultaneous reduction of Hg(II) to Hg(I) likely at the expense of Fe(II) cations at the surface of the magnetic core. Curiously, no sign of thiol oxidation was encountered. Recently, Wang *et al.* [151] tested a simpler Fe_3O_4@SiO_2-RSH nanoplatform toward Hg(II), Pb(II) and Cd(II) ions. Although no mechanism was elucidated, the fact that the adsorption capacities followed the order Hg(II) >> Cd(II) > Pb(II) might be indicative of two phenomena: (i) metal uptake is governed by soft-soft interactions between the cations and the thiolate groups and (ii) for the case of Hg(II) ions, adsorption involves reduction. A similar $CoFe_2O_4$@SiO_2-RSH nanosystem for Pb(II) removal has been employed, but no mechanism was proposed [152].

Zhu *et al.* [121] designed a Fe_3O_4/chitosan-OC(=S)SH platform for Pb(II), Cu(II) and Zn(II) adsorption. FTIR and XPS suggested that metal-ligand interactions comprise both the N atom of the residual chitosan amine groups and the sulfur groups of xanthate moieties. However, neither the specific role of each functional group nor the structure of such complexes was elucidated. Based on the adsorption capacity order Cu(II) >> Pb(II) ≈ Zn(II), it is likely that relative hard chitosan N groups play an important role in the overall performance of the adsorbent. Alternatively, Zhang et al. [72] chose nanocomposites of Fe_3O_4 and thiolated CNTs as sorbents for Pb(II) and Hg(II) ions. Thiol grafting material proved to be a better adsorbent than Fe_3O_4/CNT composites. Adsorption capacity is larger for softer Lewis acid Hg(II), which could imply the occurrence of metal-thiol complexes. A similar trend was found in another report [106].

Fe_3O_4 NPs functionalized with a copolymer obtained by the partial modification of PAA with thio-salicyl-hydrazide were tested for several divalent cations [153]. This system contains both soft (thiol) and hard (carboxyl and amine) moieties, which might explain the good adsorption properties toward soft Cd(II) and hard Co(II) cations. Regarding Pb(II) uptake, XPS studies confirmed the presence of Pb-S interactions; it is interesting that only one contribution was proposed for the deconvolution of the Pb 4f spectrum, which implies that there is only one coordination environment for Pb(II) cations. The prevalence of Pb-S interactions is coherent with the small interference effect produced by alkaline/earth metals, since these hard cations largely prefer hard ligands.

Surface ion imprinting techniques can also be used for efficient and selective sequestration of heavy metal cations. Guo *et al.* [154] added Fe_3O_4@SiO_2 NPs to a solution containing the Pb-MPTS complex as a template. Condensation of silane groups followed by Pb(II) removal with HCl results in imprinted cavities with the proper thiol configuration (**Figure 6**). This nanosystem was shown to be a good adsorbent toward Pb(II) ions with excellent selectivity over other heavy metals like Cu(II), Zn(II) and Co(II). In this case, selectivity is not only ruled by the chemical affinity between cation and thiols but it also depends on the ionic radius, coordination number and coordination geometry.

Yantasee *et al.* [155] employed Fe_3O_4@DMSA NPs for the removal of several cations, suggesting that metal adsorption was driven by free –SH groups, while –COOH groups were anchored to the magnetic core. However, no evidence regarding either the state of the sulfur before and after metal incorporation or the nature of metal-S complex was presented. Afterward, Odio *et al.* [24] tested the adsorption capabilities of the same nanoplatform aiming at Au (III). By means of XPS and UV-vis analysis, they found that before adsorption, DMSA ligands are mostly oxidized to disulfide bridges and Au(III) could adsorb onto Fe_3O_4@DMSA NPs by three possible ways: (i) chelation with free –COOH moieties; (ii) reduction to Au^0 sub-nanometer clusters triggered by surface Fe(II) oxidation in the bare sectors of the NPs and (iii) extensive reduction to Au^0 nanoclusters in the region covered by the organic ligand, which is caused likely due to oxidation of disulfide bridges to –SOx species. The process is depicted in the upper part of **Figure 7**. On the contrary, when Pb(II) was tested with the same material, the results indicated that neither Pb reduction nor Pb-S interactions contribute to the adsorption process, but it is solely caused by the occurrence of chelating carboxylates and oxo-complex at bare sectors [34]. In order to unravel the actual role of both –SH and –COOH functions in the adsorption of hardly reducible divalent cations, the same group studied the adsorption of Pb (II) and Cd(II) by a novel nanoplatform based on Fe_3O_4 NPs capped with a copolymer with pendant free –COOH and –SH groups [34]; thiol moieties were protected during the synthesis to avoid oxidation and were regenerated just before the adsorption experiments. A detailed XPS analysis indicated that both metal-carboxylate and metal-thiolate interactions are verified during adsorption. In addition, it was shown that although both cations showed higher affinity to thiols, this tendency was more pronounced for Cd(II), that is, Pb(II) is less selective, in agreement with its borderline softness characteristics. Nevertheless, the actual structure of the surface metal complexes is still elusive.

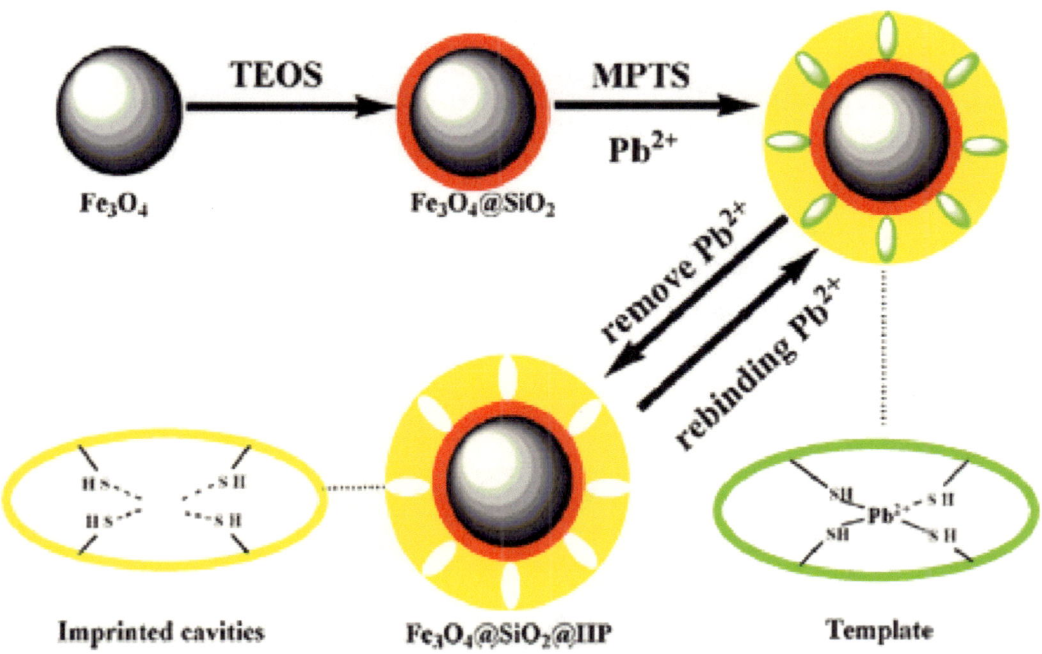

Figure 6. Surface ion imprinting technique for selective adsorption of Pb(II) ions onto magnetic NPs (Adapted from Ref. [154] with permission of Elsevier).

6.1.1.4. Other functional groups and particles with a bare surface

Fe_3O_4/PAM nanocomposites functionalized with hydroxamic acid moieties were shown to adsorb Pb(II), Cd(II), Co(II) and Ni(II) ions by forming bidentate chelating complexes [116].

The key role of hydroxamic groups was demonstrated, which agrees with the fact that the stability constant of metal-hydroxamic complexes follows the same order as the maximum adsorption capacity. The structure of these surface complexes was determined from IR and DFT studies and the system was selective toward Pb(II) uptake.

Rutledge *et al.* [122] designed a nanoplatform decorated with diphosphonic acid and thiol moieties for Pb(II), Hg(II) and f-block elements like La(III) and Eu(III) adsorption. Comparison between different adsorption sites suggested that both hard Lewis acid cations La(III) and Eu (III) are efficiently adsorbed by a hard Lewis base as diphosphonate groups, but they are inert toward thiol-mediated binding, in accordance with the soft nature of this Lewis base. However, Hg(II) uptake showed the inverse tendency, being very sensitive to –SH moieties, but a kind of unreactive toward diphosphonate basic functions. On the contrary, Pb(II) did not display such selectivity toward a particular functional group, but it showed a synergic effect upon ion uptake. The same behavior for Pb(II) was mentioned above: the fact that Pb(II) cation

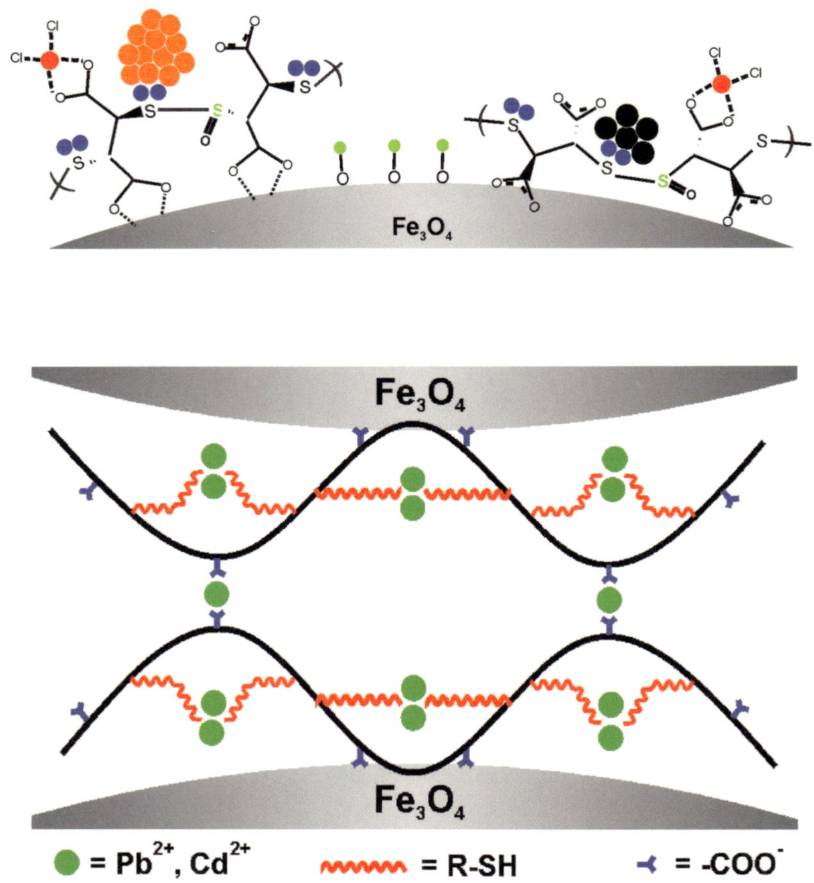

Figure 7. *Top:* Likely distribution of Au atoms during adsorption onto Fe_3O_4@DMSA NPs; different colors correspond to distinctive Au 4f XPS signals. *Bottom:* Likely distribution of Pb(II) and Cd(II) cations during adsorption onto the thiol- and carboxyl-containing Fe_3O_4 NPs; note the preponderance of metal-thiol interactions over metal-carboxylate ones (Adapted from Ref. [34] with permission of Elsevier).

http://dx.doi.org/10.5772/67513

can bind a wide variety of hard and soft ligands is related with unique electronic properties steaming from relativistic effects [156]. Further theoretical and experimental investigations on this issue are still needed.

Given that surface magnetite NPs biosynthesized by microorganisms are richer in Fe(II) content with respect to stoichiometric Fe_3O_4, this biomaterial has been tested for the adscrption and reduction of toxic oxyanions containing Cr(VI) and $m^{99}Tc(VII)$ [85]. Results confirm that bio-magnetite is a better absorber compared to a commercial magnetite of similar size, and the removal capacity changes with the particular iron substrate that was used for bacteria culture. The adsorption-reduction mechanism of chromate anions was studied by means of XPS and X-ray magnetic circular dichroism (XMCD). Authors suggested that after the fast electron transfer reactions between Cr(VI) and surface Fe(II), Cr(III) ions are incorporated into the spinel structure and occupy octahedral interstices, thus forming a layer of ferrimagnetic $CrFe_2O_4$ spinel.

6.1.2. Arsenic, phosphorous and fluoride

In a careful spectroscopic study, Liu *et al.* [157] studied As(V) and As(III) adsorption onto 34 nm magnetite NPs, avoiding the presence of oxygen during the adsorption procedure. After EXAF and XANES analysis, they confirmed that arsenate is adsorbed as a bidentate binuclear corner-sharing complex, while arsenite binds to the surface through a tridentate hexanuclear corner-sharing complex (see the left panel of **Figure 8**). Also important, they noted, based on XPS analysis, that if anoxic conditions are fulfilled, no redox reactions involving arsenic species are verified. This result disagrees with other reports [158, 159] that claimed for redox reactions

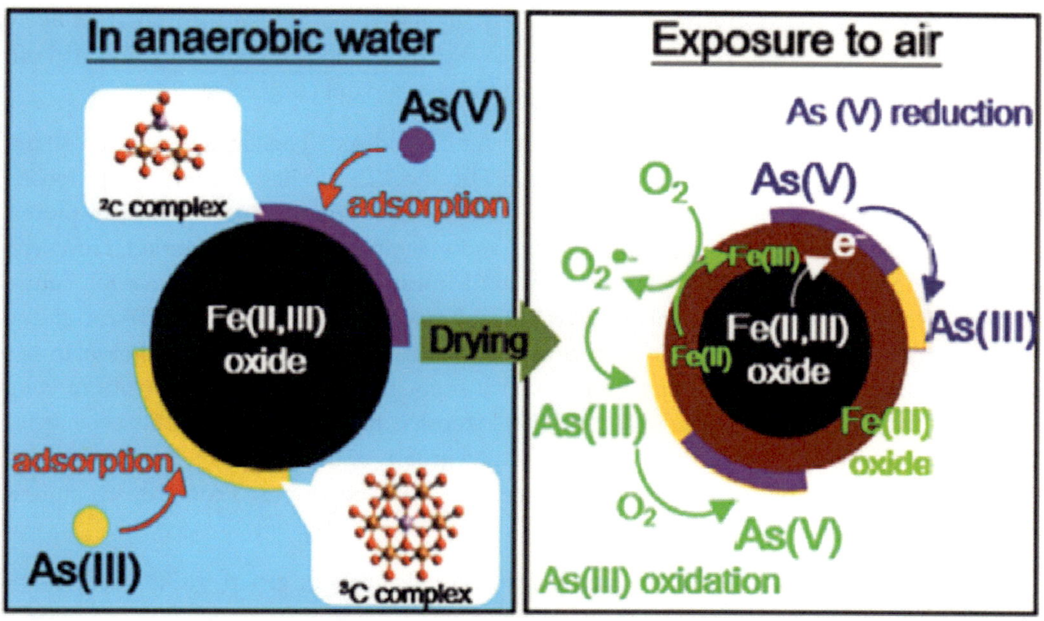

Figure 8. *Left:* Geometry of surface complexes during As(V) and As(III) adsorption onto magnetite surface in anaerobic aqueous medium. *Right:* possible redox reactions occurring at the As-Fe_3O_4 interface when exposed to air (Reproduced from Ref. [157] with permission of the American Chemical Society).

during adsorption and generates a reasonable doubt about the role of the magnetite surface in the arsenic redox processes. On the contrary, after exposure of the adsorbed material to aerobic conditions, XPS analysis showed substantial amounts of As(V) in the As(III)-treated NPs and As (III) in the As(V)-treated NPs. In both samples, the Fe(II)/Fe(III) molar ratio was less than 0.5, denoting Fe_3O_4 oxidation to γ-Fe_2O_3. As(III) oxidation to As(V) can be caused by oxygen contact. On the other hand, As(V) reduction to As(III) was explained assuming that during magnetite oxidation, an electron flow from the core to the surface takes place; eventually, these electrons could cause As(V) reduction. Both processes are depicted in the right panel of **Figure 8**.

Zhang *et al.* [160] employed Fe_3O_4/activated carbon fiber nanocomposites for As(V) removal. Based on XPS studies, authors claimed the occurrence of inner-sphere bidentate complexes between the mono-protonated anion and surface oxygen anions from either the magnetite phase or the carbon fibers. In contrast, no redox reactions were claimed. Another carbonaceous material in conjunction with ferrites has also been tested for arsenic removal. Lingamdinne *et al.* [55] chose 30 nm porous $NiFe_2O_4$/rGO nanocomposites and concluded that the high removal efficiency is due to the extended porous structure of the composite, which favors adsorption. The authors claimed that adsorption involves both electrostatic attraction and surface-complexation reactions, but to our understanding, this issue was not totally clarified.

Another recent report uses Fe_3O_4@ZIF-8 as a sorbent [123]. In this case, arsenic adsorption is entirely caused by the ZIF-8 shell, while magnetite core only acts as a magnetic device to remove the contaminant in a facile and efficient way. Alternatively, magnetite particles encapsulated with calcium alginate were tested as an adsorbent for inorganic and organic As(V) species [159]. The authors found that inorganic species are better adsorbed than monomethyl arsenate. Based on IR and XPS measurements, they suggested that arsenic incorporation likely occurs through the partial reduction of As(V) to As(III) species and the oxidation of both alginate and magnetite. However, spectroscopic studies were not conclusive.

A recent report from Penke *et al.* [161] deals with As(III) and As(V) adsorption onto 20–30 nm Al-substituted $NiFe_2O_4$ NPs. Based on Raman, FTIR and XPS studies, the authors proposed that both species are adsorbed onto the ternary oxide surface through inner-sphere complexes. In addition, redox reactions between the spinel and arsenic species were claimed. However, this report does not clarify either the geometry of the formed surface complexes or the nature of the implicated redox reactions, although it showed that the replacement of Fe(III) cations by Al(III) cations enhances the adsorption properties of the oxide due to an increase in the number of surface hydroxyl groups. Similar conclusions were drawn from the work of Peng *et al.* [162]. In this case, Fe_3O_4@$Cu(OH)_2$ core-shell nanostructures were tested as adsorbents for As(V). They stressed the key role of surface hydroxyl groups of the copper shell and suggested that arsenate complexes are formed at the surface. However, the proposed complex structure is not rigorously justified.

The use of metal hydroxides can be extended to other elements of group V like phosphorus. Thus, Lai *et al.* [125] employed a shell of hydrous lanthanum oxide incorporated into Fe_3O_4@SiO_2 NPs to drive the removal of phosphate anions from the water medium. The adsorption was fast and efficient given the high affinity of La(III) species toward phosphate ligands. This methodology overcomes the use of bare Fe_3O_4. Fe_3O_4/polypyrrole nanocomposites were employed for

http://dx.doi.org/10.5772/67513

removing fluoride anions without interference effects [163]. Based on thermodynamic and kinetic data, authors postulated an ion exchange mechanism.

6.1.3. *Dyes*

Extensive use of organic dyes has become a serious environmental problem since this family of organic compounds is difficult to decompose and transforms to carcinogenic amines. A series of ferrite MFe_2O_4/rGO (M = Mn^{2+}, Ni^{2+}, Zn^{2+}, Co^{2+}) nanocomposites were tested as combined magnetic materials for adsorption and photocatalytic degradation of Methylene Blue (MB) and Rhodamine B (RhB) under visible light [81] (see Section 6.2). Authors devoted the high adsorption capacity and fast removal rate to the large surface area of the material. For this system, though electrostatic interactions cannot be ruled out, dye retention is mainly caused by the rGO sheets, comprising π-π stacking interactions between the aromatic moieties of the dyes and the extended π-conjugated regions in the graphene structure. The same mechanism was claimed earlier using Fe_3O_4/rGO nanocomposites for MB adsorption [164]; this report also tested other materials like activated carbon and multi-walled carbon nanotubes (MWCN).

Cobalt ferrites covered by PEG chains were shown to be good adsorbents for several dyes such as methyl orange (MO), MB and Congo red (CR) [75]. Adsorption data indicate that electrostatic interactions are not the prominent cause for adsorption; instead, H-bonding interactions between –OH groups of PEG and functional groups in the dyes seem to be the responsible cause. The interactions are depicted in **Figure 9**. H-bonding has also been claimed as the main interaction of several dyes with naked $MnFe_2O_4$ NPs [43].

In a recent work, Dolatkhah and Wilson [114] functionalized Fe_3O_4 NPs with chitosan grafted with PAA and poly(itaconic) acid (PIA) chains. This polymeric material displays reversible pH-responsive behavior, which was tested for MB adsorption. As the pH increases, the ionization of the chitosan-grafted acid groups also increases, favoring the expansion of the grafted chains and the ionic interactions with MB, since this dye is cationic. Hence, adsorption is favored. Afterward, the desorption of MB is accomplished simply by acidification until dye-sorbent interactions become very weak and the polymeric chains no longer stabilize the colloid, leading to the collapse of the dye-free NPs. The process is represented in **Figure 10**.

6.1.4. *Aromatic compounds and other organic pollutants*

Rodovalho *et al.* [98] employed mixed Mn and Co ferrite functionalized with carboxyl-terminated polydimethylsiloxane brushes for the adsorption of toluene in water. Significant hydrophobic interactions between toluene and polymeric chains lead to high adsorption capacity of the NPs. Moreover, authors exploited the suitable magnetic characteristic of this mixed ferrite for fast and efficient toluene desorption through hyperthermia treatment of the toluene-loaded nanoadsorbent. McCormick and Adriaens [86] employed ultra-small biogenic magnetite in the reductive transformation of CCl_4, stressing the key role of octahedral Fe(II) cations and the importance of the electron hopping between Fe(II) and Fe(III) cations in the **B** sites, which enables a good electron mobility for the surface reduction of CCl_4. Efficient adsorption of tetracycline and diclofenac by proper functionalization of magnetic nanostructures has also been reported [165, 166].

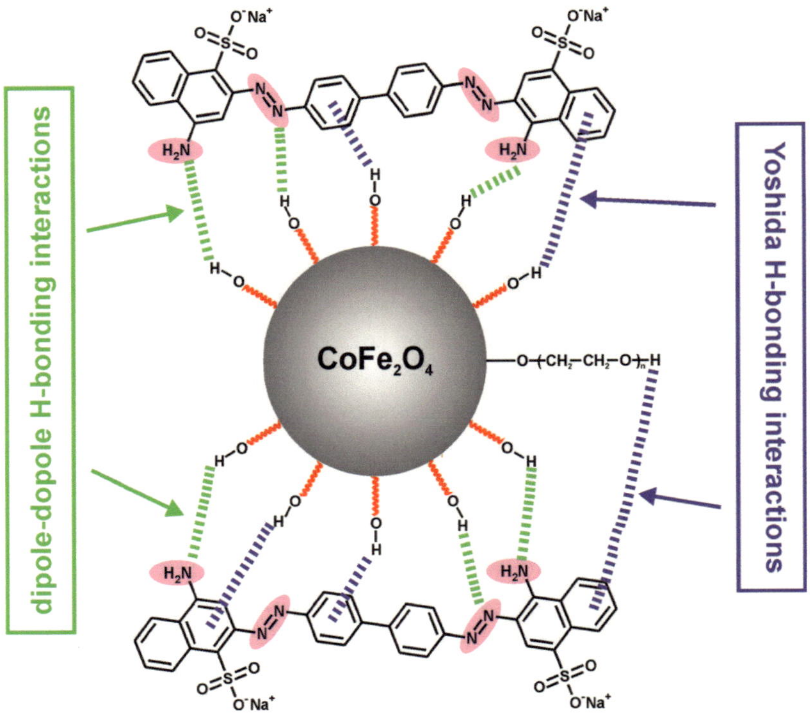

Figure 9. Proposed mechanism for CR adsorption onto PEG-functionalized MFe_2O_4 NPs (Adapted from Ref. [75] with permission of Elsevier).

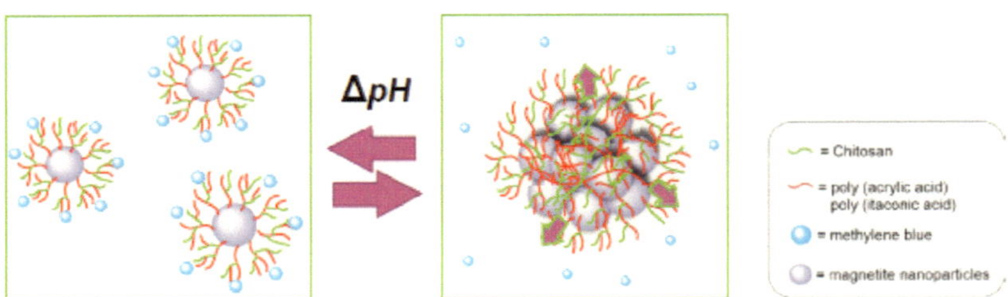

Figure 10. Mechanism of reversible pH-responsive behavior (Adapted from Ref. [114] with permission of the American Chemical Society).

6.2. Advanced oxidation technology

Advanced oxidation technologies consist of the assisted degradation of a given pollutant by using a source of highly oxidizing transient species. Such species are generally activated by the action of another substance that acts as a catalyst. Since the removal, reuse and toxicity of catalysts are major concerns, investigations have focused on the development of heterogeneous magnetic materials that can activate efficiently the oxidative degradation of the pollutants and at the same time minimize secondary contamination events. Taking into account these requirements, it is not surprising that the growing interest in ferrite NPs is due to the

http://dx.doi.org/10.5772/67513

following reasons: (i) large surface area enhances the catalytic activity; (ii) the onset of superparamagnetism enables facile removal of the catalyst; (iii) versatility of ferrite compositions makes feasible the tuning of the optical band gap of the material enabling photo-degradation approaches and (iv) chemical stability of the ferrite structure avoids metal leaking to the environment.

Nanostructured $CoFe_2O_4$ is shown to be a promising material for heterogeneous peroximonosulfate (HSO_5^-) activation in order to generate sulfate radicals ($SO_4^{-\bullet}$) that promote the oxidative decomposition of organic pollutants like 2,4-dichlorophenol [167]. The convenience of this ferrite over other cobalt oxides stems from the fact that Fe(III) cations easily cause the hydrolysis of water, leading to surface Fe(III)-OH species that can be further converted into surface Co(II)-OH complexes. In turn, such complexes are the key for the reaction of HSO_5^- to yield the $SO_4^{-\bullet}$ radicals that promote pollutant decomposition. Besides, $CoFe_2O_4$ NPs present other advantages such as no Co(II) leaching and suitable magnetic properties for the easy recovery of the catalyst. $CuFe_2O_4$ NPs have also been employed for the HSO_5^- activation in the catalytic degradation of atrazine [168] and tetrabromobisphenol [82]. In both cases, the HSO_5^- decomposition was claimed to be triggered by the cycle Cu^{2+}/Cu^+ [168]. This is in disagreement with the report of Zhang *et al.* [169], which assigned the main role to the redox pair Cu^{2+}/Cu^{3+}. Other oxidation technologies with the aid of ferrite NPs entail the persulfate ($S_2O_8^{2-}$) [170] and H_2O_2 [12, 44, 81] heterogeneous activation for the decomposition of a wide range of organic pollutants.

An improvement in the catalytic properties of ferrites can be assessed by using composites with rGO [53, 54, 81] and MWCNs [83]. Such a synergic effect is attributed to the large surface area of the composites and to the electronic properties of these carbon-based functional materials. The proposed mechanism is outlined below [171]. It comprises the initial formation of the electron-hole pair in the ferrite phase by photon absorption (I), followed by the rapid electron transfer reaction from the ferrite conduction band to the rGO sheets (II). H_2O_2 is then decomposed in the vicinity of the rGO producing highly oxidative •OH radicals (III), which are also formed from the remaining holes in the ferrite (IV). As can be seen, step (II) is crucial for the efficient separation of photo-generated carriers, which is facilitated by the high electron conductivity of the conjugated π structure of the rGO sheets, which inhibits electron-hole recombination [172]. Moreover, •OH radicals are generated close to the rGO-adsorbed target organic pollutants, thus enhancing the decomposition rate.

$$MFe_2O_4 + h\nu \rightarrow MFe_2O_4\ (h + e) \qquad \text{(I)}$$

$$MFe_2O_4\ (e) + rGO \rightarrow MFe_2O_4 + rGO\ (e) \qquad \text{(II)}$$

$$rGO\ (e) + H_2O_2 \rightarrow rGO + \bullet OH + OH^- \qquad \text{(III)}$$

$$MFe_2O_4\ (h) + OH^- \rightarrow MFe_2O_4 + \bullet OH \qquad \text{(IV)}$$

Along the same lines, Fu *et al.* [172] noted that in the case of $CoFe_2O_4$/rGO nanocomposites with 40% of GO, there is no need for H_2O_2 to achieve efficient catalytic degradation of several dyes. Other materials for dye degradation involving ferrites are $CoFe_2O_4/TiO_2$ nanocomposites [173].

7. Concluding remarks and perspectives

- Synthesis techniques for nanostructured spinel ferrites are available to tune their magnetic properties.
- The nanoparticle surface is able to bind a wide variety of molecules with distinct functional groups that not only contribute to colloidal stabilization but also serve as the starting point for further conjugation steps. Many organic and inorganic reactions can be driven at the surface of ferrites, which allow for the tailoring of specific ligands with the desired binding affinity.
- The combination of these two advantages—tuning of magnetic properties and surface versatility—makes ferrites useful and promising materials for applications where superparamagnetic behavior is required.
- Functionalized ferrite NPs, especially Fe_3O_4, are useful for removing a wide variety of heavy metals. In the case of cations, amino, carboxyl and thiol functional groups prevail as preferred candidates for metal uptake, although phosphonic and hydroxamic acids constitute promising ligands. Multifunctional ligands (synthetic and natural polymers) contribute to increase the stability and the adsorption capacity of the sorbent. At intermediate pH values, the tendency between metal-ligand affinities shows that for carboxyl and amino groups, the NPs are more selective toward hard Lewis acids, while for softer ligands like sulfur groups, the tendency is inverted.
- For removing arsenic, additional studies are warranted since controversy exists about the structure of the inner-sphere complexes and the nature of redox reactions at the interface. Also, the use of organic ligands to drive arsenic removal has not been exhausted yet.
- Most adsorption studies are limited to thermodynamic and kinetic analysis and the investigations of metal-binding interactions are supported by phenomenological models. But the mode of coordination and the geometry of the surface complexes are not clear and so detailed spectroscopic studies are still needed. Since this is a tough task due to the inherent difficulties for the achievement of a rigorous surface picture, the use of theoretical calculations could help in this regard.
- Organic dyes are preferentially adsorbed by ligand-decorated magnetic NPs. Composites with functional carbonaceous materials and grafting of smart polymers are promising lines of development.
- Spinel ferrites are useful materials for different advanced oxidation technologies, especially as composites with graphene-based materials due to the electronic and adsorptive properties of these carbon-based functional materials, which enhance the overall efficiency of the process.

Acknowledgements

The preparation of this chapter was partially supported by the CONACyT (Mexico) Projects 2013-05-231461, CB-2014-01-235840 and 2015-270810.

http://dx.doi.org/10.5772/67513

Author details

Oscar F. Odio[1,2] and Edilso Reguera[2*]

*Address all correspondence to: edilso.reguera@gmail.com

1 Universidad de La Habana, Instituto de Ciencia y Tecnología de Materiales, La Habana, Cuba

2 Instituto Politécnico Nacional, Centro de Investigación en Ciencia Aplicada y Tecnología Avanzada-Unidad Legaria, México

References

[1] Kalia S, Kango S, Kumar A, Haldorai Y, Kumari B, Kumar R. Colloid Polym. Sci. 2014;292(9):2025–2052.

[2] Wu L, Mendoza-Garcia A, Li Q, Sun S. Chem. Rev. 2016;116(18):10473–10512.

[3] Kharisov BI, Dias HVR, Kharissova OV. Arab. J. Chem. 2014. DOI: 10.1016/j.arabjc.2014.10.049.

[4] Su C. J. Hazard. Journal of hazardous materials. 2017;322(Pt A):48–84.

[5] Reddy DH, Lee SM. Adv. Colloid Interface Sci. 2013;201–202:68–93.

[6] Reddy DHK, Yun Y-S. Coordin. Chem. Rev. 2016;315:90–111.

[7] Mehta D, Mazumdar S, Singh SK. J. Water Process Eng. 2015;7:244–265.

[8] Laurent S, Forge D, Port M, Roch A, Robic C, Vander Elst L, Muller RN. Chem. Rev. 2008;108(6):2064–2110.

[9] Lu AH, Salabas EL, Schüth F. Angew. Chem. Int. Ed. 2007;46(8):1222–1244.

[10] Mathew DS, Juang R-S. Chem. Eng. J. 2007;129(1–3):51–65.

[11] Jang JT, Nah H, Lee JH, Moon SH, Kim MG, Cheon J. Angew. Chem. 2009;48(7):1234–1238.

[12] Albuquerque AS, Tolentino MVC, Ardisson JD, Moura FCC, De Mendona R, MacEdo WAA. Ceram. Int. 2012;38:2225–2231.

[13] Tahar LB, Basti H, Herbst F, Smiri LS, Quisefit JP, Yaacoub N, Grenèche JM, Ammar S. Mater. Res. Bull. 2012;47(9):2590–2598.

[14] Sorensen CM. Magnetism. In: Klabunde KJ, editor. Nanoscale Materials in Chemistry. New York: John Wiley & Sons; 2001. p. 169–221.

[15] Néel L. Ann. Geophys. 1949;5:99–136.

[16] Mohapatra J, Mitra A, Bahadur D, Aslam M. Cryst. Eng. Comm. 2013;15(3):524–532.

[17] Fernandes C, Pereira C, Fernández-García MP, Pereira AM, Guedes A, Fernández-Pacheco R, Ibarra A, Ibarra MR, Araújo JP, Freire C. J. Mater. Chem. C. 2014;2(29):5818.

[18] Artus M, Tahar LB, Herbst F, Smiri L, Villain F, Yaacoub N, Grenèche J-M, Ammar S, Fiévet F. J. Phys. Condens. Mat. 2011;23:506001.

[19] Dutta P, Pal S, Seehra MS, Shah N, Huffman GP. J. Appl. Phys. 2009;105:07B501.

[20] Yuan Y, Rende D, Altan CL, Bucak S, Ozisik R, Borca-Tasciuc D-A. Langmuir. 2012;28:13051–13059.

[21] Costo R, Morales MP, Veintemillas-Verdaguer S. J. Appl. Phys. 2015;117(6):064311.

[22] Jovanović S, Spreitzer M, Tramšek M, Trontelj Z, Suvorov D. J. Phys. Chem. C. 2014;118 (25):13844–13856.

[23] Daou TJ, Grenèche JM, Pourroy G, Buathong S, Derory A, Ulhaq-Bouillet C, Donnio B, Guillon D, Begin-Colin S. Chem. Mater. 2008;20(18):5869–5875.

[24] Odio OF, Lartundo-Rojas L, Santiago-Jacinto P, Martínez R, Reguera E. J. Phys. Chem. C. 2014;118:2776–2791.

[25] Aslibeiki B, Kameli P, Ehsani MH, Salamati H, Muscas G, Agostinelli E, Foglietti V, Casciardi S, Peddis D. J. Magn. Magn. Mater. 2016;399:236–244.

[26] Jia X, Chen D, Jiao X, He T, Wang H, Jiang W. J. Phys. Chem. C. 2008;112(4):911–917.

[27] Topkaya R, Kurtan U, Baykal A, Toprak MS. Ceram. Int. 2013;39(5):5651–5658.

[28] Vestal CR, Zhang ZJ. J. Am. Chem. Soc. 2003;125:9828–9833.

[29] Li Z, Chen H, Bao H, Gao M. Chem. Mater. 2004;16(8):1391–1393.

[30] Vestal CR, Zhang ZJ. Nano Lett. 2003;3:1739–1743.

[31] Batlle X, Pérez N, Guardia P, Iglesias O, Labarta A, Bartolomé F, García L, Bartolomé J, Roca A, Morales M. J. Appl. Phys. 2011;109(7):07B524-1–07B524-6.

[32] Pereira AM, Pereira C, Silva AS, Schmool DS, Freire C, Grenèche J-M, Araújo JP. J. Appl. Phys. 2011;109(11):114319–114324.

[33] Yang H, Hasegawa D, Takahashi M, Ogawa T. Appl. Phys. Lett. 2009;94(1):013103-1–013103-3.

[34] Odio OF, Lartundo-Rojas L, Palacios EG, Martínez R, Reguera E. Appl. Surf. Sci. 2016;386:160–177.

[35] Bishop KJ, Wilmer CE, Soh S, Grzybowski BA. Small. 2009;5(14):1600–1630.

[36] Peddis D, Cannas C, Musinu A, Ardu A, Orrù F, Fiorani D, Laureti S, Rinaldi D, Muscas G, Concas G, Piccaluga G. Chem. Mater. 2013;25:2–10.

[37] Dormann JL, Fiorani D, Tronc E. J. Magn. Magn. Mater. 1999;202:251–267.

[38] Tholence JL, Solid St. Commun. 1980;35(2):113–117.

[39] Shtrikman S, Wohlfarth EP, Phys. Lett. 1981;85(8–9):467–470.

[40] Singh V, Seehra MS, Bonevich J, J. Appl. Phys. 2009;105(7): 07B518.

[41] Dormann JL, Bessais L, Fiorani D, J. Phys. C. 1998;21(10):2015–2034.

[42] Seehra MS, Pisane KL, J. Phys. Chem. Solids. 2016;93:79–81.

[43] Yang L, Zhang Y, Liu X, Jiang X, Zhang Z, Zhang T, Zhang L. Chem. Eng. J. 2014;246:88–96.

[44] Roonasi P, Nezhad AY. Mater. Chem. Phys. 2015;172:143–149.

[45] Massart R, Cabuil V. J. Chim. Phys. PCB. 1987;84(7–8):967–973.

[46] Mascolo M, Pei Y, Ring T. Materials. 2013;6(12):5549–5567.

[47] Vadivel M, Babu RR, Arivanandhan M, Ramamurthi K, Hayakawa Y. RSC Adv. 2015;5 (34):27060–27068.

[48] Krishna Surendra M, Annapoorani S, Ansar EB, Harikrishna Varma PR, Ramachandra Rao MS. J. Nanopart. Res. 2014;16(12):2773.

[49] Zhang Y, Nan Z. Mater. Lett. 2015;149:22–24.

[50] Anirudhan T, Shainy F. J. Ind. Eng. Chem. 2015;32:157–166.

[51] Garza-Navarro MA, Torres-Castro A, García-Gutiérrez DI, Ortiz-Rivera L, Wang YC, González-González VA. J. Phys. Chem. C. 2010;114(41):17574–17579.

[52] Tran HV, Tran LD, Nguyen TN. Mater. Sci. Eng.: C. 2010;30(2):304–310.

[53] Yao Y, Yang Z, Zhang D, Peng W, Sun H, Wang S. Ind. Eng. Chem. Res. 2012;51:6044–6051.

[54] Yao Y, Cai Y, Lu F, Wei F, Wang X, Wang S. J. Hazard. Mater. 2014;270:61–70.

[55] Lingamdinne LP, Choi Y-L, Kim I-S, Chang Y-Y, Koduru JR, Yang J-K. RSC Adv. 2016;6:73776–73789.

[56] Pereira C, Pereira AM, Fernandes C, Rocha M, Mendes R, Fernández-García MP, Guedes A, Tavares PB, Grenèche J-M, Araújo JP, Freire C. Chem. Mater. 2012;24(8):1496–1504.

[57] Zhang Y, Shi Q, Schliesser J, Woodfield BF, Nan Z. Inorg. Chem. 2014;53(19):10463–10470.

[58] Verma S, Pravarthana D. Langmuir. 2011;27(21):13189–13197.

[59] Park J, An K, Hwang Y, Park JG, Noh HJ, Kim JY, Park JH, Hwang NM, Hyeon T. Nat. Mater. 2004;3(12):891–895.

[60] Park J, Lee E, Hwang NM, Kang M, Kim SC, Hwang Y, Park JG, Noh HJ, Kim JY, Park JH. Angew. Chem. 2005;117(19):2932–2937.

[61] Sun S, Zeng H, Robinson DB, Raoux S, Rice PM, Wang SX, Li G. J. Am. Chem. Soc. 2004;126(1):273–279.

[62] Yang C, Wu J, Hou Y. Chem. Comm. 2011;47(18):5130–5141.

[63] Silvestri A, Mondini S, Marelli M, Pifferi V, Falciola L, Ponti A, Ferretti AM, Polito L. Langmuir. 2016;32(28):7117–7126.

[64] Moriya M, Ito M, Sakamoto W, Yogo T. Cryst. Growth Des. 2009;9(4):1889–1893.

[65] Mondini S, Cenedese S, Marinoni G, Molteni G, Santo N, Bianchi CL, Ponti A. J. Colloid Interface Sci. 2008;322(1):173–179.

[66] Cai W, Wan J. J. Colloid Interface Sci. 2007;305(2):366–370.

[67] Zhang B, Tu Z, Zhao F, Wang J. Appl. Surf. Sci. 2013;266:375–379.

[68] Maity D, Chandrasekharan P, Si-Shen F, Xue J-M, Ding J. J. Appl. Phys. 2010;107 (9):09B310.

[69] Maity D, Kale SN, Kaul-Ghanekar R, Xue J-M, Ding J. J. Magn. Magn. Mater. 2009;321 (19):3093–3098.

[70] Deligöz H, Baykal A, Tanrıverdi EE, Durmus Z, Toprak MS. Mater. Res. Bull. 2012;47 (3):537–543.

[71] Gonçalves RH, Cardoso CA, Leite ER. J Mater Chem. 2010;20(6):1167–1172.

[72] Zhang C, Sui J, Li J, Tang Y, Cai W. Chem. Eng. J. 2012;210:45–52.

[73] Baldi G, Bonacchi D, Franchini MC, Gentili D, Lorenzi G, Ricci A, Ravagli C. Langmuir. 2007;23(7):4026–4028.

[74] Ji GB, Tang SL, Ren SK, Zhang FM, Gu BX, Du YW. J. Cryst. Growth. 2004;270(1–2): 156–161.

[75] Wu X, Wang W, Li F, Khaimanov S, Tsidaeva N, Lahoubi M. Appl. Surf. Sci. 2016;389: 1003–1011.

[76] Zong M, Huang Y, Ding X, Zhang N, Qu C, Wang Y. Ceram. Int. 2014;40(5):6821–6828.

[77] Meidanchi A, Akhavan O. Carbon. 2014;69:230–238.

[78] Rahman MM, Khan SB, Faisal M, Asiri AM, Alamry KA. Sensors Actuators B. 2012;171–172:932–937.

[79] Komarneni S, D'Arrigo MC, Leonelli C, Pellacani GC, Katsuki H. J. Am. Ceram. Soc. 1998;81(11):3041–3043.

[80] Bastami TR, Entezari MH, Kwong C, Qiao S. Front. Chem. Sci. Eng. 2014;8(3):378–385.

[81] Bai S, Shen X, Zhong X, Liu Y, Zhu G, Xu X, Chen K. Carbon. 2012;50:2337–2346.

[82] Ding Y, Zhu L, Wang N, Tang H. Appl. Catal. B. 2013;129:153–162.

[83] Zhang X, Feng M, Qu R, Liu H, Wang L, Wang Z. Chem. Eng. J. 2016;301:1–11.

[84] Aubery C, Solans C, Sanchez-Dominguez M. Langmuir. 2011;27(23):14005–14013.

[85] Cutting RS, Coker VS, Telling ND, Kimber RL, Pearce CI, Ellis BL, Lawson RS, van der Laan G, Pattrick RAD, Vaughan DJ, Arenholz E, Lloyd JR. Environ. Sci. Technol. 2010;44 (7):2577–2584.

[86] McCormick ML, Adriaens P. Environ. Sci. Technol. 2004;38(4):1045–1053.

[87] Kikukawa N, Takemori M, Nagano Y, Sugasawa M, Kobayashi S. J. Magn. Magn. Mater. 2004;284:206–214.

[88] Choodamani C, Nagabhushana GP, Ashoka S, Daruka Prasad B, Rudraswamy B, Chandrappa GT. J. Alloys Compd. 2013;578:103–109.

[89] Liu B, Li Q, Zhang B, Cui Y, Chen H, Chen G, Tang D. Nanoscale. 2011;3(5):2220–2226.

[90] Marin T, Montoya P, Arnache O, Calderon J. J. Phys. Chem. B. 2016;120(27):6634–6645.

[91] Bellusci M, Aliotta C, Fiorani D, La Barbera A, Padella F, Peddis D, Pilloni M, Secci D. J. Nanopart. Res. 2012;14(6):904.

[92] Duan H, Kuang M, Wang X, Wang YA, Mao H, Nie S. J. Phys. Chem. C. 2008;112 (22):8127–8131.

[93] Dai Q, Lam M, Swanson S, Yu RH, Milliron DJ, Topuria T, Jubert PO, Nelson A. Langmuir. 2010;26(22):17546–17551.

[94] Marcelo G, Pérez E, Corrales T, Peinado C. J. Phys. Chem. C. 2011;115(51):25247–25256.

[95] Tombácz E, Tóth IY, Nesztor D, Illés E, Hajdú A, Szekeres M, Vékás L. Colloids Surf. A. 2013;435:91–96.

[96] Cheng K, Peng S, Xu C, Sun S. J. Am. Chem. Soc. 2009;131(30):10637–10644.

[97] Das M, Mishra D, Maiti TK, Basak A, Pramanik P. Nanotechnology. 2008;19(41):415101.

[98] Rodovalho FL, Capistrano G, Gomes JA, Sodré FF, Chaker JA, Campos AFC, Bakuzis AF, Sousa MH. Chem. Eng. J. 2016;302:725–732.

[99] Dong A, Ye X, Chen J, Kang Y, Gordon T, Kikkawa JM, Murray CB. J. Am. Chem. Soc. 2010;133(4):998–1006.

[100] Wu Y, Guo J, Yang W, Wang C, Fu S. Polymer. 2006;47(15):5287–5294.

[101] Ge J, Hu Y, Biasini M, Dong C, Guo J, Beyermann WP, Yin Y. Chemistry. 2007;13 (25):7153–7161.

[102] Stöber W, Fink A, Bohn E. J. Colloid Interface Sci. 1968;26(1):62–69.

[103] Graf C, Vossen DLJ, Imhof A, van Blaaderen A. Langmuir. 2003;19(17):6693–6700.

[104] Gill CS, Price BA, Jones CW. J. Catal. 2007;251:145–152.

[105] Rocha M, Fernandes C, Pereira C, Rebelo SLH, Pereira MFR, Freire C. RSC Adv. 2015;5:5131–5141.

[106] Li G, Zhao Z, Liu J, Jiang G. J. Hazard. Mater. 2011;192(1):277–283.

[107] Lattuada M, Hatton TA. Langmuir. 2007; 23(4):2158–2168.

[108] Hood M, Mari M, Muñoz-Espí R. Materials. 2014;7(5):4057–4087.

[109] Gelbrich T, Feyen M, Schmidt AM. Macromolecules. 2006;39(9):3469–3472.

[110] Sun Y, Ding X, Zheng Z, Cheng X, Hu X, Peng Y. Eur. Polym. J. 2007;43(3):762–772.

[111] Li G-Y, Huang K-L, Jiang Y-R, Ding P, Yang D-L. Biochem. Eng. J. 2008;40(3):408–414.

[112] Zhang T, Ge J, Hu Y, Yin Y. Nano Lett. 2007;7(10):3203–3207.

[113] Huang SH, Chen DH. J. Hazard. Mater. 2009;163(1):174–179.

[114] Dolatkhah A, Wilson LD. ACS Appl. Mater. Interfaces. 2016;8(8):5595–5607.

[115] Wu W, He Q, Jiang C. Nanoscale Res. Lett. 2008;3(11):397–415.

[116] Zhao F, Tang WZ, Zhao D, Meng Y, Yin D, Sillanpää M. J. Water Proc. Eng. 2014;4:47–57.

[117] Zhao YG, Shen HY, Pan SD, Hu MQ. J. Hazard. Mater. 2010;182(1–3):295–302.

[118] Pan S, Shen H, Xu Q, Luo J, Hu M. J. Colloid Interface Sci. 2012;365(1):204–212.

[119] Ren Y, Abbood HA, He F, Peng H, Huang K. Chem. Eng. J. 2013;226:300–311.

[120] Ge F, Li MM, Ye H, Zhao BX. J. Hazard. Mater. 2012;211–212:366–372.

[121] Zhu Y, Hu J, Wang J. J. Hazard. Mater. 2012;221–222:155–161.

[122] Rutledge RD, Warner CL, Pittman JW, Addleman RS, Engelhard M, Chouyyok W, Warner MG. Langmuir. 2010;26(14):12285–12292.

[123] Zou Z, Wang S, Jia J, Xu F, Long Z, Hou X. Microchem. J. 2016;124:578–583.

[124] Zheng J, Cheng C, Fang W-J, Chen C, Yan R-W, Huai H-X, Wang C-C. Cryst. Eng. Comm. 2014;16(19):3960.

[125] Lai L, Xie Q, Chi L, Gu W, Wu D. J. Colloid Interface Sci. 2016;465:76–82.

[126] Xu Z, Hou Y, Sun S. J. Am. Chem. Soc. 2007;129(28):8698–8699.

[127] Dolci S, Ierardi V, Remskar M, Jagličić Z, Pineider F, Boni A, Pampaloni G, Veracini CA, Domenici V. J. Mater. Sci. 2013;48(3):1283–1291.

[128] Nishio K, Gokon N, Tsubouchi S, Ikeda M, Narimatsu H, Sakamoto S, Izumi Y, Abe M, Handa H. Chem. Lett. 2006;35(8):974–975.

[129] Soler MA, Lima EC, Nunes ES, Silva FL, Oliveira AC, Azevedo RB, Morais PC. J. Phys. Chem. A. 2011;115(6):1003–1008.

[130] Maurizi L, Bisht H, Bouyer F, Millot N. Langmuir. 2009;25(16):8857–8859.

[131] Daou TJ, Begin-Colin S, Grenèche JM, Thomas F, Derory A, Bernhardt P, Legaré P, Pourroy G. Chem. Mater. 2007;19(18):4494–4505.

[132] Hatakeyama M, Kishi H, Kita Y, Imai K, Nishio K, Karasawa S, Masaike Y, Sakamoto S, Sandhu A, Tanimoto A, Gomi T, Kohda E, Abe M, Handa H. J. Mater. Chem. 2011;21 (16):5959.

[133] Wilson D, Langell MA. Appl. Surf. Sci. 2014;303:6–13.

[134] Palchoudhury S, An W, Xu Y, Qin Y, Zhang Z, Chopra N, Holler RA, Turner CH, Bao Y. Nano Lett. 2011;11(3):1141–1146.

[135] Rath SS, Sinha N, Sahoo H, Das B, Mishra BK. Appl. Surf. Sci. 2014;295:115–122.

[136] Lin CL, Lee CF, Chiu WY. J. Colloid Interface Sci. 2005;291(2):411–420.

[137] Aslam M, Schultz EA, Sun T, Meade T, Dravid VP. Cryst. Growth Des. 2007;7(3):471–475.

[138] Sathish S, Balakumar S. Mater. Chem. Phys. 2016;173:364–371.

[139] Hua M, Zhang S, Pan B, Zhang W, Lv L, Zhang Q. J. Hazard. Mater. 2012;211–212:317–331.

[140] Gómez-Pastora J, Bringas E, Ortiz I. Chem. Eng. J. 2014;256:187–204.

[141] Järup L. Br. Med. Bull. 2003;68(1):167–182.

[142] Shen H, Pan S, Zhang Y, Huang X, Gong H. Chem. Eng. J. 2012;183:180–191.

[143] Zhao D, Gao X, Wu C, Xie R, Feng S, Chen C. Appl. Surf. Sci. 2016;384:1–9.

[144] Xin X, Wei Q, Yang J, Yan L, Feng R, Chen G, Du B, Li H. Chem. Eng. J. 2012;184:132–140.

[145] Tan Y, Chen M, Hao Y. Chem. Eng. J. 2012;191:104–111.

[146] Jainae K, Sukpirom N, Fuangswasdi S, Unob F. J. Ind. Eng. Chem. 2015;23:273–278.

[147] Zhao F, Repo E, Sillanpää M, Meng Y, Yin D, Tang WZ. Ind. Eng. Chem. Res. 2015;54:1271–1281.

[148] Liu Y, Fu R, Sun Y, Zhou X, Baig SA, Xu X. Appl. Surf. Sci. 2016;369:267–276.

[149] Mahdavian AR, Mirrahimi MA-S. Chem. Eng. J. 2010;159(1–3):264–271.

[150] Wang H, Chen QW, Chen J, Yu BX, Hu XY. Nanoscale. 2011;3(11):4600–4603.

[151] Wang Z, Xu J, Hu Y, Zhao H, Zhou J, Liu Y, Lou Z, Xu X. J. Taiwan Inst. Chem. Eng. 2016;60:394–402.

[152] Viltužnik B, Košak A, Zub YL, Lobnik A. J. Sol-Gel Sci. Technol. 2013;68(3):365–373.

[153] Zargoosh K, Abedini H, Abdolmaleki A, Molavian MR. Ind. Eng. Chem. Res. 2013;52:14944–14954.

[154] Guo B, Deng F, Zhao Y, Luo X, Luo S, Au C. Appl. Surf. Sci. 2014;292:438–446.

[155] Yantasee W, Warner CL, Sangvanich T, Addleman RS, Carter TG, Wiacek RJ, Fryxell GE, Timchalk C, Warner MG. Environ. Sci. Technol. 2007;41:5114–5119.

[156] Claudio ES, Godwin HA, Magyar JS. Fundamental Coordination Chemistry, Environmental Chemistry, and Biochemistry of Lead(II). In: Karlin KD, editor. Progress in Inorganic Chemistry, Volume 51. New York: John Wiley & Sons; 2003. p. 1–144.

[157] Liu C-H, Chuang Y-H, Chen T-Y, Tian Y, Li H, Wang M-K, Zhang W. Environ. Sci. Technol. 2015;49(13):7726–7734.

[158] Su C, Puls RW. Water Air Soil Poll. 2008;193(1–4):65–78.

[159] Lim SF, Zheng YM, Chen JP. Langmuir. 2009;25(9):4973–4978.

[160] Zhang S, Li XY, Chen JP. J. Colloid Interface Sci. 2010;343(1):232–238.

[161] Penke YK, Anantharaman G, Ramkumar J, Kar KK. RSC Adv. 2016;6:55608–55617.

[162] Peng B, Song T, Wang T, Chai L, Yang W, Li X, Li C, Wang H. Chem. Eng. J. 2016;299:15–22.

[163] Bhaumik M, Leswifi TY, Maity A, Srinivasu VV, Onyango MS. J. Hazard. Mater. 2011;186(1):150–159.

[164] Ai L, Zhang C, Chen Z. J. Hazard. Mater. 2011;192(3):1515–1524.

[165] Ou J, Mei M, Xu X. J. Solid State Chem. 2016;238:182–188.

[166] Zhang S, Dong Y, Yang Z, Yang W, Wu J, Dong C. Chem. Eng. J. 2016;304:325–334.

[167] Yang Q, Choi H, Al-Abed SR, Dionysiou DD. Appl. Catal. B. 2009;88:462–469.

[168] Guan YH, Ma J, Ren YM, Liu YL, Xiao JY, Lin Lq, Zhang C. Water Res. 2013;47:5431–5438.

[169] Zhang T, Zhu H, Croué J-P. Environ. Sci. Technol. 2013;47(6):2784–2791.

[170] Yan J, Lei M, Zhu L, Anjum MN, Zou J, Tang H. J. Hazard. Mater. 2011;186:1398–1404.

[171] Jumeri FA, Lim HN, Ariffin SN, Huang NM, Teo PS, Fatin SO, Chia CH, Harrison I. Ceram. Int. 2014;40:7057–7065.

[172] Fu Y, Chen H, Sun X, Wang X. Appl. Catal. B. 2012;111–112:280–287.

[173] Haw C, Chiu W, Abdul Rahman S, Khiew P, Radiman S, Abdul Shukor R, Hamid MAA, Ghazali N. New J. Chem. 2016;40:1124–1136.

Chapter 10

CVD-Made Spinels: Synthesis, Characterization and Applications for Clean Energy

Patrick Mountapmbeme Kouotou, Guan-Fu Pan and Zhen-Yu Tian

Additional information is available at the end of the chapter

http://dx.doi.org/10.5772/66285

Abstract

To reduce emissions and protect environment from pollution caused by volatile organic compounds (VOCs) and CO, catalytic oxidation can be applied as an efficient and promising technique. This review provides a novel and facile strategy to synthesize spinel-type and non-spinel-type transition metal oxides (TMOs). Specifically, single (Co_3O_4, α-Fe_2O_3, Mn_3O_4, CuO, Cu_2O and Cr_2O_3) and binary ($Co_{3-x}Cu_xO_4$, $Co_{3-x}Mn_xO_4$ and $Co_{3-x}Fe_xO_4$) TMOs have been prepared using pulsed spray evaporation chemical vapor deposition approach (PSE-CVD). PSE-CVD offers several advantages over conventional methods, such as relatively low cost, simplicity and high throughput, which makes it a promising strategy. Moreover, the PSE delivery system allows using less stable precursors and permits improving the reproducibility of the film properties with tailored compositions. The above listed TMOs prepared by PSE-CVD were successfully tested as catalysts toward the complete oxidation of some real fuels such as CO, C_2H_2, C_3H_6, n-C_4H_8 and C_2H_6O as representatives of VOCs and industrial exhaust streams. The active TMOs explored in this review could be potential catalysts candidates in one of the research areas that are currently under scrutiny, as the battle for the future of energy and environment involves the generation and application of clean energy.

Keywords: clean energy, VOCs, biofuels, catalytic oxidation, PSE-CVD, TMOs, in-situ diagnostic; mechanism

1. Introduction

Volatile organic compounds (VOCs) are widely recognized as the major contributors to the global air pollution [1]. VOCs are composed of a variety of substances, which may be either natural or of anthropogenic origin from different human activities such as transportation and many factories or industrial processes including chemical, power and pharmaceutical plants, gas stations, petroleum refining, printing, food processing, automobile, as well as textile manufacturing [2]. The volatility of the emission from the above-listed sources enables them to diffuse more or less away from their place of issue, thus causing direct and indirect impacts on human health, animals and nature. In recent years, with the rapid increase of population, industrialization, transportation and urbanization, extremely severe and persistent haze pollution has been frequently observed in developing countries. For example, by 2020, VOCs emissions are predicted to increase by 49% relative to 2005 levels in China [3]. Therefore, in addition to the increasingly stringent controls for VOCs emissions level, it is urgent to develop and apply approaches to accelerate the reduction in VOCs emissions which is also vital in the context of climate change.

VOCs include a wide range of compounds such as aromatic and aliphatic hydrocarbons, alcohols, ketones, aldehydes, which are not easy to be oxidized. Abatement by catalytic oxidation appears to be a preferable technique compared to the thermal incineration in reducing VOCs. In fact, thermal oxidation is known to be expensive since it requires a substantial energy input to destroy dilute gas phase containing VOCs at relatively high temperature (750°C) which favors the formation of toxic by-products. In contrast, catalytic oxidation allows operating at much lower temperatures (200–500°C) and leads to none or negligible NO_x formation in the combustion chamber. More importantly, catalytic oxidation can destroy VOCs and convert them into harmless CO_2 and H_2O [4]. In addition, the selectivity of catalytic oxidation could be well controlled. However, to achieve deep oxidation at mild temperature, highly active, nonselective and stable catalysts for extended periods of time are required.

Nowadays, the selection of catalyst for various organic pollutants abatement has been the subject of many studies, although the optimization of catalyst formulation does not appear to be an easy task. Noble metals and transition metal oxides (TMOs) have been widely explored in most commercial applications [5–11]. Noble metals are very active at low temperature, but their use is limited due to the high price, low thermal stability and tendency to poisoning [12]. In contrary, TMOs are considered as suitable alternatives because of higher thermal stability and lower price [13]. Among TMOs, single and mixed oxides, such as manganese and cobalt oxides, perovskites, zirconia-based catalysts, have been claimed for their effectiveness in VOCs oxidation [14–19]. In particular, Co_3O_4-based catalysts, which have been studied several decades ago regarding the high activity for CO and VOCs oxidation [20–22], have received again considerable attention in the recent years [23–28]. However, the physico-chemical and catalytic properties of TMOs thin films can be modulated with respect to the morphology, surface and bulk composition as well as metallic ratio, which are strongly dependent upon the

http://dx.doi.org/10.5772/66285

preparation approaches and experimental conditions [29–32]. Thus, suitable synthesis route for the deposition of thin films of high purity and crystallinity is urgently needed.

In recent years, great efforts have been made to the development of efficient TMOs synthesis methods, including sol-gel [33], thermal decomposition [34], hydrothermal synthesis, electrodeposition [35] and pulsed spray evaporation chemical vapor deposition (PSE-CVD) [23–28]. Among these techniques, PSE-CVD shows the benefit of being a reliable one-pot method for the growth of complex oxides with controlled composition, since most of the functional oxides contain more than two elements in their structures and further tuning of their properties requires controlled doping ratio. The strategy of using multiple precursors in a single liquid feedstock and its combination with PSE-CVD have been proved to be a rationally controllable route for the growth of functionally mixed oxides such as spinels and perovskites [36]. In addition to complex oxides structures, PSE-CVD synthesis route offers also the potential to produce nano-scale layers of pure metals, metal carbides as well as alloys, which presents a large variety of potential applications.

In this review, we mainly focus on the progress made in the deposition of single and binary metal oxides-containing thin films using gas-phase processes namely PSE-CVD for complete catalytic oxidation of CO and VOCs operating at low temperature generally below 500°C or even at much lower temperature. Following a general introduction, a brief recall of the mechanisms involved in the catalytic oxidation over TMOs is described. The main sections deal with catalytic oxidation of VOCs over single and mixed TMOs followed by remarks and perspectives. We examined several typical metal oxides that are widely studied as the essential components for catalytic oxidation of CO and VOCs and explored the effect of some important influencing factors such as the redox properties, composition, doping, film morphology and the particle size of the metals oxides. The specific mechanisms involved in the catalytic activity process toward low-temperature VOC oxidation are discussed.

2. Reaction mechanisms with TMOs

Oxides-type catalysts made of transition metals are well known to selectively catalyze a large number of chemical processes. Most often, these oxides are used in the form of powder or supported thin film for oxidation reactions in the chemical industry or in automotive emission control. Although low-temperature catalytic oxidation of CO was intensively studied and the mechanism has been well addressed, it is still difficult to extend the results obtained from this reaction to catalytic oxidation of VOCs due to the different properties of pollutants and reaction conditions [37]. Depending on the partial reaction order, different reaction mechanisms have been proposed for CO and VOCs oxidation. First, the Langmuir mechanism states that the reaction occurs via the so-called Eley-Rideal (ER) process in which the reaction proceeds via a collision between an impinging gas-phase molecule and an adsorbed species, the controlling step being the reaction between an adsorbed molecule and a molecule from the gas phase [38]. The ER process for simple molecule such as CO can be schematically represented in **Scheme 1**.

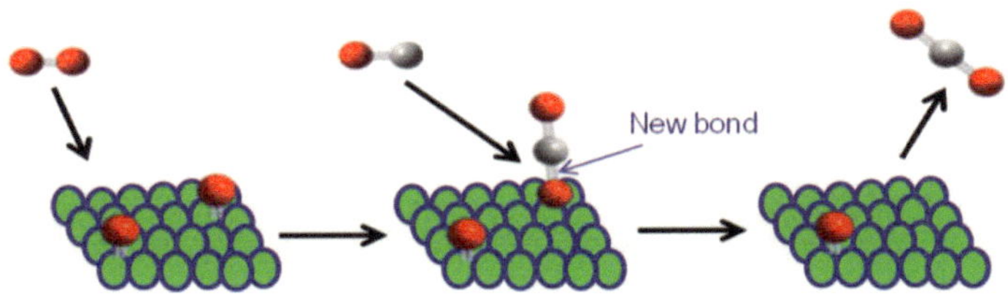

Scheme 1. Schematic illustration of the ER process for CO oxidation.

Second, the Langmuir-Hinshelwood (LH) mechanism indicates the reaction happens through interactions among the adsorbed molecules, radicals or fragments of the reactant molecules [39] (see **Scheme 2**). According to LH, the controlling step is the surface reaction between two adsorbed molecules on analogous active sites.

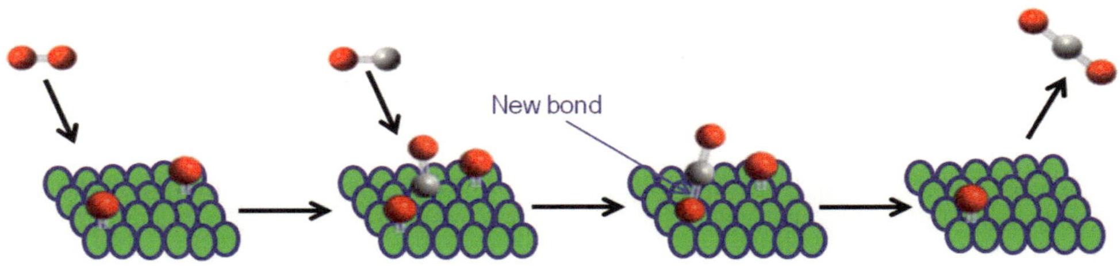

Scheme 2. Schematic illustration of the LH process for CO oxidation.

Finally, the Mars-van Krevelen (MvK) mechanism [40] points out the lattice oxygen enters the reaction sequences, caused an oxidation-reduction sequence in the reaction of the reactant molecules and oxygen on different redox sites as schematically represented in **Scheme 3**. This mechanism has been widely accepted and used for the oxidation of a series of organic compounds.

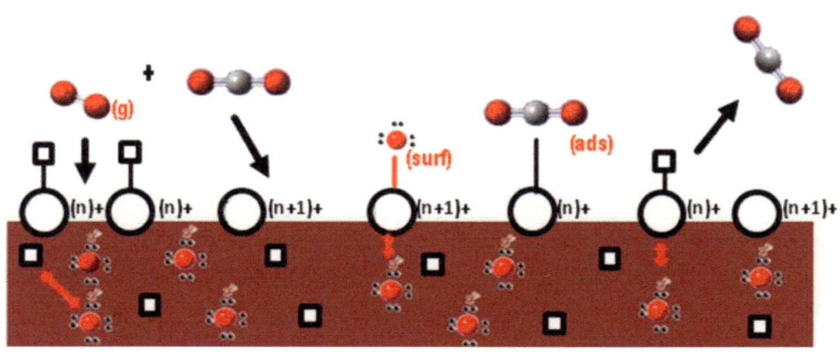

Scheme 3. Illustration of the MvK mechanism for CO oxidation.

For the transition metal catalysts either in single oxide (Co_3O_4, CuO/Cu_2O, Fe_2O_3, Cr_2O_3 as well as Mn_3O_4) or in binary oxide phase (Co-Cu, Co-Fe, Co-Mn and so on), it has been widely

http://dx.doi.org/10.5772/66285

demonstrated and recognized that the determinant factors affecting their activities and performances toward complete oxidation of CO and VOCs are the formation of highly active oxygen species activated by the oxygen vacancies in addition to the close relationship between redox properties, oxygen vacancies and the bulk oxygen mobility through the Mars-van Krevelen (MvK) mechanism. In general, the MvK mechanism can proceed in two successive steps in terms of the cyclic reaction, namely the transfer of the bulk oxygen ions to the surface sites being the first step for the decomposition of bulk oxides and then the recovering of the bulk oxide. In this first step, oxygen vacancies on the catalyst surface are reduced as they react with the organic molecules. Therefore, the presence of surface oxygen vacancies plays a crucial role in the decomposition of bulk oxide. In the second step, the preformed reduced site is immediately regenerated through the consumption of gaseous oxygen or the transfer of oxygen atoms from the bulk to the surface, that is, the oxygen molecules compete with the bulk lattice oxygen for the surface oxygen vacancies, resulting in the inhibition of the bulk oxide decomposition process [41]. Since the catalyst is reduced in the first step and then reoxidized in the second step, this mechanism is also known as the redox mechanism.

3. Tailored synthesis, characterization and application of single and binary oxides

The most active oxides frequently used are made of Ag, V, Cr, Mn, Fe, Co, Ni and Cu [42]. In fact, these *n*-type and *p*-type metal oxides are generally active catalysts (particularly *p*-type) for deep oxidation since they are electron-deficient in the lattice and conduct electrons by means of positive "holes" [37]. In addition, the adsorbed oxygen species generally observed at their surfaces might participate in the reaction sequences together with the lattice oxide ions to improve the catalysts' performances. Therefore, numerous investigations on the development of TMOs catalysts are mainly focused on this type of oxides. Several *n*-type and *p*-type single oxides such as Co_3O_4 [23], α-Fe_2O_3, Mn_3O_4 [43], CuO [44], Cu_2O [45] and Cr_2O_3 [46] have been prepared via PSE-CVD and successfully tested as active catalysts toward total oxidation of some real fuels such as CO, C_2H_2, C_3H_6, n-C_4H_8 and C_2H_6O as representatives of industrial exhaust stream or VOCs. Among these oxides mentioned above, a few of them seem to be particularly promising.

3.1. CO and VOCs oxidation over PSE-CVD made single oxide catalysts

3.1.1. Cobalt oxide with spinel structure

Cobalt oxide has a spinel structure and its formula can be written as CoO-Co_2O_3, or Co_3O_4. It presents an ideal spinel structure in which Co^{2+} cations occupy one-eighth of the tetrahedral sites, Co^{3+} cations occupy half of the octahedral sites and 32 sites are occupied by O^{2-} ions [47]. Spinel Co_3O_4 has been used broadly and successfully for the destruction of CO and hydrocarbons compounds [48] and is claimed to be one of the most active catalysts in destruction of these compounds. The performances of some Co-based catalysts seem to be comparable to some noble metal catalysts. The high activity of Co_3O_4 is likely to be related to the relatively low enthalpy of vaporization (ΔH_{vap}) of O_2 [47]. Therefore, the Co–O bond strength of Co_3O_4 can affect desorption of lattice oxygen [49]. For example, CO frequently seems to react with pre-adsorbed or lattice oxygen to give CO_2, which may further react to form surface carbonate species.

Kouotou et al. have studied the total oxidation of C_3H_6 and CO over PSE-CVD made spinel Co_3O_4 deposited on stainless steel grid mesh [23] and the as-deposited Co_3O_4 catalysts exhibited good activity compared with the reaction over non-coated mesh (NCM) as blank experiment (see **Figure 1**). The total conversion of the investigated compounds to CO_2 was obtained respectively at around 380°C for C_3H_6 and 350°C for CO, which gives an obvious temperature shift relative to the NCM. This result shows that Co_3O_4 prepared by PSE-CVD was active for total oxidation of CO and C_3H_6. The catalytic performance of Co_3O_4 toward the oxidation of CO and C_3H_6 was attributed to the abundance of active Co^{3+}/Co^{2+} cations and oxygen vacancies generally present at the surface of such materials. Co^{3+}/Co^{2+} cations and oxygen vacancies were suggested to act as active sites for the oxidation process and are key parameters governing catalytic process during the total conversion rate of CO and C_3H_6.

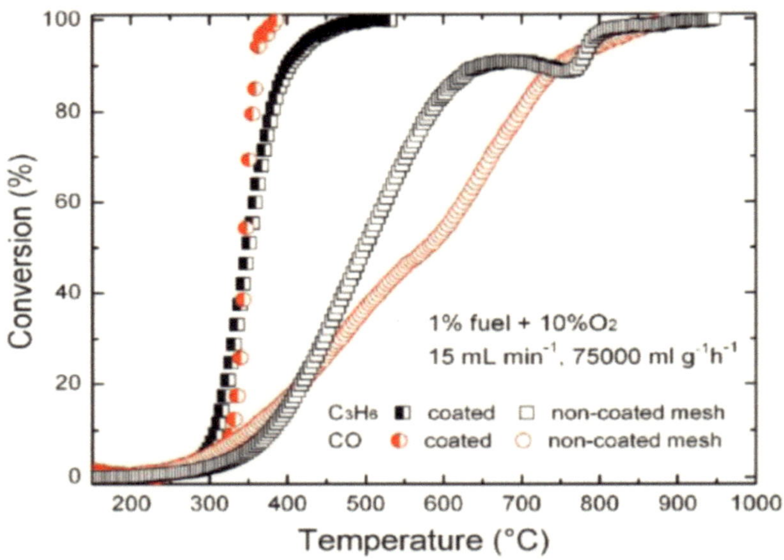

Figure 1. Light-off curves of C_3H_6 and CO catalytic conversion over grid-mesh of stainless steel coated with Co_3O_4 and non-coated mesh as reference. Reproduced from [23] with permission. Copyright 2013, the Royal Society of Chemistry.

3.1.2. Hematite

Hematite, known as iron oxide (α-Fe_2O_3), has been extensively studied because of its excellent chemical stability, natural abundance, low cost, relatively nontoxic and environmentally benign [50]. As an important precursor, α-Fe_2O_3 is the most stable iron oxide phase, featuring some unique properties such as *n*-type semiconductor and magnetic as well as corrosion-

http://dx.doi.org/10.5772/66285

resistant properties [51]. α-Fe_2O_3 can be converted into other functional materials such as maghemite (γ-Fe_2O_3) and magnetite (Fe_3O_4) [52]. These properties have driven α-Fe_2O_3 in numerous promising applications. Among them, application of α-Fe_2O_3 as catalysts caught wide attention. A systematic study was carried out by Walker et al. [53] who evaluated the possible application of iron catalysts for automotive emission control. Despite the promising results, α-Fe_2O_3 has been scarcely tested as catalyst for the abatement of CO and VOCs. Recently, α-Fe_2O_3 thin films have been selectively prepared by PSE-CVD approach and successfully tested against the low-temperature catalytic oxidation of CO and propene. As a reference, the reaction was first performed on NCM in the temperature range of 180–950°C. An identical mesh coated with 20 mg α-Fe_2O_3 was then tested under the same inlet gas conditions, but raising the temperature only to 500°C, which is the lattice stability temperature limit of α-Fe_2O_3 [54]. In the presence of α-Fe_2O_3, the conversion of CO started at ~220°C and complete conversion occurred at 398°C, while these values were shifted to ~300 and 930°C on NCM, respectively (see **Figure 2**). It should be mentioned that the reusability of the obtained α-Fe_2O_3 films and reproducibility of their catalytic performance were satisfactory within experimental uncertainty. The obtained results were compared to results reported by Walker et al. [53] who used unsupported Fe_2O_3 and several supported catalysts for CO oxidation. The temperature at 50% of CO conversion (defined as T_{50}) was found to be 398°C with Fe_2O_3 [53], while PSE-CVD made α-Fe_2O_3 thin films [55] enabled 50% conversion of CO to CO_2 at 320°C, 78°C lower than Fe_2O_3 from Walker et al.'s work, revealing the better catalytic performance of the PSE-CVD deposited thin films. More details regarding the comparison can be found in **Table 2** (see Section 2.1.6).

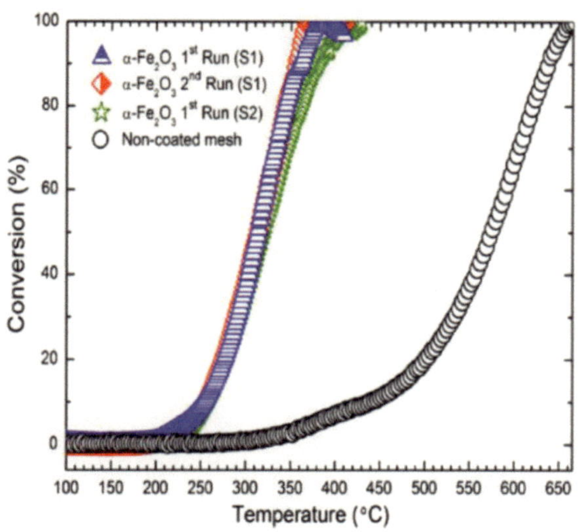

Figure 2. CO conversion profiles over stainless steel grid meshes coated with α-Fe_2O_3 films at a WHSV of 45,000 g_{cat}^{-1} h^{-1}. Two samples (S1 and S2) prepared at the same conditions were tested. First run and 2nd run are the catalytic tests performed for the first and second time, respectively. The performance for NCM is included as a reference. Reproduced from [55] with permission. Copyright 2013, Elsevier.

The reaction mechanism involved in CO oxidation over various Fe_2O_3 surfaces has been widely discussed in the literature [56–58]. As an example, a suprafacial mechanism describing the oxidation of CO over hematite has been proposed by Kandalam et al. [57] and was suggested to be applicable to the catalytic reaction with PSE-CVD made hematite [55]. According to Kandalam et al. [57], CO molecule first adsorbs onto hematite, weakening a Fe-O bond near the crystal surface and then, a second CO molecule adsorbs and forms CO_2 by breaking the weakened Fe–O bond [57]. As hematite films are composed of a bulk and surface region on which active adsorption sites exist, part of these sites could be occupied by trapped oxygen atoms, which can originate either via dissociation of adsorbed oxygen during the oxidation reaction or via diffusion from the lattice to the surface. In addition, the number of surface iron atoms with which the adsorbed O_2 can interact was earlier reported to play a significant role [59].

Besides CO oxidation, deep oxidations of C_3H_6 have been achieved at low temperature over a series of α-Fe_2O_3 thin films coated at different temperature on stainless steel. The effect of the deposition temperature on the film morphology and redox properties has been systematically investigated and their influences on the catalytic oxidation of C_3H_6 have been clearly demonstrated [54]. Since the samples presented different redox behavior that is the temperature-programmed reduction (TPR), oxygen states distribution (lattice and adsorbed) obtained by the X-ray photoelectron spectroscopy (XPS) and morphology by helium ion microscopy (HIM, see **Figure 3**) [54], the catalytic tests were performed for all samples. The objective was to investigate these effects on the catalytic properties. To analyze the catalytic behavior after reoxidation as a prerequisite for application in consecutive cycles, the catalytic conversion of C_3H_6 was carried out both before and after pretreatment of the catalyst under oxygen flow. This was done because of the large amount of adventitious carbon, C–O and O–C=O type moieties detected at the surface of the samples prepared at 400 and 450°C (see **Table 1** and **Figure 3**) would be expected to limit their performance by forming a barrier layer between the active sites of the catalysts and the reactant molecules. The catalytic tests on fresh samples were performed over catalysts prepared at 350 and 450°C which present the lowest and the highest concentration of carbonaceous species, respectively. In addition, all pretreated samples were used at least twice in the catalytic tests to assess the reproducibility, as shown in **Figure 4**. **Figure 4** displays the light-off curves of C_3H_6 conversion obtained with α-Fe_2O_3 fresh samples (**Figure 4a**) and with α-Fe_2O_3 pretreated samples (**Figure 4b**). With fresh samples, catalyst (CAT) prepared at 350°C (CAT_{350}) with lower adsorbed species presents better performance than $CAT_{450°C}$. For each sample, the conversion begins at around 250°C. Temperatures for 10, 50 and 90% C_3H_6 conversion were presented for fresh and pretreated samples. The film prepared at 350°C was the most active one regarding the conversion profile as a function of temperature, followed by the films obtained at 400 and 450°C, respectively (**Table 1**). The experiments were repeated and the results were reproducible (**Figure 4b**). α-Fe_2O_3 thin films present competitive activity to that reported for supported noble metals. As an example, with only 20 mg of α-Fe_2O_3 deposited at 350°C, the oxidation of 50% of propene was reached at 331°C, whereas that of 200 mg Au/Al_2O_3 and $La_{1.7}Sr_{0.3}CuO_4S_{0.2}$ was obtained at 365 and 419°C, respectively. More details regarding the comparison can be found in **Table 2** (see Section 2.1.6).

http://dx.doi.org/10.5772/66285

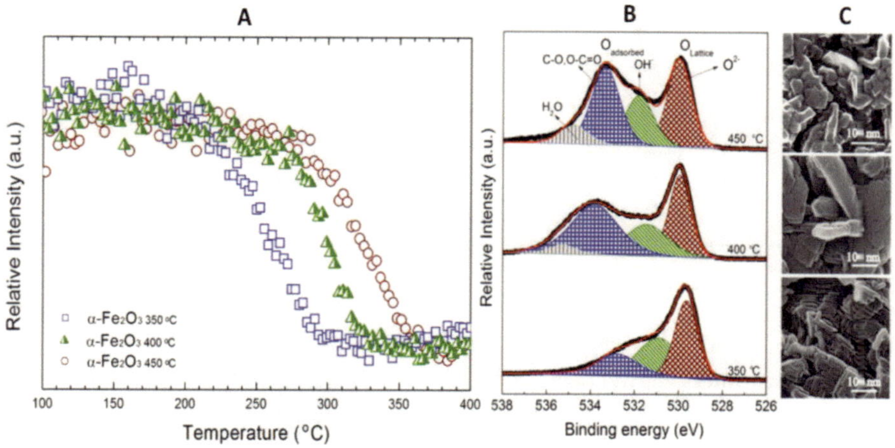

Figure 3. TPR profiles obtained for α-Fe_2O_3 thin films, showing variation of the reduction properties in function of deposition temperature (A); XPS O 1s core-shell for α-Fe_2O_3 thin films, thin and thick lines are fitted and experimental results respectively (B) and HIM images of α-Fe_2O_3 thin films coated on stainless steel at different temperatures (C). Reproduced from [54] with permission. Copyright 2013, the Royal Society of Chemistry.

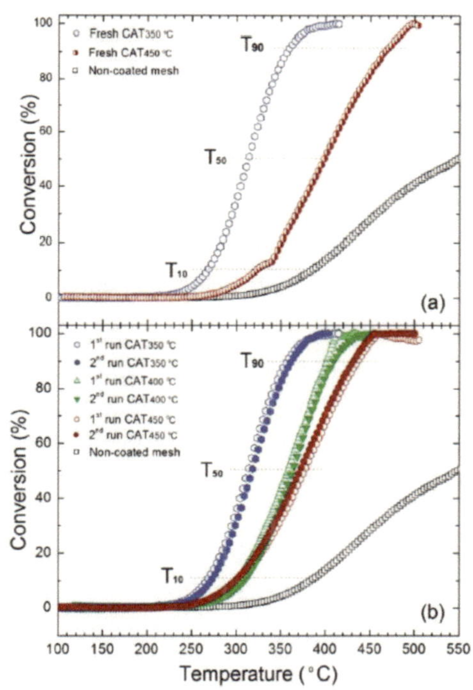

Figure 4. Light-off-curves of C_3H_6 conversion over a series of α-Fe_2O_3 coated on the mesh of stainless steel at different temperature and NCM. CAT_{350}, CAT_{400} and CAT_{450} represent α-Fe_2O_3 deposited at 350, 400 and 450°C, respectively. The total flow rate was kept at 15 ml min^{-1} with 1% of C_3H_6 and 10% of O_2 diluted in Ar at the WHSV of 45,000 g_{cat}^{-1} h^{-1}. Reproduced with permission from [54]. Copyright 2013, the Royal Society of Chemistry.

	OL (%)	OH⁻ (%)	O–C=O, C=O (%)	H_2O (%)	Oads/OL	T10 (°C)		T50 (°C)		T90 (°C)	
						FS	PS	FS	PS	FS	PS
CAT350	46.52	25.87	22.76	–	1.05	264	260	326	315	405	350
CAT400	31.19	23.69	41.20	3.92	2.21	–	295	–	355	–	400
CAT450	27.31	25.47	43.82	3.41	2.66	326	302	405	380	470	435

Table 1. XPS peak deconvolution result in percentage O_{abs}/O_L ratio, T_{10}, T_{50}, T_{90} of α-Fe_2O_3 at different deposition temperature.

To identify the phenomenon governing the difference in activity between the three catalysts, a correlation between the catalytic behavior and the α-Fe_2O_3 characterization results was made [54]. XPS results revealed the presence of both adsorbed and lattice oxygen (**Figure 3B**). The Oadsorbed/Olattice ratio of CAT_{350} was the lowest among the three samples [54]. The catalytic tests over fresh and pretreated samples reveal the negligible role of the adsorbed oxygen, suggesting that the lattice oxygen plays a key role in the reaction sequence. This was corroborated by the TPR experiments, as shown in **Figure 3A**. In addition, it has been observed by Xie et al. [60] and Wang et al. [61] that the morphology or the crystal plane of metal oxides nano-crystal can remarkably alter their catalytic performance. As revealed by the HIM analysis, the film morphology of PSE-CVD made hematite was found to be significantly dependent on the deposition temperature (see **Figure 3C**). The increase of the preparation temperature leads to a variation of film morphology, which is responsible of the increase of the grain size. The grain size is one of the most crucial factors that determine the catalytic performances of catalysts. It is known that the smaller the grain size, the larger the specific surface area. Therefore, it was assumed that CAT_{350} with the smallest grain size and fine crystal structure possesses the largest specific surface area. This argument was consistent with the experimental observation that CAT_{350} presents the best performance.

Based on the TPR, XPS and HIM results, the possible oxidation mechanism of C_3H_6 over the as-prepared catalysts was suggested to follow a Mars-van Krevelen (MvK) mechanism, an intrafacial mechanism which involves migration of bulk oxygen to the surface, where it participates in the reaction with the reactant and replacement of bulk oxygen by oxygen from the gas phase [62]. Previous investigations have pointed out that this mechanism is valid in the combustion of hydrocarbons over transition metal oxide catalysts [63]. Also, catalytic combustion of C_3H_6 over Co_3O_4 has been reported by Liotta et al. [10] to proceed with the MvK mechanism.

3.1.3. Manganese oxides

Considering the low toxicity and availability, manganese oxides have attracted great attention among the various different transition metal oxides. Besides its unique physic-chemical properties, high activity and durability, they were extensively studied as catalysts [64–66]. Manganese oxides possess a wide range of crystal phases (β-MnO_2, γ-MnO_2, α-Mn_2O_3,

γ-Mn_2O_3, α-Mn_3O_4 and Mn_5O_8) as well as variable oxidations states (+II, +III, +IV), which confer strong ability to switch from one oxidation state to another one and enable the formation of defects in the lattice, beneficial to the high oxygen mobility and oxygen storage [67]. Catalytic oxidation of VOCs (benzene and toluene) was investigated over a series of manganese oxide catalysts (Mn_3O_4, Mn_2O_3 and MnO_2). The sequence of catalytic activity was found as follows: $Mn_3O_4 > Mn_2O_3 > MnO_2$, which was closely correlated with the oxygen mobility on the catalyst [68]. Following the same logic, an investigation of the role of lattice oxygen in catalytic activity of manganese oxides toward the oxidation of ethanol, ethyl acetate and toluene has been performed by Santos et al. [69]. The results indicate that Mn_3O_4 improves catalytic performance due to the increased reactivity and mobility of lattice oxygen. In a recent paper, Tian et al. [43] have reported that Mn_3O_4 is a highly stable and active catalyst in many respects comparable to conventional catalysts based on noble metals. The catalytic performance was investigated with respect to the total oxidation of CO and C_3H_6 at atmospheric pressure. The full comparison of the reactants and products is given in **Figure 5**. The oxidation of CO over Mn_3O_4 (**Figure 5a$_1$, a$_3$**) becomes observable at 190°C and complete conversion occurs at 343°C with a temperatures shift to 280 and 820°C for an experiment without Mn_3O_4 (**Figure 5b$_1$, b$_3$**). Compared to the CO oxidation over manganese oxides (MnO_x and Mn_2O_3) prepared by precipitation [70], T_{25} was observed at 298°C, which is higher than the value (250°C) obtained with PSE-CVD Mn_3O_4 [43]. Moreover, the temperature at 50% (T_{50}) of CO conversion was observed at 271°C, demonstrating that Mn_3O_4 prepared by PSE-CVD is highly active in the deep oxidation of CO.

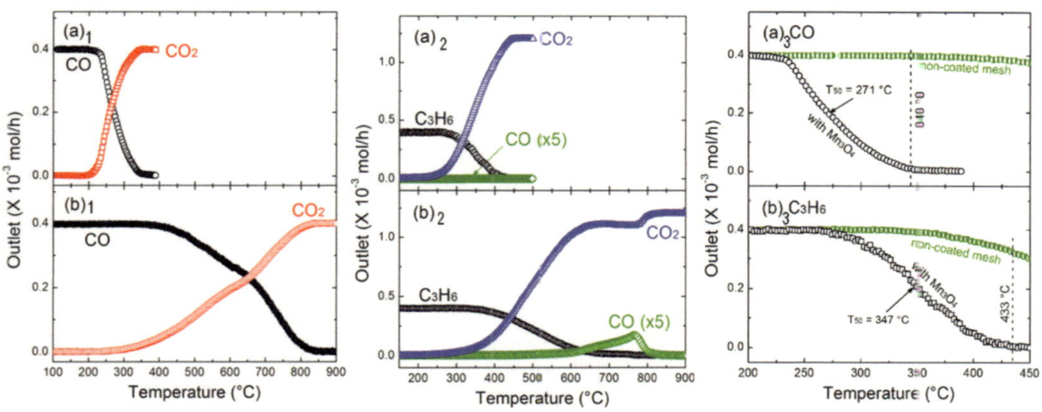

Figure 5. Production and light-off curves of CO and C_3H_6 oxidation over Mn_3O_4-coated mesh (a_1, a_2 and a_3) and NCM (b_1, b_2 and b_3), respectively. Light-off curve of CO (a_3) and C_3H_6 (b_3) oxidation over Mn_3O_4-coated mesh and NCM. The reaction was carried at the total flow rate was kept at 15 ml min^{-1} with 1% of C_3H_6 and 10% of O_2 diluted in Ar corresponding the WHSV of 45,000 g_{cat}^{-1} h^{-1}. Reproduced with permission from [43]. Copyright 2013, American Chemical Society.

It was also observed that the conversion of C_3H_6 on Mn_3O_4 (**Figure 5, a$_3$**) becomes detectable at 220°C and conversion of C_3H_6 approaches 100% at 433°C. However, the oxidation of C_3H_6 on NCM starts at ~290°C and a temperature as high as 821°C (**Figure 5b$_2$, b$_3$**) was required for complete conversion of C_3H_6. More importantly, at low C_3H_6 conversions, the main product is

CO_2, but a trace of CO peaking at 769°C (**Figure 5b$_2$**) was formed on NCM, which results from partial oxidation. However, except CO_2 and water, no other by-products were detected in the presence of Mn_3O_4, indicating that Mn_3O_4 is effective to reduce CO emission.

The catalytic oxidation of CO and C_3H_6 over Mn_3O_4 was suggested to likely follow the MvK mechanism which involves reversible reduction/reoxidation steps of the lattice oxygen atoms. As reported in the above-mentioned text, trapped oxygen was apparently released with the increase of the temperature and Mn_3O_4 tends to be reduced through the subsurface oxidation of CO or C_3H_6 with the lattice or surface oxygen. Subsequently, the reduced metal oxide site is reoxidized by the atmospheric O_2. According to the redox results [43], Mn_3O_4 was easier to be reoxidized than to be reduced, demonstrating that the reduction step could play a crucial role in the kinetics of the catalytic oxidation process. The abundance of adsorbed oxygen, revealed in the XPS results in addition to the excellent redox behavior [43], can play a key role in the oxidation process.

3.1.4. Copper oxide

Copper oxide is a promising semiconductor material which is widely exploited for a broad field of applications. As CuO is nontoxic and its constituents are abundantly available, the synthesis of CuO is considered as an important research topic for catalytic processes. Copper oxide is unique as it has square planar coordination of copper by oxygen in the monoclinic structure. TMOs, such as Co_3O_4 and Mn_3O_4, are effective catalysts for the oxidative destruction of VOCs [24]. However, not much is known concerning the abatement of CO and VOCs with CuO [71–73]. The selective oxidation of propene with O_2 to propylene oxide and acrolein has been study by Hua et al. [74] and the results indicate that Cu_2O nanocrystals control the catalytic selectivity as well as the activity in propylene oxidation with O_2. In addition, the authors also reveal the underlying structure-activity relationships of this complex heterogeneous catalytic reaction at the molecular level and identify the catalytically active sites. Most recently, Tian et al. [44] and Pan et al. [45] have studied the low-temperature complete oxidation of VOCs of olefins type such as C_2H_2 and C_3H_6 over CuO_x. They have demonstrated that CuO and Cu_2O are active and promising catalyst for the abatement of VOCs.

The catalytic performance of the CuO was evaluated for the complete oxidation of C_3H_6 by Tian et al. [44]. The background effect of the mesh on the combustion process was examined by carrying out the oxidation of C_3H_6 on NCM under the same gas inlet conditions. **Figure 6** compares the conversion temperature of C_3H_6 production over CuO films and NCM. The conversion plots show clearly that CuO favors the total oxidation C_3H_6 (**Figure 6a**) at lower temperatures relative to NCM. In the presence of CuO, the consumption of C_3H_6 was observable at about 190°C and complete conversion was reached within 310°C, while these two values shift toward higher temperatures for the reaction on NCM. No trace of CO was detected in the oxidation process with CuO (**Figure 6b**). However, a significant amount of CO was formed in the reaction without catalyst, which was assigned to come from the partial oxidation reaction. In the two cases, CO_2 was observed to be the final product. The temperatures *T10, T50,* and *T90,* corresponding to the 10, 50 and 90% C_3H_6 conversion during the temperature-programmed reaction, were selected to compare the catalytic performance of the deposited CuO as well as some representative catalysts available in the literature (see **Table 2** in Section 2.1.6) toward C_3H_6 oxidation.

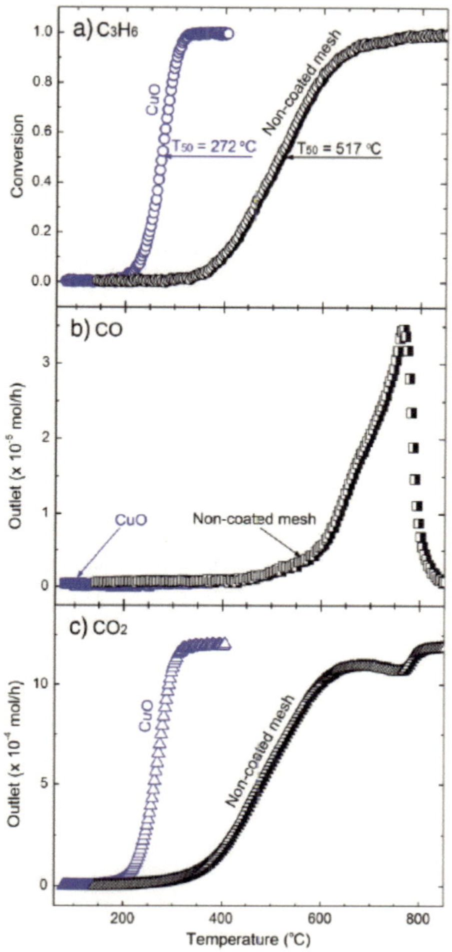

Figure 6. Light-off curves for C_3H_6 oxidation: conversion of C_3H_6 (a); associated CO production (b) and CO_2 production (c) with CuO-coated and NCM. The reaction was carried at the total flow rate was kept at 15 ml min^{-1} with 1% of C_3H_6 and 10% of O_2 diluted in Ar corresponding the WHSV of 45,000 g_{cat}^{-1} h^{-1}. Reproduced with permission from [44]. Copyright 2013, Elsevier.

Cu_2O was also tested by Pan et al. [45] as catalyst for the deep oxidation C_2H_2 and C_3H_6. The catalytic tests were carried out three times for the same sample and the results were quite close, demonstrating that the prepared Cu_2O has good reusability with reproduced results. **Figure 7A** compares the temperature-dependent conversion ratio of C_2H_2 and C_3H_6 with Cu_2O-coated mesh and NCM. Compared to the NCM condition, the complete oxidation of C_2H_2 decreased from 450 to 300°C and for C_3H_6 decreased from 675 to 425°C over Cu_2O. It should be mentioned that during the oxidation of C_2H_2 and C_3H_6 over Cu_2O-coated mesh, CO was not detected. However, CO was detected abundantly over NCM. Compared to other TMOs such as Mn_3O_4 [43] and Co_3O_4 [23] for the oxidation of C_2H_2 and C_3H_6, Cu_2O exhibits much better catalytic performances. With an Arrhenius expression, apparent activation energies (E_a) of <15% of C_2H_2 and C_3H_6 conversion were deduced. The E_a of C_2H_2 and C_3H_6 oxidation over NCM was 93.5 and 92.0 kJ mol^{-1}, while these values shift to 51.7 and 57.0 kJ mol^{-1} with Cu_2O-coated samples (average values for three times), respectively. Compared to E_a obtained with other TMOs, such as Co_3O_4 (128.9 kJ mol^{-1} for C_2H_2, 127.1 kJ mol^{-1} for C_3H_6) [48] and Mn_3O_4 (84.7 kJ mol^{-1} for C_3H_6) [43], the reaction with Cu_2O also shows lower E_a. Thus, the relatively low E_a was suggested to contribute in the acceleration of the oxidation processes and enhances the catalytic performance for the oxidation of C_2H_2 and C_3H_6. As redox mechanism is generally accepted to be the dominant mechanism in the oxidation of low-rank hydrocarbons over TMOs, both authors also suggested CuO and Cu_2O follow a redox process. According to Pan et al. [45], Cu_2O is suggested to react first with oxygen, giving rise to CuO. Second, the reaction of C_2H_2 and C_3H_6 with the trapped or lattice oxygen occurs, leading to CuO reduction and release of oxygen to form Cu_2O. From the XPS results (**Figure 7B**), O 1s core shell shows mainly O_2^{2-} and O^- species. Both O_2^{2-} and O^- are known as strongly electrophilic reactants capable to attack an organic molecule in the region of its highest electron density and result in the oxidation of the carbon skeleton. As the electrophilic oxygen species such as lattice and adsorbed oxygen generally participates in the total oxidation of hydrocarbons to CO_2, these electrophilic oxygen species (O_2^{2-} or O^-) presented at the surface of Cu_2O were suggested to be also benefit for the complete conversion of C_2H_2 and C_3H_6.

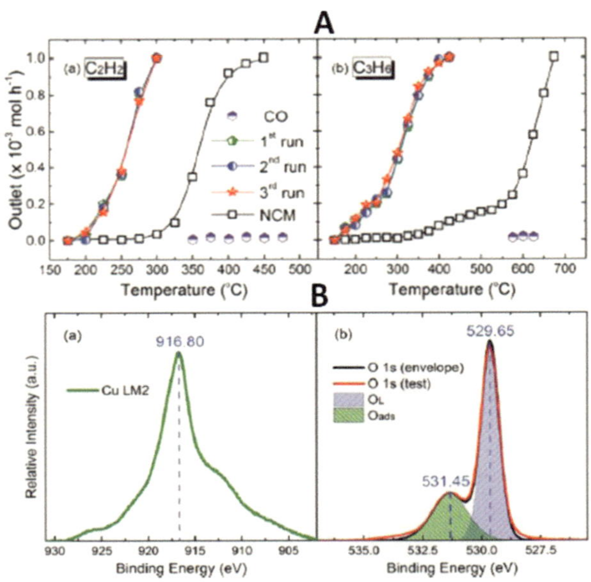

Figure 7. Outlet profiles of C_2H_2 A (a) and C_3H_6 A (b) oxidation over NCM and mesh grid of stainless steel coated with Cu_2O. The reaction was carried at the total flow rate was kept at 15 ml min^{-1} with 1% of fuel and 10% O_2 diluted in Ar, corresponding the WHSV of 45,000 $g_{cat}^{-1}\,h^{-1}$. Cu LM2 B (a) and O 1s signals B (b) of representative Cu_2O thin film. Reproduced with permission from [45]. Copyright 2015, the Royal Society of Chemistry.

3.1.5. Chromium oxide

Chromium oxide (Cr_2O_3) has been broadly and successfully used for numerous applications such as wear resistance, corrosion protection, optics and electronics due to its high melting point, heat resistance, mechanical strength, chemical inertness, optical characteristics, high hardness and low friction coefficient [75–79]. It has been proved to be active when used as catalyst for the destruction of halogenated compounds. In addition to its good thermal stability, Cr_2O_3 exhibited attractive performance in the low-temperature abatement of VOCs. Recently, Liang et al. [46] reported the synthesis of Cr_2O_3 by PSE-CVD for the total oxidation of propene. **Figure 8** compares the temperature-dependent conversion of C_3H_6 to CO_2 as the final product and the associated CO production over Cr_2O_3 films and NCM. The conversion plots show clearly that Cr_2O_3 enables the oxidation of C_3H_6 at lower temperatures relative to the NCM. In the presence of Cr_2O_3, the consumption of C_3H_6 becomes observable at about 400°C and complete conversion occurs around 550°C, while these temperatures were observed to shift, respectively, to 600 and 775°C for the reaction on NCM. Compared to the T_{50} of C_3H_6 conversion, T_{50} obtained with Cr_2O_3 was quite close to the values reported for Au/Al_2O_3 and $La_{1.7}Sr_{0.3}CuO_4S_{0.2}$, which indicates that the PSE-CVD made Cr_2O_3 exhibits similar catalytic performance to the noble metal and perovskite. During the oxidation of C_3H_6 over Cr_2O_3-coated mesh, only a small amount of CO was detected. However, considerable quantities of CO (**Figure 8b**) were formed in the reaction without catalyst, which could originate from the partial oxidation reaction.

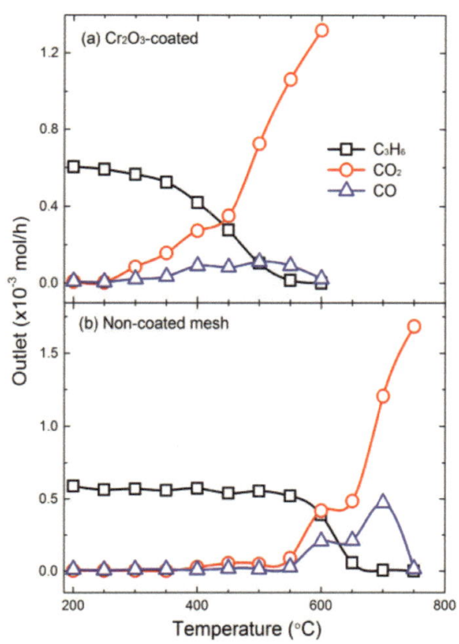

Figure 8. Outlet profiles of C_3H_6 oxidation over Cr_2O_3-coated and NCM. The reaction was carried at the total flow rate was kept at 15 ml min^{-1} with 1% of fuel and 10% O_2 diluted in Ar, corresponding the WHSV of 62,500 g_{cat}^{-1} h^{-1}. Reproduced with permission from [46]. Copyright 2015, Wiley-VCH.

Diffuse reflectance infrared Fourier transformed spectroscopy (DRIFTS) was used to study the possible catalytic reaction mechanism of Cr_2O_3. A feed of argon gas containing 1% of C_3H_6 and 10% of O_2 was introduced into the reactor which was equipped with a small piece of stainless steel grid mesh as the catalyst support. The IR spectra at different temperatures are presented in **Figure 9**. Surface adsorption species with IR bands at 729 and 913 cm^{-1} are clearly observed at temperature lower than 100°C, which were assigned to adsorption of C_3H_6 on the deposited film. These bands were still detectable below 300°C. When the temperature increases to 500°C, the coordinated C_3H_6 vanishes gradually and CO_2 was detected. The temperature for the complete conversion of C_3H_6 to CO_2 was around 500°C, in good agreement with the catalytic tests.

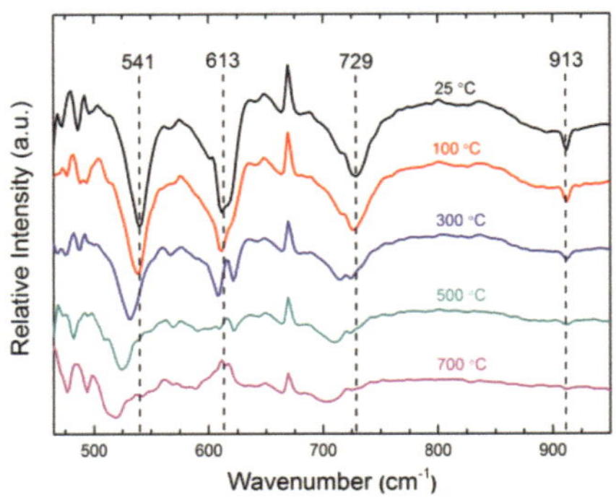

Figure 9. In-situ DRIFTS spectra of Cr_2O_3 during C_3H_6 adsorption at different temperatures. Reproduced with permission from [46]. Copyright 2015, Wiley-VCH.

3.1.6. Comparison of the catalytic performances of single TMOs with literature

Table 2 summarizes and compares the catalytic performance of the PSE-CVD made single TMOs with that of the selected catalysts in the literature. Even though the experimental condition was not exactly the same, regarding the weight of the catalyst and the WHSV, our catalysts present comparable activity to the systems presented in the table. Special attention is paid to the comparison of the catalytic performance with noble metals and transition metal oxides. As it can be seen from **Table 2**, noble metals and TMOs (single) present attractive results in terms of CO and propene total oxidation.

Type of catalyst	Material	Weight (mg)	Gas composition	WHSV [a] (ml g^{-1} h^{-1})	T_{10} [b] (°C)	T_{50} [b] (°C)	T_{90} [b] (°C)	Refs.
Single TMOs	α-Fe_2O_3	20	1% C_3H_6/10% O_2 in Ar	45,000	260	313	350	[54]
	α-Fe_2O_3 400°C				295	355	400	
	α-Fe_2O_3 450°C				302	380	435	
	Co_3O_4	12	1% C_3H_6/10% O_2 in Ar	75,000	306	347	396	[23]

http://dx.doi.org/10.5772/66285

Type of catalyst	Material	Weight (mg)	Gas composition	WHSV [a] (ml g^{-1} h^{-1})	T_{10} [b] (°C)	T_{50} [b] (°C)	T_{90} [b] (°C)	Refs.
	Co_3O_4	41.5	2% C_3H_6/20%O_2 in Ar	73,000	293	327	356	[48]
	Co_3O_4	12		75,000	325	354	385	[24]
	CuO	12	1% C_3H_6/10% O_2 in Ar		–	272	–	[44]
	Cu_2O	20		45,000	–	300	–	[45]
	Cr_2O_3	10			–		–	[46]
	Mn_3O_4	12		75,000		347		
Precious metal	Au/Al_2O_3	200	1.5% C_3H_6/4% O_2 He	22,500	288	349	410	[80]
	Au/Al_2O_3		1% C_3H_6/9% O_2 in He	219,512	–	365	–	[91]
	$Au/BaO/Al_2O_3$			250,000	–	290	–	
	$Au/Rb_2O/Al_2O_3$			257,143	–	307	–	
	$Au/Li_2O/Al_2O_3$			225,000	–	327	–	
	$Au/MgO/Al_2O_3$			214,286	–	359	–	
	Ag/Al_2O_3	50	3% C_3H_6/10% O_2 in N_2	12,000	–	–	420	[95]
Single TMOs	α-Fe_2O_3 thin film	20	1% CO/10% O_2/89% Ar	45,000	–	319	364	[25]
	Fe_2O_3	50	3.44% CO/20.6 O_2% in He	1,200,000	–	300	330	[96]
	Fe_2O_3	1000	2.5% CO/1.7% O_2/0.5% H_2 in He	60,000	–	367	397	[53]
	Fe_2O_3/TiO_2				–	333	412	
	Fe_2O_3/Al_2O_3		2.5% CO/1.7% O_2/0.5% H_2/He		–	327	377	
	Co_3O_4 thin film	12	1% CO/10% O_2/89% Ar	75,000	–	335	350	[23]
	Co_3O_4 bulk	50	1% CO/8% O_2 in He	44,400	–	–	350	[90]
	Co_3O_4 thin film	12	1% CO/10% O_2 in Ar	75,000	306	347	396	[23]
	CeO_2	100	2% CO/2% O_2 in N_2	60,000	–	–	374	[97]
	Al_2O_3	100	2% CO/2% O_2 in N_2		–	–	394	
	Mn_3O_4	12	1% CO/10% O_2 in Ar	75,000	–	271		[43]
Precious metals	$Pt/H_2SO_4/ZrO_2$	50	3.5% CO/4% O_2 in N_2	120,000	–	–	290	[98]
	Pt/Al_2O_3	2/200	1% CO/1.38% O_2 in N_2	90,000	–	333	430	[89]
	Au/SiO_2	100	1% CO/99% dry air	12,000	–	337	423	[99]

[a]: *Weight hourly space velocity*
[b]: *Temperature at X% conversion (X=10, 50 and 90 %)*

Table 2. Overview of the catalytic performances of PSE-CVD single made TMOs compared with the literature.

3.2. CO and VOCs oxidation over PSE-CVD made binary TMOs catalysts

Among the different mixed-oxide structures, numerous references can be found, dealing with the reactivity of perovskites such as compounds with A_2BO_4 structure or spinel-type mixed oxide (AB_2O_3) for the CO and VOCs oxidation reaction. Since the 1970s, some spinel-type mixed oxides are known to exhibit activity for CO and VOCs oxidation. In order to further improve the catalytic performance of TMOs, some combinations of oxides have been formulated and the obtained binary or mixed TMOs have exhibited better activities than the single or mixed components or even comparable with that of noble metals. Such catalysts include Co-Mn [24], Cu-Co [28] and Co-Fe [25–27]. It is important to note that, because of the weak performances of Cr_2O_3 catalyst compared to other single oxide presented in this review, in addition to highly toxicity, we have restricted the application of Cr-based catalysts to low operation temperatures.

3.2.1. Catalytic oxidation of VOCs over mixed Co-Mn oxides made by PSE-CVD

Tian et al. have studied the catalytic oxidation of VOCs over spinels $Co_{3-x}Mn_xO_4$ ($0 \leq x \leq 0.34$) binary oxides obtained by PSE-CVD. The grown Co-Mn binary oxides were tested toward the total oxidation of C_2H_2 and C_3H_6 as illustrative examples [24]. The TPR/TPO results and the light-off curves of the samples are presented in **Figures 10** and **11** and summarized in **Table 3. Figure 10A** compares the temperature-dependent conversion of C_2H_2 and C_3H_6 over $Co_{3-x}Mn_xO_4$ films and NCM. The detailed outlet profiles of the fuels and products are given in **Figure 11**. The temperatures T_{50}, T_{90}, $T_{50,}$ and T_{90} of CO_2, corresponding to the temperatures of CO_2 production, were selected as parameters to indicate the catalytic activity of the deposited samples toward the deep oxidation of hydrocarbons, as shown in **Table 3**. The catalytic performances were improved from Co_3O_4 to $Co_{2.66}Mn_{0.34}O_4$, both being superior to the NCM.

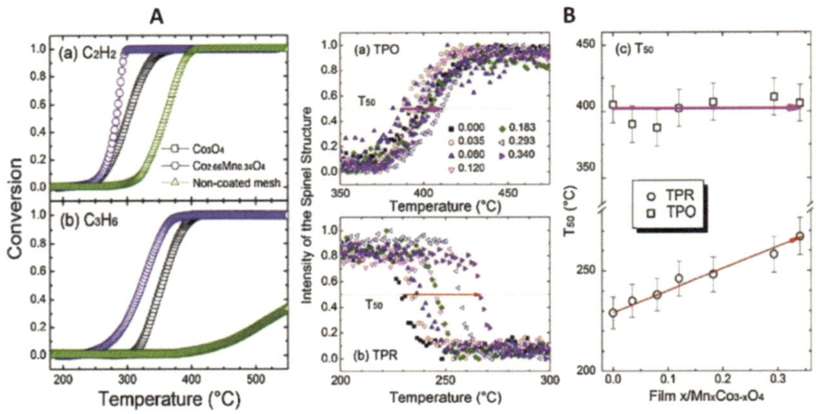

Figure 10. (A) Light-off curves of C_2H_2 and C_3H_6 over $Co_{3-x}Mn_xO_4$ (x = 0 and 0.34) spinel structures grown on mesh of stainless steel substrates and NCM. The reaction was carried at the total flow rate was kept at 15 ml min^{-1} with 2% of fuel and 20% O_2 diluted in Ar at a total flow rate of 0.015 L/min; (B) TPO/TPR patterns of Co-Mn oxide. Reproduced with permission from [45]. Copyright 2015, Elsevier.

http://dx.doi.org/10.5772/66285

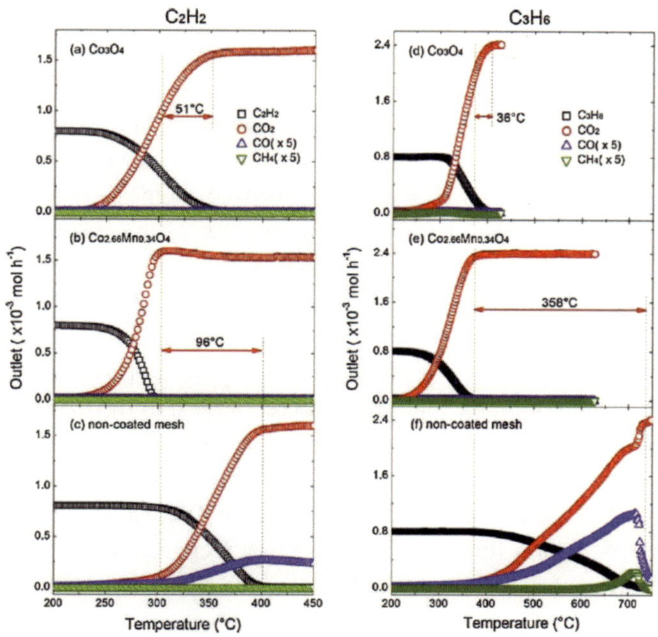

Figure 11. Production profiles of C_2H_2/C_3H_6, CO, CO_2 and CH_4 in the oxidation of C_2H_2 and C_3H_6 over $Co_{3-x}Mn_xO_4$ oxides (x = 0 and 0.34) grown on mesh of stainless steel substrates and NCM. Reproduced with permission from [24]. Copyright 2015, Elsevier.

Fuel	Catalyst	T_{50} (°C)[a]	T_{90} (°C)[a]	T_{50} CO_2 (°C)[b]	T_{90} CO_2(°C)[b]	*Ea* (kJ mol^{-1})[c]
C_2H_2	NCM	355	386	356	387	168.97
	Co_3O_4	299	335	301	337	139.77
	$Co_{2.66}Mn_{0.34}O_4$	282	294	282	295	131.86
C_3H_6	NCM	603	692	614	723	121.98
	Co_3O_4	354	385	356	387	158.32
	$Co_{2.66}Mn_{0.34}O_4$	321	356	323	357	114.59

Reproduced with permission from [24]. Copyright 2012, Elsevier.

[a]T_{50} and T_{90} represent the temperatures at which the conversion of C_2H_2/C_3H_6 reaches 50 and 90%, respectively.

[b]T_{50} CO_2 and T_{90} CO_2 represent the temperatures at which 50 and 90% conversion to CO_2 is reached.

[c]Ea is the apparent activation energy for the hydrocarbon activation at atmospheric pressure.

Table 3. Catalytic oxidation of C_2H_2/C_3H_6 over Co-Mn oxides catalysts deposited on grid mesh of stainless steel substrates and NCM.

The selectivity-conversion plots show clearly that the investigated coatings favor the total oxidation of both unsaturated hydrocarbons at lower temperatures relative to the NCM. For

C_2H_2, T_{50} and T_{90} were observed at 355 and 386°C over NCM, whereas these values decreased by respective 56 and 51°C over cobalt oxide and further decreased by 17 and 41°C over $Co_{2.66}Mn_{0.34}O_4$, which reveals that the catalyst with small amount of manganese content shows the highest activity (see **Figure 10** and **Table 3**). Tian et al. [48] reported that T_{50} of C_2H_2 oxidation with Co_3O_4 deposited on monolithic cordierite support was 297°C, which was in good agreement with the current Co_3O_4 sample prepared on mesh. Note that the T_{50} of C_2H_2 (282°C, **Table 3**) with $Co_{2.66}Mn_{0.34}O_4$ [24] was quite close to the value (280°C) reported by Ivanova et al. [80] who used Au/Al_2O_3, revealing that $Co_{2.66}Mn_{0.34}O_4$ prepared by CVD features a competitive activity. The low-temperature shift of the light-off curve was more pronounced for C_3H_6 since the T_{90} difference between the NCM and $Co_{2.66}Mn_{0.34}O_4$ is 336°C, demonstrating that $Co_{2.66}Mn_{0.34}O_4$ was very active for the total oxidation of propene and supporting the conclusion of Liang et al. [81] who reported that the manganese insertion into the cobalt oxide spinel could enhance the catalytic activity of the oxidation of light olefins. Considering the higher activity, Mn-doped spinel of cobalt oxide could be a potential system for catalytic oxidation of hydrocarbons. As presented in **Table 3**, the values of T_{50} and T_{90} of hydrocarbons conversion and those of CO_2 selectivity were found to be almost identical for Co_3O_4. However, differences were observed relative to NCM especially in the oxidation of C_3H_6, illustrating an occurrence of partial oxidation by generating products other than CO_2 (see **Figure 11f**). With Co_3O_4 and $Co_{2.66}Mn_{0.34}O_4$, CO_2 was observed to be the only carbonaceous product and no secondary products were formed all over the entire oxidation process of C_3H_6, while CO and CH_4, coming from incomplete oxidation, were detected in addition to the production of CO_2 on NCM. Even though some by-products were formed at certain temperatures with NCM, all indicate that CO_2 was the final product. Taking the temperature at which hydrocarbons were completely converted to CO_2 on $Co_{2.66}Mn_{0.34}O_4$ sample as a reference, Co_3O_4 and NCM show a respective increase of 51 and 96°C for C_2H_2 and 36 and 358°C for C_3H_6, which indicate that the use of cobalt together with manganese significantly affects the catalytic activity. By applying the Arrhenius equation in the conversion range within 15% [48], the *Ea* was calculated (see **Table 3**). In general, the manganese-doped cobalt oxide exhibits lower *Ea*. Compared to those activation energies obtained in the oxidation of C_2H_2 and C_3H_6 over Co_3O_4-coated monolith with a large flow rate (500 sccm) [48], the values obtained on Co_3O_4-coated mesh were about 10 and 30 kJ mol^{-1} larger, the difference of which may come from the comprehensive effect of the support and inlet condition. Zhang et al. [82] and Aguilera et al. [83] also reported that the modification of Co_3O_4 by MnO_x could promote the preferential oxidation of CO, toluene and alcohols at lower temperatures. The catalytic oxidation of the two hydrocarbons employing Co-Mn oxides systems consists of two irreversible steps, namely the reaction of the hydrocarbon with the lattice or trapped oxygen leading to its reduction and release of oxygen from the surface of the metal oxide and the reoxidation of the partly reduced metal oxide site by means of oxygen in a subsequent step. The TPO analysis indicates that the manganese introduction does not really influence the bulk reoxidation behavior of Co-Mn oxides [24]. Considering the TPR (**Figure 10B**) and catalytic results which reveal that Co_3O_4 has higher reducibility but lower activity than $Co_{2.66}Mn_{0.34}O_4$, the formation of oxygen vacant site plays a key role in the redox mechanism, the importance of which was pointed out previously by Noller and Vinek [84]. According to the enhancement of the thermal stability,

http://dx.doi.org/10.5772/66285

the amount of oxygen vacancy was increased upon manganese incorporation, which accelerates the oxidation process. Furthermore, the catalytic activity could also benefit from the substitution of cobalt with manganese that was more active and from the cooperative effect among metallic species by increasing the oxide reduction sites.

3.2.2. *PSE-CVD made cobalt ferrite for low-temperature oxidation of CO and VOCs*

Thin films of cobalt ferrite binary oxides (with the general formula $Co_{3-x}Fe_xO_4$) at different composition ($Co_{0.9}Fe_{2.1}O_4$, $Co_{1.8}Fe_{1.2}O_4$ and $Co_{2.1}Fe_{0.9}O_4$) were prepared by PSE-CVD. The systematic characterization of their properties and their potential application as catalysts for low-temperature CO, C_3H_6, n-C_4H_8 and C_2H_6O oxidation has been reported by Kouotou et al. and Tian et al. [27] The effect of iron substitution by cobalt in the structure on the optical and redox properties was investigated. The catalytic performance of the Co-Fe oxides was discussed with respect to the participation of surface and lattice oxygen in the oxidation process. According to XPS and temperature-programmed reduction/oxidation (TPR/TPO) results, a suprafacial mechanism was the dominant mechanism for CO oxidation to CO_2, while C_3H_6, n-C_4H_8 and C_2H_6O were oxidized through an intrafacial process (MvK mechanism).

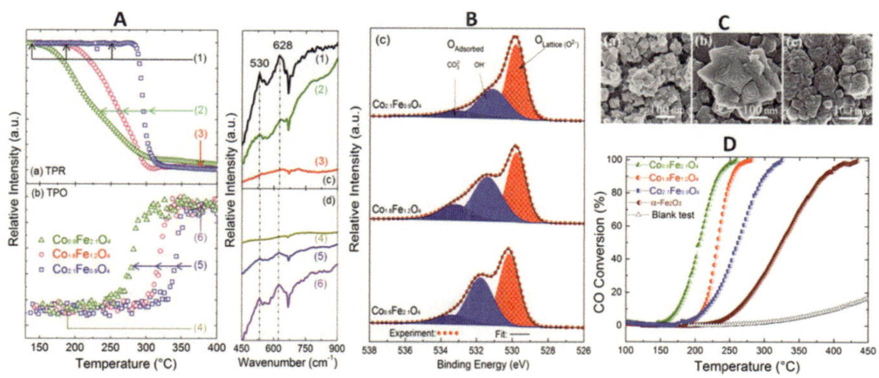

Figure 12. (A) Redox behaviors of Co-Fe oxides: (a) TPR, (b) TPO, (c) progressive loss of the spinel structure and (d) recover of Co-Fe-O IR vibration; (B) O1s XPS spectra of the Co-Fe-O samples; (C) HIM image displaying films morphology; (D) light-off curves of CO conversion with the Co-Fe-O samples, αFe_2O_3 and NCM. The results obtained over αFe_2O_3 [55]. The reaction was carried at the total flow rate was kept at 15 ml min^{-1} with 1% of CO and 10% of O_2 diluted in Ar, corresponding the WHSV of 45,000 g_{cat}^{-1} h^{-1}. Reproduced with permission from [25]. Copyright 2014, the Royal Society of Chemistry.

CO oxidation was performed at atmospheric pressure over all samples. The results were compared with those obtained with pure α-Fe_2O_3 [55] and a blank sample of NCM, as shown in **Figure 12D**. Single α-Fe_2O_3 becomes active in CO oxidation at around 230°C and achieves complete CO conversion to CO_2 above 400°C. Co-Fe-O composites exhibit lower temperatures for the initiation of CO oxidation than the single α-Fe_2O_3. The complete CO oxidation over $Co_{0.9}Fe_{2.1}O_4$, $Co_{1.8}Fe_{1.2}O_4$ and $Co_{2.1}Fe_{0.9}O_4$ occurs at 255, 275 and 325°C, respectively. The results indicate that cobalt ferrites were more catalytically active than single α-Fe_2O_3. The obtained performance order was the following: α-$Fe_2O_3 < Co_{2.1}Fe_{0.9}O_4 < Co_{1.8}Fe_{1.2}O_4 < Co_{0.9}Fe_{2.1}O_4$. It was observed that the catalytic performance was decreased with the increase of the Co content in the matrix of Co-Fe-O [25]. The sample with low Co content ($Co_{0.9}Fe_{2.1}O_4$) exhibits the highest catalytic performance. An attempt to explain such behavior was made with respect to the difference in the film morphology, redox property and the chemical composition as well as the ionic state at the surface of the material as clearly displayed in **Figure 12** [25].

The TPR and TPO result shows that Co-Fe-O samples were reduced at higher temperature, increasing the Co content (**Figure 12A**). Generally, Co_3O_4, Co^{2+}-Co^{3+} ion pairs are known to be very active in low-temperature CO oxidation [22]. Therefore, the presence in Co-Fe-O composites of both Co^{2+} and Co^{3+} together with Fe^{3+} in the octahedral and tetrahedral sites should enable a decrease of the reduction temperature and an improvement of the catalytic performance of CO conversion to CO_2 over samples with higher Co content ($Co_{2.1}Fe_{0.9}O_4$ and $Co_{1.8}Fe_{1.2}O_4$), which own the $[Co^{2+}Fe^{3+}Co^{3+}]_{O\text{-site}}O_4$ cationic distribution in the O-site. Surprisingly, the opposite behavior was observed with $Co_{0.9}Fe_{2.1}O_4$ (the most active sample) in which only Fe^{3+} and Co^{2+} were present in the octahedral sites ($[Fe^{3+}Co^{2+}]_{O\text{-site}}O_4$). It was thus suggested that the CO oxidation over Co-Fe-O catalyst does not proceed with the redox mechanism, even if $Co_{0.9}Fe_{2.1}O_4$ presents the lowest reduction temperature (**Figure 12A**). The redox mechanism hypothesis was strongly supported by the fact that CO oxidation of $Co_{2.1}Fe_{0.9}O_4$ was initiated at ~200°C while the reduction started at ~280°C (**Figure 12A, D**). Since the formation of carbonates on the cobalt surface has been suggested by Thormählen et al. to play an important role in the low-temperature oxidation of CO [85], the earlier initiation of the reaction at low temperature was therefore assigned to the surface-adsorbed oxygen revealed by XPS analysis (**Figure 12B**). It was thus proposed that CO oxidation over Co-Fe-O follows a suprafacial mechanism where CO molecules react with adsorbed oxygen, mainly as Co_3^{2-} and OH^-, giving rise to form CO_2.

In order to study further the catalytic activity of cobalt ferrite thin films, Tian et al. have investigated the catalytic performance of cobalt ferrite ($Co_{2.1}Fe_{0.9}O_4$) with respect to the total oxidation of propene, *n*-butene and DME at atmospheric pressure referring to NCM [27]. The catalytic effect of the mesh has been excluded by the observation that there is no significant difference between the oxidation over NCM and in a blank system (**Figure 13**). The results show that the cobalt ferrite films favor the complete conversion of the reactants at much lower temperatures relative to NCM. Besides the reactant gas, CO_2 was detected as the unique product in the oxidation processes over cobalt ferrite, while additional CO was observed in the reaction on NCM. T_{50} and T_{90}, corresponding to respective 50 and 90% conversion of the reactant gas, were used as parameters to compare the performance of the deposited samples. With cobalt ferrite, T_{50} and T_{90} of propene oxidation were 348 and 382°C. These values shifted to 578 and 691°C for the experiment carried out with NCM (see **Figure 13a, b**), respectively. Compared to the reaction over Co_3O_4 with *Ea* of 158.32 kJ/mol [24], the introduction of iron tends to initiate the oxidation of propene with lower *Ea*, which makes the cobalt ferrite more suitable for the catalytic applications. **Figure 13c** and **d** compares the results of *n*-C_4H_8 oxidation with and without cobalt ferrite. T_{50} and T_{90} of *n*-C_4H_8 oxidation over cobalt ferrite were observed at 358 and 402°C, whereas these values shifted by respective 100 and 135°C toward higher temperatures over NCM. It has been reported that a small quantity of 1,3-

http://dx.doi.org/10.5772/66285

butadiene was selectively formed in the oxidation of n-C_4H_8 at temperature higher than 350°C over $MnMoO_4$-based catalysts [86]. It should be noted that 1,3-butadiene was not detected in the reaction with the PSE-CVD made $Co_{2.1}Fe_{0.9}O_4$ indicating that the material presents higher catalytic activity.

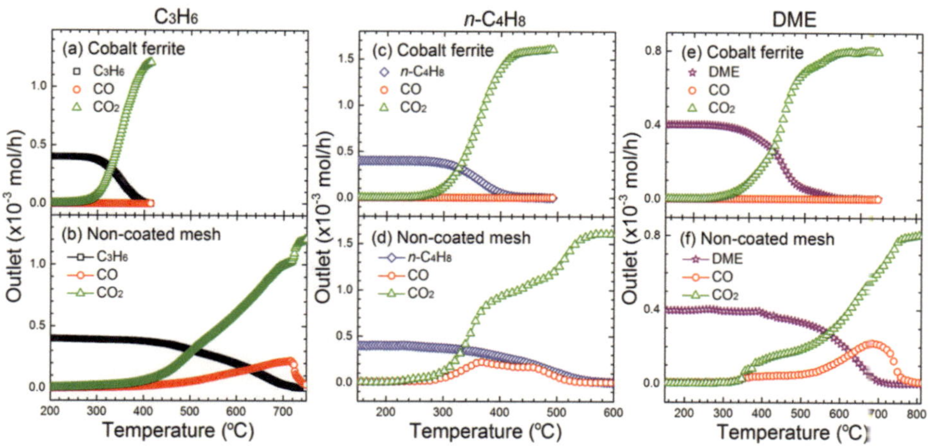

Figure 13. Outlet profiles of C_3H_6, n-C_4H_8 and DME oxidation over cobalt ferrite-coated and non-coated meshes. The reaction was carried at the total flow rate was kept at 15 ml min^{-1} with 1% of Fuel and 10% of O_2 diluted in Ar, corresponding the WHSV of 45,000 g_{cat}^{-1} h^{-1}. Reproduced with permission from [27]. Copyright 2015, Elsevier.

For DME, cobalt ferrite also exhibits much better performance than NCM, as revealed in **Figure 13e** and **f.** T_{50} and T_{90} of the reaction over cobalt ferrite are 356 and 409°C. These values were observed to be 613 and 682°C for the reaction on NCM, respectively. According to Liu et al. [87], the reaction network for DME conversion contains four pathways giving rise to CH_3OH, HCHO, $HCOOCH_3$ and CO_x. Low selectivity of CH_3OH and $HCOOCH_3$ was reported [87]. HCHO was not detected, which could be resulted either from the low concentration or fast conversion to CO_x.

The good catalytic performance of cobalt ferrite could be correlated with the $O_{adsorbed}$ on the surface and attractive redox properties (**Figure 12A**, **B**). It is widely accepted that the catalytic performance of ferrite-type catalysts depends on its oxygen mobility since the reaction follows MvK mechanism [88]. The abundance of $O_{Adsorbed}$ could participate in the catalytic oxidation process, as proposed by Veleva and Trifirò [86]. The good reducibility and reoxidability tend to enable the catalytic reactions at relatively low temperatures by involving various cations distributed in the octahedral and tetrahedral sites. Moreover, the slightly low band gap energy of cobalt ferrite could also indicate that the migration of $O_{lattice}$ or O^{2-} from the bulk to the surface becomes easier and leads to good reducibility [25].

Because cobalt ferrite binary oxide with high Co concentration ($Co_{2.1}Fe_{0.9}O_4$) showed very interesting catalytic performance against the deep oxidation of C_3H_6, n-C_4H_8 and DME, Kouotou et al. [26] explored the catalytic activity of the same material, but with low Co concentration ($CoFe_2O_4$). Authors aimed to compare the performances of $CoFe_2O_4$ (low cobalt content) versus $Co_{2.1}Fe_{0.9}O_4$ (high cobalt content) from recent investigation toward catalytic oxidation of CO, DME, C_3H_6 and n-C_4H_8. The role of Co or Fe atom in the matrix of cobalt ferrite mixed oxide and adsorbed/lattice oxygen at the surface of the deposited material in the catalytic reaction was identified. The light-off curves of CO, C_3H_6, n-C_4H_8 and DME oxidation over $CoFe_2O_4$ are shown in **Figure 14**. The results indicate that $CoFe_2O_4$ is a very active catalyst for the total oxidation of CO, C_3H_6, n-C_4H_8 and DME at atmospheric pressure. Total destruction of such pollutants at low temperature enables the production of CO_2 as the only detectable C-containing product. In the presence of $CoFe_2O_4$, the oxidation becomes detectable at around 150, 200, 230 and 270°C for CO, C_3H_6, n-C_4H_8 and DME, respectively and complete conversion occurs at 253°C for CO, 336°C for C_3H_6 and n-C_4H_8 and 502°C for DME.

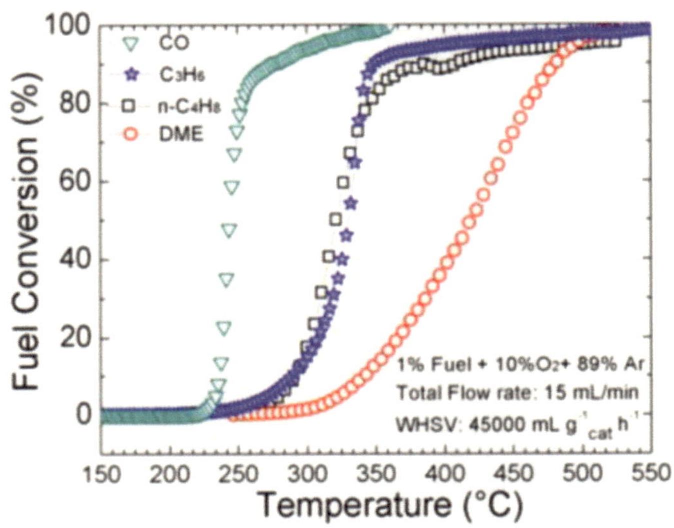

Figure 14. Light-off curves of CO, C_3H_6, n-C_4H_8 and DME oxidation over the PSE-CVD made $CoFe_2O_4$ sample. Reproduced with permission from [26]. Copyright 2015, Wiley-VCH.

For the CO oxidation, the comparison of the temperature at 90% conversion (T_{90}) over $CoFe_2O_4$ (260°C) as catalyst with that obtained with $Co_{0.9}Fe_{2.1}O_4$ (230°C), $Co_{1.8}Fe_{1.2}O_4$ (250°C) and $Co_{2.1}Fe_{0.9}O_4$ (298°C) from our previous work has been made. The results indicate the following performance order: $Co_{0.9}Fe_{2.1}O_4 > Co_{1.8}Fe_{1.2}O_4 > CoFe_2O_4 > Co_{2.1}Fe_{0.9}O_4$. $CoFe_2O_4$ owns similar composition as $Co_{0.9}Fe_{2.1}O_4$, but it exhibits less activity for CO oxidation. By the way, the comparison of the T_{50} of CO over $CoFe_2O_4$ (243°C) with results obtained with Pt/Al_2O_3 (304°C) and $Pt/CoO_x/Al_2O_3$ (340°C) catalysts reported by Torncrona et al. [89] indicates that $CoFe_2O_4$ is more active at low temperature.

In the investigation of the CO oxidation over cobalt ferrite oxides with different composition ($Co_{0.9}Fe_{2.1}O_4$, $Co_{1.8}Fe_{1.2}O_4$ and $Co_{2.1}Fe_{0.9}O_4$) [25], it was established that the performance of cobalt ferrite oxide was not dependent on the variation of the Co content in the mixed oxides. Therefore, these results suggest that CO oxidation over $CoFe_2O_4$ follows a suprafacial mechanism where CO molecules react with adsorbed oxygen at the surface of the catalyst to form CO_2. Considering that the difference in the catalytic performance of CO over $CoFe_2O_4$ with seemingly the same composition as $Co_{2.1}Fe_{0.9}O_4$ [25], it was therefore evident that the process is independent on the Co population in the cobalt ferrite oxide, but on the lattice and adsorbed oxygen, mainly adsorbed oxygen at the surface of the material.

The oxidation of C_3H_6 and n-C_4H_8 on $CoFe_2O_4$ starts at temperatures lower than that of DME and the trend for the rate of oxidation is $C_3H_6 \approx n\text{-}C_4H_8 >$ DME [26]. In the presence of $CoFe_2O_4$ nanoparticles, the oxidation becomes detectable at around 250°C for C_3H_6 and n-C_4H_8 and 285°C for DME and the T_{90} is 334, 371 and 470°C for C_3H_6, n-C_4H_8 and DME, respectively. The temperature at the complete oxidation was 475°C for C_3H_6, 485°C for n-C_4H_8 and 506°C for DME. These values appear to be higher than those for $Co_{2.1}Fe_{0.9}O_4$ catalysts, where complete oxidation was observed at 400, 425 and 446°C for C_3H_6, n-C_4H_6 and DME, respectively [27]. However, they were lower than those observed for C_3H_6 on Au/Al_2O_3 [80] and $La_{1.7}Sr_{0.3}CuO_4S_{0.2}$ [90]. The activity in n-C_4H_8 oxidation was comparable to the measurement on Co_3O_4 [48]. In general, for the oxidation reactions over $CoFe_2O_4$ at low temperatures, no toxic or partial oxidation products are formed when complete conversion of VOCs is achieved. For DME oxidation, it has been established that the reaction network consists of four pathways giving rise to CH_3OH, HCHO, $HCOOCH_3$ and COx [87]. Kouotou et al. [26] detected only CO_2, indicating the higher activity of $CoFe_2O_4$ catalyst for total oxidation of DME at low temperature.

3.2.3. *PSE-CVD made $CuCo_2O_4$ for low-temperature catalytic combustion of CO and VOCs*

To investigate the potential application of the grown Co-Cu oxides, Tian et al. [28] investigated the catalytic performance of the prepared samples with respect to the total oxidation of CO and C_3H_6. The authors carried out the same tests on bare meshes to exclude the background effect of the mesh structures on the oxidation processes. **Figure 15** compares the light-off curves of CO and C_3H_6 by using Co_3O_4 and Co-Cu oxide referring to the bare mesh. With Co_3O_4, the conversion of CO occurs at 270°C and complete conversion was achieved at 362°C. For Co-Cu oxide, these values shift to 125 and 221°C, respectively. CO oxidation over bare mesh was observed to begin at around 395°C and finish at 833°C. Under similar conditions, CO oxidation on other TMOs such as Mn_3O_4 was reported to happen at 190°C and complete at 343°C [43]. Compared to Mn_3O_4, Co-Cu oxide exhibits better performance. This could be explained by the abundance of adsorbed oxygen, Co^{3+} and Cu^{2+} revealed in the XPS analysis. By applying the Arrhenius equation at low conversion profiles (within 5%), the *Ea* values were calculated to be 68.2, 77.3 and 82.9 kJ mol^{-1} for the CO oxidation with Co_3O_4, Co-Cu oxide and bare mesh, respectively. As shown in **Figure 15b**, both Co-Cu oxide and Co_3O_4 improve the catalytic oxidation of C_3H_6 compared to the bare mesh. In the presence of Co-Cu oxide, the propene conversion becomes detectable at 220°C. This value was 30 and 120°C lower than that with Co_3O_4 and bare mesh, respectively. With Co-Cu oxide and Co_3O_4, complete conversion of C_3H_6 was reached within 412 and 454°C, while this value was observed to be 772°C for the reaction on NCM. It should be mentioned that CO_2 is the unique measurable product and no trace of CO was detected in the oxidation process with the prepared $CuCo_2O_4$ and Co_3O_4, as depicted in **Figure 16**. This result agrees well that the conclusion drawn in our recent work that the active transition metal oxides prevent the formation of CO in the oxidation of low-rank hydrocarbons. In the reaction on bare mesh, a large amount of CO peaking at 758°C was produced, which comes from partial oxidation.

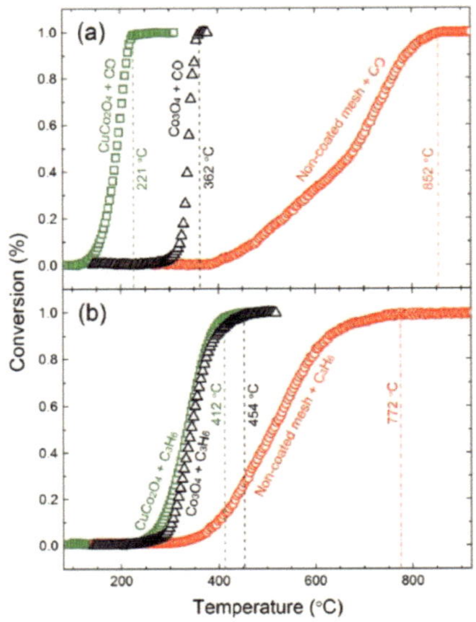

Figure 15. Catalytic performance of Co-Cu oxides: oxidation of CO (a) and propene (b). The reaction was carried at the total flow rate was kept at 15 ml min^{-1} with 1% of fuel and 10% of O_2 diluted in Ar, corresponding the WHSV of 75,000 g_{cat}^{-1} h^{-1}. Reproduced with permission from [28]. Copyright 2014, the Royal Society of Chemistry.

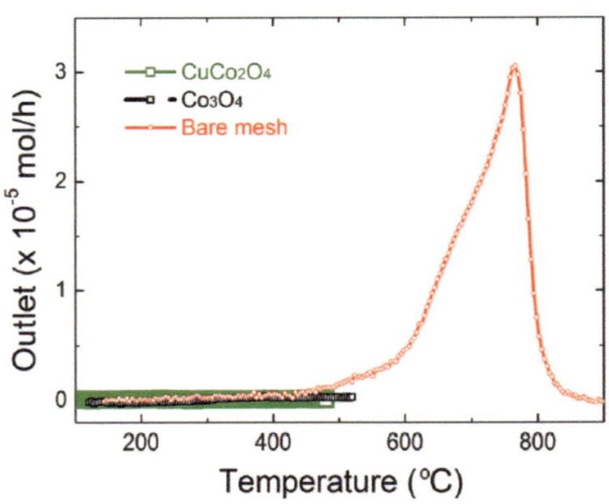

Figure 16. Outlet profiles of CO in the propene oxidation over $CuCo_2O_4$, Co_3O_4 and NCM. Reproduced with permission from [28]. Copyright 2014, the Royal Society of Chemistry.

Compared to the reported T_{50} values in the temperature-programmed reaction, 330°C obtained with $CuCo_2O_4$ was lower than that measured with Au/Al_2O_3 (349°C) [80], Co_3O_4 (354°C) [48],

$Au/MgO/Al_2O_3$ (359°C) [91], $La_{1.7}Sr_{0.3}CuO_4S_{0.2}$ (419°C) [90], (327°C) [91] and $Co_{2.66}Mn_{0.34}O_4$ (321°C) [24]. Although CuO shows better performance with the T_{50} at 272°C [44], $CuCo_2O_4$ has more application prospects than CuO by considering the higher thermal stability. It is worth mentioning that the magnitude order of T_{50} for CuO, $CuCo_2O_4$ and Co_3O_4 follows the order of the band gaps [28]. It can be explained that the migration of O^{2-} from the bulk to the surface gets easier for samples with lower band gaps which have better electron mobility. Moreover, the *Ea* for propene oxidation with $CuCo_2O_4$ is 82.6 kJ mol^{-1}, which is lower than that obtained with Co_3O_4 (130 kJ mol^{-1}) [24], CuO (109.5 kJ mol^{-1}) [44], $Co_{2.66}Mn_{0.34}O_4$ (115.5 kJ mol^{-1}) [24] as well as the non-catalyzed reaction (138.3 kJ mol^{-1}). The MvK mechanism has been established for CO and propene oxidation over CuO [44], Co–Mn oxides and Co–Fe oxides [25–27]. The same redox behavior could exist in the reaction with Co-Cu oxides.

3.2.4. *Comparison of the catalytic performances of binary TMOs with literature*

Table 4 compares the catalytic performance of the PSE-CVD made mixed TMOs with that of the selected catalysts in the literature. It can be noted that our catalysts present comparable activity to the other systems. Special attention is paid to the comparison of the catalytic performance with noble metals and transition metal oxides. As can be seen from **Table 4**, noble metals and TMOs (mixed) present attractive results in terms of CO, propene, *n*-butene and DME total oxidation.

Type of catalyst	Material	Weight (mg)	Gas composition	WHSV [a] (ml g^{-1} h^{-1})	T_{10} [b] (°C)	T_{50} [b] (°C)	T_{90} [b] (°C)	Refs.
Mixed TMOs	$CoFe_2O_4$	20	1% C_3H_6/10% O_2 in Ar	45,000	–	334	–	[26]
	$Co_{2.1}Fe_{0.9}O_4$				–	348	–	[27]
	$CuCo_2O_4$	12	1% C_3H_6/10% O_2 in Ar	75,000	–	330	–	[28]
	$Co_{2.66}Mn_{0.34}O_4$		2% C_3H_6/20%O_2 in Ar		277	321	356	[24]
	$Cu_{0.72}Co_{2.28}O_4$	40	13% C_3H_6/52% O_2 in N_2	15,000	230	275	>400	[100]
	$La_{1.7}Sr_{0.3}CuO_4S_{0.2}$	200	0.1% C_3H_6/5% O_2 in N_2	30,000	368	419	500	[90]
Precious metal	Au/Al_2O_3	200	1.5% C_3H_6/4% O_2 in He	22,500	288	349	410	[80]
	Au/Al_2O_3		1% C_3H_6/9% O_2 in He	219,512	–	365	–	[91]
	$Au/BaO/Al_2O_3$			250,000	–	290	–	
	$Au/Rb_2O/Al_2O_3$	200	1% C_3H_6/9% O_2 in He	257,143	–	307	–	
	$Au/Li_2O/Al_2O_3$			225,000	–	327	–	
	$Au/MgO/Al_2O_3$			214,286	–	359	–	
	Ag/Al_2O_3	50	3% C_3H_6/10% O_2 in N_2	12,000	–	–	420	[95]
Mixed TMOs	$CeAlO_3$	100	2% CO/2% O_2 in N_2	60,000	–	–	465	[97]

Type of catalyst	Material	Weight (mg)	Gas composition	WHSV [a] (ml g^{-1} h^{-1})	T_{10} [b] (°C)	T_{50} [b] (°C)	T_{90} [b] (°C)	Refs.
	$CuCo_2O_4$	12	1% CO/10% O_2 89%Ar	75,000	–	190		[28]
	$CoFe_2O_4$	20		45,000	–	–	260	[23]
	$Co_{0.9}Fe_{2.1}O_4$				–	–	230	[25]
	$Co_{2.1}Fe_{0.9}O_4$				–	–	298	
	$Co_{1.8}Fe_{1.2}O_4$				–	–	250	
Precious metals	$Pt/H_2SO_4/ZrO_2$	50	3.5% CO/4%O_2 in N_2	120,000	–	–	290	[98]
	Pt/Al_2O_3	2/200	1% CO/1.38%O_2 in N_2	90,000	–	333	430	[89]
	Au/SiO_2	100	1% CO/99% dry air	12,000	–	337	423	[99]
Mixed TMOs	$CoFe_2O_4$	20	1% *n*-C_4H_8/10% O_2 in Ar	45,000	–		371	[23]
	$Co_{2.1}Fe_{0.9}O_4$		1% *n*C_4H_8/10%O_2 in Ar		–	358	402	[27]
Mixed TMOs	$CoFe_2O_4$	20	1% C_2H_6O/10%O_2 in Ar	45,000	–	–	470	[26]
	$Co_{2.1}Fe_{0.9}O_4$		–		356	409	[27]	

[a]: *Weight hourly space velocity*
[b]: *Temperature at X% conversion (X = 10, 50 and 90 %)*

Table 4. Overview of the catalytic performances of PSE-CVD made binary TMOs compared with the literature.

4. Remarks and perspectives

This review highlights the newly developed CVD approach called PSE-CVD for single and binary TMOs thin films coating and their performances in the catalytic oxidation of CO and VOCs. The advances in the synthesis and properties of single and binary oxides thin films have shown potential applications in several catalytic processes. In this review, we first presented different single oxides fabricated by using a more elaborated CVD method. In particular, the complete oxidation of CO and VOCs over Co_3O_4, CuO/Cu_2O, Fe_2O_3, Cr_2O_3 as well as Mn_3O_4 is comprehensively summarized. In addition, the catalytic activity in CO and VOCs oxidation is systematically compared and it can be well-documented that the CO catalytic oxidation activity is related to either the abundance of active metallic cation sites (Co^{3+}, Cu^{2+}, Fe^{3+}, Cr^{3+} and Mn^{3+}) present in Co_3O_4, CuO/Cu_2O, Fe_2O_3, Cr_2O_3 as well as Mn_3O_4, respectively, or the weakly bonded molecular oxygen species at their surface. On the basis of the present results, we could conclude that the VOCs catalytic oxidation activity of single TMOs is related with either the reactive surface oxygen species or bulk oxygen mobility, reoxidation of metallic cations species and active oxygen vacancies of single oxides.

Regarding binary systems, Co-based binary oxides generally prepared by co-precipitation, impregnation, surfactant-template methods and by combined impregnation and combustion

http://dx.doi.org/10.5772/66285

synthesis were prepared by PSE-CVD and summarized in the present review. The idea was to formulate active and stable mixed oxide based on the performance obtained with single oxide. In fact, Co_3O_4 is known as the most active single oxides for CO and VOCs deep oxidation and the activity strongly depends on Co^{3+}/Co^{2+} redox couple. Compared with pure Co_3O_4 catalysts, the binary oxide catalysts are generally composed of one or more catalytically active components and a functional support, in which the interaction between the catalytic components and the supports can provide improved redox properties and enhanced catalytic behavior [92–94]. As for single oxide catalysts, the active component alone can catalyze the various reactions at a certain kinetic rate, which is usually relatively low. The synergistic effect between binary oxide catalysts leads to the improved catalytic activity. Therefore, the combination of Co and individual metal of single oxide such as Fe_2O_3, Mn_3O_4 and CuO enables the obtaining of binary oxide like Co-Mn-O, Co-Fe-O and Co-Cu-O with new chemical and cationic distribution benefits to the obtained catalytic performances summarized in this review. From the obtained results, it seems like combining two transition metals influence the interaction of active cobalt oxide with the second oxide, affecting the textural and structural properties of the binary oxides. In particular, the investigations of the interaction mechanisms as a function of the surface redox properties of the binary oxides are comprehensively summarized. The catalytic activity of the binary oxides in CO and VOC oxidation is compared and different reaction mechanisms occurring in CO and VOC oxidation are presented. Probably, the surface adsorbed oxygen species over the surface oxygen vacancies of the catalysts are mainly attributed to the CO oxidation activity at low temperatures. The total oxidations of VOCs over transition metal oxides are suggested to follow the MvK redox process.

Nevertheless, in the case of VOCs oxidation, the catalytic activity depends on the nature of catalysts and VOCs. While CO oxidation over single oxide seems to be mainly governed by the amount of available surface oxygen species, VOCs oxidation activity of PSE-CVD made single oxide in contrary is scarcely determined by the surface oxygen species. In the case of CO oxidation, PSE-CVD made single and binary oxides show relevant concentration of surface oxygen species which might be involved in the oxidation at low temperatures. Moreover, the higher activity of Co_3O_4 and $Co_{0.9}Fe_{2.1}O_4$ binary oxides with respect to the Fe_2O_3 or CuO with respect to $CuCO_2O_4$ and other binary oxides is mainly attributed to their higher mobility of lattice oxygen species.

The investigation of the catalytic properties of single and binary TMOs made by PSE-CVD has provide valuable results in the literature, about the employ of a well-established nonconventional CVD approach for the selective synthesis of oxide made of Co, Cu, Fe, Cr and Mn and their performances toward low temptation conversion of CO, C_2H_2, C_3H_6, n-C_4H_8 and C_2H_6O in harmless CO_2 and H_2O. As perspectives, we think that the following aspects deserving further investigations are: (1) further research study of the interactions between different kinds of binary oxide catalytic systems and related corrections between their redox properties and catalytic activities; (2) investigation of the interactions among three or more component catalytic systems, such as noble metal doped or noble metal alloys doped Co-Mn, Co-Fe and Co-Cu binary oxides. It is expected that the novel structures matrix can offer new opportunities to expand our understanding of this kind of interaction as well as relationship between structure and property; (3) because of his lattice oxygen reservoir character when used as supporter, single Co_3O_4, CuO, Fe_2O_3 crystals structure needs to be deposited on CeO_2 nanoparticles in order to control this kind of interaction between these two oxides and to enhance application potentials of these multifunctional materials. It is expected that such an integration of the current investigations on different systems will bring a more comprehensive understanding of the interactions during the catalytic process and therefore will be of great significance in searching for and find novel multi-oxide catalysts of further enhanced catalytic activity, selectivity as well as durability.

5. Summary

The catalysis science of non-precious metals has significantly improved recently, due to development of more elaborated synthesis approaches for their controlled synthesis and advanced characterization techniques that allow more fundamental insights into the reaction mechanisms. As summary of the present review on the low-temperature oxidation of CO and VOCs over selected non-precious metal such as TMOs catalytic materials either in their single and binary or mixed phases prepared using PSE-CVD, the following main results can be summarized:

1. The catalytic combustion has become the most popular method for the environmental emission control. Thus, catalytic oxidation of CO and VOCs is highly desirable to proceed at low temperature for the consideration of energy savings, low cost, operation safety and environmental friendliness.

2. To reduce the temperature of VOCs catalytic oxidation, great efforts have been made to develop efficient and low cost catalysts via an elaborated synthesis method, namely pulsed spray evaporation chemical vapor deposition. The grid mesh of stainless steel as inert substrates was used instead of catalytically active support to enable the real evaluation of the catalytic performance of the as-deposited thin films layer toward CO and VOCs oxidation, offering the major advantage of low-pressure drop.

3. TMOs exhibited comparative activity versus noble metals toward the catalytic oxidation of VOCs at low temperature. The catalytic performance of TMOs was found to be generally affected by many factors such as the composition, the valence of metallic particles, doping, the film morphology and the particle size of the metals oxide.

4. Single oxides systems composed of transition metals (Co, Cu, Cr, Fe and Mn) are efficient for the abatement of CO and VOCs. In addition, the effect of doping should be taken into consideration since they greatly improve the thermal properties and the catalytic performances of the as-prepared materials. For instance, the combination of two transition metals or the doping that is the introduction of another transition metal ion in the matrix of single oxides to form a mixed oxide catalysts such as Co-Mn, Co-Fe and Co-Cu oxides also contributes to stabilize the as-deposited catalysts, enable the total oxidation of VOCs and exhaust stream at low temperature.

http://dx.doi.org/10.5772/66285

5. The favored mechanism for CO total oxidation over TMOs catalysts is the suprafacial mechanism, involving reaction between adsorbed molecules on metal sites, while the occurrence of intrafacial mechanism, also known as Mark-van Krevelen mechanism, has been demonstrated for the total oxidation of the representative VOCs (C_2H_2, C_3H_6, *n*-C_4H_8, C_2H_6O) over TMOs catalysts at low temperature. Therefore, it can be concluded that the validity of each mechanism strongly depends on the redox properties of the catalysts as well as the adsorbed species at the catalyst surface.
6. The comparison of the catalytic performance of PSE-CVD made TMOs with other catalysts reported in the literature indicated that TMOs present comparative activity with some noble metals and mixed oxides like perovskites. Such performances confirm that TMOs should be really considered as potential candidates for the replacement of limited available and high-cost precious metal catalysts.
7. This review validates PSE-CVD as low cost and promising synthesis approach for the control synthesis of active single or mixed TMOs for low-temperature catalytic oxidation of CO and VOCs.

Conflict of interest

The authors confirm that this review article has no conflict of interest.

Acknowledgements

Prof. Dr. Tian and Dr. Mountapmbeme Kouotou are grateful for the support from the Recruitment Program of Global Youth Experts and the Chinese Academy of Sciences visiting Professorship for Senior International scientist (grant no. 2015PT016). The authors are grateful to Prof. Dr Katharina Kohse-Höinghaus for her support and discussions and for allowing us to perform part of this research in her laboratory in Bielefeld.

Author details

Patrick Mountapmbeme Kouotou[1,3*], Guan-Fu Pan[1,2] and Zhen-Yu Tian[1,2*]

*Address all correspondence to: patrick@iet.cn and tianzhenyu@iet.cn

1 Intistitute of Engineering Thermophysics, Chinese Academy of Sciences, Beijing, China

2 University of Chinese Academy of Sciences, Beijing, China

3 Higher Institute of the Sahel, University of Maroua, Maroua, Cameroon

References

[1] R. M. Heck, R. J. Farrauto and S. T. Gulati, Catalytic air pollution control: commercial technology, John Wiley & Sons, 2009.

[2] S. Scire and L. F. Liotta, *Appl. Catal. B: Environ.*, 2012, 125, 222–246.

[3] H. Huang, Y. Xu, Q. Fenga and D. Y. C. Leung, *Catal. Sci. Technol.*, 2015, 5, 2649–2669.

[4] C. He, J. Li, J. Cheng, L. Li, P. Li, Z. Hao and Z. P. Xu, *Ind. Eng. Chem. Res.*, 2009, 48, 6930–6936.

[5] J. C.-S. Wu and T.-Y. Chang, *Catal. Today*, 1998, 44, 111–118.

[6] J. Carpentier, S. Siffert, J.-F. Lamonier, H. Laversin and A. Aboukais, *J. Porous Mater.*, 2007, 14, 103–110.

[7] M. Guillemot, J. Mijoin, S. Mignard and P.Magnoux, *Appl. Catal. B: Environ.*, 2007, 75, 249–255.

[8] J.-F.Lamonier, A.-B. Boutoundou, C. Gennequin, M. Pérez-Zurita, S. Siffert and A. Aboukais, *Catal. Lett.*, 2007, 118, 165–172.

[9] J.-M. Giraudon, A. Elhachimi and G. Leclercq, Appl. Catal. B: Environ., 2008, 84, 251–261.

[10] L. F. Liotta, M. Ousmane, G. D. Carlo, G. Pantaleo, G. Deganello, G. Marci, L. Retailleau and A. Giroir-Fendler, *Appl. Catal. A: Gen.*, 2008, 347, 81–88.

[11] S. C. Kim and W. G. Shim, *Appl. Catal. B: Environ.*, 2009, 92, 429–436.

[12] P. O. Thevenin, A. G. Ersson, H. M. J. Kusar, P. G. Menon and S. G. Järas, *Appl. Catal. A: Gen.*, 2001, 212, 189–197.

[13] M. F. M. Zwinkels, S. G. Järas, P. G. Menon and T. A. Griffin, *Catal. Rev. Sci. Eng.*, 1993, 35, 319–358.

[14] M. Baldi, E. Finocchio, F. Milella and G. Busca, *Appl. Catal. B: Environ.*, 1989, 16, 43–51.

[15] M. Labari, S. Siffert, J.-F. Lamonier, E. A. Zhilinskaya and A. Aboukaıs, *Appl. Catal. B: Environ.*, 2003, 43, 261–271.

[16] R. Spinicci, M. Faticanti, P. Marini, S. D. Rossi and P. Porta, *J. Mol. Catal. A: Chem.*, 2003, 197, 147–155.

[17] M. Alifanti, M. Florea, S. Somacescu and V. I. Parvulescu, *Appl. Catal. B: Environ.*, 2005, 60, 33–39.

[18] T. Ataloglou, J. Vakros, K. Bourikas, C. Fountzoula, C. Kordulis and A. Lycourghiotis, *Appl. Catal. B: Environ.*, 2005, 57, 299–312.

[19] B. P. Barbero, J. A. Gamboa and L. E. Cadus, *Appl. Catal. B: Environ.*, 2006, 65, 21–30.

[20] G. K. Boreskov, In: J.R. Anderson, M. Boudart (Eds.), *Catalysis, Science and Technology, vol. 3*, Springer Verlag, New York, 1982.

[21] M. Shelef, K. Otto and H. Gandhi, *J. Catal.*, 1968, 12, 361–375.

[22] Y.-F. Y. Yao, *J. Catal.*, 1974, 33, 108–122.

[23] P. M. Kouotou, Z. Y. Tian, U. Mundloch, N. Bahlawane and K. Kohse-Höinghaus, *RSC Adv.*, 2012, 23, 10809–10812.

[24] Z. Y. Tian, P. H. T. Ngamou, V. Vannier, K. Kohse-Höinghaus and N. Bahlawane, *Appl. Catal. B: Environ.*, 2012, 117–118, 125–134.

[25] P. M. Kouotou, H. Vieker, Z. Y. Tian, P. H. T. Ngamou, A. E. Kasmi, A. Beyer, A. Gölzhäuser and K. Kohse-Höinghaus, *Catal. Sci. Technol.*, 2014, 4, 3359–3367.

[26] P. M. Kouotou and Z. Y. Tian, *Phys. Status Solidi A*, 2015, 212, 1508–1513.

[27] Z. Y. Tian, P. M. Kouotou, A. E. Kasmi, P. H. T. Ngamou, K. Kohse-Höinghaus, H. Vieker, A. Beyer and A. Gölzhäuser, *Proc. Combust. Inst.*, 2015, 35, 2207–2214.

[28] Z. Y. Tian, H. Vieker, P. M. Kouotou and A. Beyer, *Faraday Discuss.*, 2015, 177, 249–262.

[29] M. Kang, M. W. Song and C. H. Lee, *Appl. Catal. A: Gen.*, 2003, 251, 143–156.

[30] M. M. Natile and A. Glisenti, *Chem. Mater.*, 2005, 17, 3403–3414.

[31] L. Xue, C.-B. Zhang, H. He and Y. Teraoka, *Appl. Catal. B: Environ.*, 2007, 75, 167–174.

[32] J. Y. Luo, M. Meng, X. Li, X. G. Li, Y. Q. Zha, T. D. Hu, Y. N. Xie and J. Zhang, *J. Catal.*, 2008, 254, 310–324.

[33] G. Spinolo, S. Ardizzone and S. Trasatti, *J. Electroanal. Chem.*, 1997, 423, 49–57.

[34] J. R. A. Sietsma, J. D. Meeldijk, J. P. d. Breejen, M. Versluijs-Helder, A. J. V. Dillen, P. E. D. Jongh and K. P. D. Jong, *Angew. Chem. Int. Ed.*, 2007, 119, 4631–4633.

[35] Y. F. Yuan, X. H. Xia, J. B. Wu, X. H. Huang, Y. B. Pei, J. L. Yang and S. Y. Guo, *Electrochem. Commun.*, 2011, 13, 1123–1126.

[36] P. H. T. Ngamou and N. Bahlawane, *J. Solid State Chem.*, 2009, 182, 849–854.

[37] J. J. Spivey, *Ind. Eng. Chem. Res.*, 1987, 26, 2165–2180.

[38] M. Boudart and G. Djega-Maxiadassou, Kinetics of Heterogeneous Catalytic Reactions, Princeton University Press: Princeton, N.J.,1984.

[39] C. N. Hinshelwwod, Kinetics of Chemical Change in Gaseous Systems, Clarendon Press, Oxford, 3rd ed., 1933.

[40] P. Mars and D. W. V. Krevelen, *Spec. Suppl. Chem. Eng. Sci.*, 1954, 3, 41–59.

[41] B. Varghese, T. C. Hoong, Z. Yanwu, M. V. Reddy, B. V. R. Chowdari, A. T. S. Wee, T. B. C. Vincent, C. T. Lim and C. H. Sow, *Adv. Funct. Mater.*, 2007, 17, 1932–1939.

[42] C. M. Pradier, F. Rodriguez, P. Marcus, M. V. Landau, M. L. Kaliya, A. Gutman and M. Herskowitz, *Appl. Catal. B: Environ.*, 2000, 27, 73–85.

[43] Z. Y. Tian, P. M. Kouotou, N. Bahlawane and P. H. T. Ngamou, *J. Phys. Chem. C*, 2013, 117, 6218–6224.

[44] Z. Y. Tian, H. J. Herrenbrück, P. M. Kouotou, H. Vieker, A. Beyer, A. Gölzhäuser and K. Kohse-Höinghaus, *Surf. Coat. Technol.*, 2013, 230, 33–38.

[45] G.-F. Pan, S.-B. Fan, J. Liang, Y.-X. Liu and Z. Y. Tian, *RSC Adv.*, 2015, 5, 42477–42481.

[46] J. Liang, G.-F. Pan, S.-B. Fan, W.-L. Cheng and Z. Y. Tian, *Phys. Status Solidi C*, 2015, 12, 1001–1005.

[47] L. F. Liotta, H. Wu, G. Pantaleo and A. M. Venezia, *Catal. Sci. Technol.*, 2013, 3, 3085–3102.

[48] Z. Y. Tian, N. Bahlawane, F. Qi and K. Kohse-Höinghaus, *Catal. Commun.*, 2008, 11, 118–122.

[49] M. Haneda, Y. Kintaichi, N. Bion and H. Hamada, *Appl. Catal. B: Environ.*, 2003, 46, 473–482.

[50] X. C. Jiang and A. B. Yu, *J. Mater. Process. Technol.*, 2009, 209, 4558–4562.

[51] L. P. Zhu, N. C. Bing, L. L. Wang, H. Y. Jin, G. H. Liao and L. J. Wang, *Dalton Trans.*, 2012, 41, 2959–2965.

[52] F. Jiao, A. Harrison, J. C. Jumas, A. V. Chadwick, W. Kockelmann and P. G. Bruce, *J. Am. Chem. Soc.*, 2006, 128, 5468–5474.

[53] J. S. Walker, G. I. Straguzzi, W. H. Manogue and G. C. A. Schuit, *J. Catal.*, 1988, 110, 298–309.

[54] P. M. Kouotou, Z. Y. Tian, H. Vieker, A. Beyer, A. Gölzhäuser and K. Kohse-Höinghaus, *J. Mater. Chem. A*, 2013, 1, 10495–10504.

[55] P. M. Kouotou, Z. Y. Tian, H. Vieker and K. Kohse-Höinghaus, *Surf. Coat. Technol.*, 2013, 230, 59–65.

[56] W. Bergermayer and H. Schweiger, *Phys. Rev. B*, 2004, 69, 195409.

[57] A. Kandalam, P. Jena, S. Khanna, B. Chatterjee and B. V. Reddy, *Am. Phys. Soc. Meeting, March 13*–17 (abstract W11.003). 2007

[58] S. Wagloehner, D. Reichert, D. Leon-Sorzano, P. Balle, B. Geiger and S. Kurerti, *J. Catal.*, 2008, 260, 305–314.

[59] N. N. Bulgakov and V. A. Sadykov, *React. Kinet. Catal. Lett.*, 1996, 58, 397–402.

[60] X. W. Xie and W. J. Shen, *Nanoscale*, 2009, 1, 50–56.

[61] X. Wang, W. Tian, T. Y. Zhai, C. Y. Zhi, Y. Bando and D. Golberg, *J. Mater. Chem.*, 2012, 22, 23310–23326.

[62] R. J. H. Voorhoeve, J. P. Remeika and D. W. Johnson, *Science* 1973, 180, 62–64.

[63] C. Hu, Q. Zhu, Z. Jiang, L. Chen and R. Wu, *Chem. Eng. J.*, 2009, 152, 583–590.

[64] P. Li, C. Y. Nan, Z. Wei, J. Lu, Q. Peng and Y. D. Li, *Chem. Mater.*, 2010, 22, 4232–4236.

[65] Y. Li, H. Tan, X.-Y. Yang, B. Goris, J. Verbeeck, S. Bals, P. Colson, R. Cloots, G. V. Tendeloo and B.-L. Su, *Small*, 2011, 7, 475–483.

[66] H. Lv, M. G. Yao, Q. J. Li, Z. P. Li, B. Liu, R. Lui, S. C. Lu, D. M. Li, J. Mao, X. L. Ji, J. Liu, Z. Q. Chen, B. Zou, T. Cui and B. B. Liu, *J. Phys. Chem. C*, 2012, 116, 2165–2171

[67] Y. Chang and J. G. McCarty, *Catal. Today*, 1996, 30, 163–170.

[68] S. C. Kim and W. G. Shim, *Appl. Catal. B: Environ.*, 2010, 98, 180–185.

[69] V. P. santos, M. F. R. Pereire, J. J. M. Orfao and J. L. Figueiredo, *Top. Catal.*, 2009, 52, 470–481.

[70] V. Iablokov, K. Frey, O. Geszti and N. Kruse, *Catal. Lett.*, 2010, 134, 210–216.

[71] E. Finocchio, R. J. Willey, G. Busca and V. Lorenzelli, *J. Chem. Soc. Faraday Trans.*, 1997, 93, 175–180.

[72] P.-O. Larsson, H. Berggren, A. Andersson and O. Augustsson, *Catal. Today*, 1997, 35, 137–144.

[73] G.-H. Lee, M. S. Lee, G.-D. Lee, Y.-H. Kim and S.-S. Hong, *J. Ind. Eng. Chem.*, 2002, 6, 572–577.

[74] Q. Hua, T. Cao, X.-K. Gu, J. Lu, Z. Jiang, X. Pan, L.-F. Luo, W.-X. Li and W. Huang, *Angew. Chem. Int. Ed.*, 2014, 53, 4856–4861.

[75] F. D. Lai, C. Y. Huang, C. M. Chang, L. A. Wang and W. C. Cheng, *Microelectron. Eng.*, 2003, 67–68, 17–23.

[76] E. Sourty, J. L. Sullivan and M. D. Bijker, *Tribol. Int.*, 2003, 36, 389–396.

[77] F. Luo, X. Pang, K. Gao, H. Yang and Y. Wang, *Surf. Coat. Technol.*, 2007, 202, 58–62.

[78] A. G. Tarasov, V. A. Gorshkov and V. I. Yukhvid, *Inorg. Mater.*, 2007, 43, 724–728.

[79] A. Cellard, V. Garnier, G. Fantozzi, G. Baret and P. Fort, *Ceram. Int.*, 2009, 35, 913–916.

[80] S. Ivanova, C. Petit and V. Pitchon, *Gold Bull.*, 2006, 39, 3–8.

[81] Q. Liang, K. D. Chen, W. H. Hou and Q. J. Yan, *Appl. Catal. A: Gen.*, 1998, 166, 191–199.

[82] Q. H. Zhang, X. H. Liu, W. Q. Fan and Y. Wang, *Appl. Catal. B: Environ.*, 2011, 102, 207–214.

[83] D. A. Aguilera, A. Perez, R. Molina and S. Moreno, *Appl. Catal. B: Environ.*, 2011, 104, 144–150.

[84] H. Noller and H. Vinek, *J. Mol. Catal.*, 1989, 51, 285–294.

[85] P. Thormählen, M. Skoglundh, E. Fridell and B. Andersson, *J. Catal.*, 1999, 188, 300–310.

[86] S. Veleva and F. Trifirò, *React. Kinet. Catal. L.*, 1976, 4, 19–24.

[87] H. Liu, P. Cheung and E. Iglesia, *J. Phys. Chem. B*, 2003, 107, 4118–4127.

[88] H. Lee, J. C. Jung, H. Kim, Y. M. Chung, T. J. Kim, S. J. Lee, S. H. Oh, Y. S. Kim and I. K. Song, *Catal. Lett.*, 2009, 131, 344–349.

[89] A. Törncrona, M. Skoglundh, P. Thormählen, E. Fridell and E. Jobson, *Appl. Catal. B: Environ.*, 1997, 14, 131.

[90] M. Machida, K. Ochiai, K. Ito and K. Ikeue, *J. Catal.*, 2006, 238, 58–66.

[91] A. C. Gluhoi, N. Bogdanchikova and B. E. Nieuwenhuys, *J. Catal.*, 2005, 232, 96–101.

[92] S. Hocevar, J. Batista and J. Levec, *J. Catal.*, 1999, 184, 39–48.

[93] C. Kleinlogel and L. J. Gauckler, *Adv. Mater.*, 2001, 13, 1081–1085.

[94] W. J. Shen, Y. Ichihashi and Y. Matsumura, *Catal. Lett.*, 2002, 83, 33–35.

[95] R. Zhang, H. Alamdari and S. Kaliaguine, *Catal. Lett.*, 2007, 119, 108–119.

[96] P. Li, D. E. Miser, S. Rabiei, R. T. Yadav and M. R. Hajaligol, *Appl. Catal. B: Environ.*, 2003, 43, 151–162.

[97] P. A. Deshpande, S. T. Aruna and G. Madras, *Catal. Sci. Technol.*, 2011, 1, 1683–1691.

[98] H. C. Wu, L. C. Liu and S. M. Yang, *Appl. Catal. A: Gen.*, 2001, 211, 159–165.

[99] K. Qian, W. X. Huang, Z. Q. Jiang and H. X. Sun, *J. Catal.*, 2007, 248, 137–141.

[100] H. Baussart, R. Delobel, M. L. Bras and J. M. Leroy, *J. Chem. Soc., Faraday Trans. I*, 1979, 75, 1337–1345.

Chapter 11

Spinel-Structured Nanoparticles for Magnetic and Mechanical Applications

Malik Anjelh Baqiya, Ahmad Taufiq, Sunaryono,
Khuroti Ayun, Mochamad Zainuri, Suminar Pratapa,
Triwikantoro and Darminto

Additional information is available at the end of the chapter

http://dx.doi.org/10.5772/66293

Abstract

Nanoparticles of Fe_3O_4 have been successfully synthesized using a simple coprecipitation technique from natural iron sands, employing HNO_3 and NH_4OH as dispersing and precipitating agents, respectively. The substitution of Fe with Mn to result in $Fe_{3-x}Mn_xO_4$ ($0 \leq x \leq 3$) was conducted to control the magnetic strength of this nano-sized spinel powder. It is shown that magnetic properties depend not only on the particle size and Mn doping but also on the particles clustering. The applications for magnetic fluids, gels, and coating are extensively described. Meanwhile, the spinel $MgAl_2O_4$ nanoparticles have also been prepared by the same simple method from commercial starting materials. This powder was used as a nano-reinforcer of Al-matrix composites. In addition, $MgAl_2O_4$ micro-sized powder forming a thick layer was successfully grown by electroless plating on the interface of matrix-filler in Al/SiC composites. The strengthening of mechanical properties with respect to the varying uses of these $MgAl_2O_4$ powders is discussed.

Keywords: Fe_3O_4, $MgAl_2O_4$, Powders, Coprecipitation, Magnetic, mechanical properties

1. Introduction

Spinel structures have the general formula of AB_2X_4, where X can be oxygen (oxides) or a chalcogen element, such as sulfur (thio-spinels) and selenium (seleno-spinels). A and B in the spinel structures can be divalent, trivalent, or tetravalent cations, such as iron, magnesium, aluminum (Al), zinc, manganese, chromium, and titanium. The various compounds of spinel family including their chemical and physical properties depend not only on the arrangement of cations and anions in the structures but also on how the spinels are produced. The spinels can be magnetic or non-magnetic compounds depending on the nature of the cations among the A and B sub-lattices. Fe_3O_4 ($FeO.Fe_2O_3$) is one of the so-called spinel ferrite having magnetic properties, whereas $MgAl_2O_4$ is also spinel (should not be confused with the spinel structure described earlier) which is non-magnetic one. This chapter covers some of the physical properties of the spinel compounds in more detail.

2. Magnetite (Fe_3O_4) nanoparticles

Fe_3O_4 (magnetite) or the black iron oxide in the form of nanoparticles has huge potential for applications in many fields. However, preparing Fe_3O_4 particles in small size with high quality is enormously challenging for several reasons. Therefore, the preparation of Fe_3O_4 nanoparticles in several forms and various sizes with high purity and homogeneity becomes an essential task before applications. The following section provides more details on the preparation and applications of Fe_3O_4 nanoparticles.

Several synthesis methods to prepare high-purity Fe_3O_4 nanoparticles in various forms and sizes have been reported in the literature. In general, Fe_3O_4 nanoparticles have been successfully prepared through sonochemical [1], thermal decomposition [2], ionothermal [3], hydrothermal [4], micro-emulsion [5], sol-gel [6], modified hydrolysis [7], solvothermal [8], electrospinning [9], coprecipitation [10] methods, and so forth. However, high quality of the particles, such as good particle size distribution with high crystalline, structural and compositional homogeneity, still has become one of the main challenges in preparing Fe_3O_4 nanoparticles [11].

The improvement of synthesis methods has also been carried out for producing Fe_3O_4 nanoparticles in order to reduce cost and time of preparation. A simple method to produce Fe_3O_4 nanoparticles via coprecipitation method has been introduced using a local natural iron sand from Indonesia as a starting material. This coprecipitation method is considered to be the easiest method for preparing Fe_3O_4 nanoparticles [12] because of its simplicity, efficiency, and inexpensiveness [13]. The use of starting material from commercial product can be substituted by natural iron sand.

As for the preparation of Fe_3O_4 nanoparticles using coprecipitation method, the ratio of ferrous and ferric ions is very crucial affecting the synthesis result. The effects include the particle diameter and magnetic properties. Iida et al. [14] have reported that the valence of the metal salt is an important factor for determining the particle size of the magnetic particles. In their experiments, they have succeeded in obtaining Fe_3O_4 nanoparticles in the size range of 9–40 nm with various molar contents of ferrous ions in the total amount of the iron ions. Furthermore, Gnanaprakash et al. [15] have also reported that initial pH and temperature of the iron salt solution during coprecipitation process have strong influences for the formation of magnetite nanoparticles. They reported that the average of particle size having initial pH value

http://dx.doi.org/10.5772/66293

of 0.7, 1.5, 3, and 4.7 are 6.7, 7.6, 9.9, and 9.9 nm, respectively. Determination of particle size is very important according to how they will be used for certain applications, namely for magnetic fluids and gels.

In the magnetic nanoparticles' syntheses, the so-called templating method has been employed to form certain particle shapes and sizes, such as one dimensional (1-D) particles or nanorods. Lian et al. [16] have successfully synthesized nanorods of Fe_3O_4 through the addition of polyethylene glycol (PEG) 1000 as a template with the average particle diameter and length of 80 nm and 2 μm, respectively. Therefore, templating technique has important role on the particle growth of magnetic nanomaterials. The greater value of the PEG's molecular weight will lead to the higher possibility to obtain nanorods resulted from the synthesis [17]. Furthermore, based on the research of He et al. [18], Fe_3O_4 nanowire has successfully been prepared depending on the volume ratio of PEG and H_2O. They have found that the best formation is achieved with the ratio of 1:3; a higher ratio gives higher viscosity and inhibits the particle formation.

3. Fe_3O_4 spinels for magnetic fluids and gels

Generally, the study of Fe_3O_4 nanoparticles has attracted attention for several disciplines. Physicists investigate the physical properties and propose the theories of the Fe_3O_4 nanoparticles. Chemists focus on the synthesis methods of the Fe_3O_4 nanoparticles. Biologists concentrate on developing Fe_3O_4 nanoparticles in bioapplications. Meanwhile, engineers study Fe_3O_4 nanoparticles with a focus on technological applications. Therefore, there is a great deal of interest in Fe_3O_4 nanoparticles by scientists and engineers, starting from synthesis and characterizations and followed by their many application potentials.

Fe_3O_4 ferrofluids or magnetic fluids are colloidal suspension containing magnetite nanoparticles in single domain, which are dispersed in a liquid carrier. Ferrofluids have unique characteristics because they have both the liquid and magnetic properties. In the ferrofluids, the liquid carriers can be polar or nonpolar medium. In order to prevent agglomerations, the Fe_3O_4 nanoparticles should be layered or covered by an appropriate material as a shell [19]. In the last decades, it has been reported that the Fe_3O_4 ferrofluids can be applied in various fields, that is, optical grating, optical switching [20], thermoelectric conversion [21], quenching media [22], sensor [23], controlling lubricant migration [24], biomedical applications [25], and so forth.

Another application for the magnetite spinel is in ferrogels, namely a hydrogel containing magnetic ferrous particles. The Fe_3O_4-ferrogel can be defined as a polymer (nano)composites with polyvinyl alcohol (PVA) and Fe_3O_4 as the matrix and filler, respectively, cross-linked together with a certain amount of liquid (water). Ferrogels can be physically or chemically cross-linked without any chemical reaction between them. The physical properties of ferrogels depend on the preparation conditions, such as polymer and solvent ratio, particle size, distribution and concentration of the magnetic material. By combining the elastic medium and the magnetic properties from the particles, the elastic behavior of ferrogel can be manipulated by external magnetic field. This leads to a great opportunity for various advanced applications, namely soft actuators and artificial muscles [26].

In the development of magneto-elastic materials, Li et al. [27] have prepared a polymeric gel in variety of shapes depending on the temperature. Furthermore, Zrínyi et al. [28] have developed a magneto-sensitive gel with Fe_3O_4 as the magnetic filler. Ramanujan and Lao [26] have successfully produced magnetic gel consisting of PVA and micro-sized Fe_3O_4 particles for artificial tissues. They have demonstrated an elastic behavior of the gel controlled by an external magnetic field.

4. Spinel $MgAl_2O_4$ grown as a thick coating and fillerreinforcer in composites

Compared to the conventional aluminum alloys, aluminum metal matrix composites (Al-MMC) have been used in many kinds of applications (e.g. automobiles) due to their high specific strength and Young's modulus, improved yield and creep strengths, light weight, and excellent properties of wear resistance. The important parameter for improving their performances lies in the structure and bonding properties between the matrix and fillers. Some materials in any form of particles have been widely used as the particle-matrix interfaces in order to improve the wettability [29], namely MgO, SiC, and Al_2O_3 in the micro- or nanostructures. It has been noted that the interface, particles size, and its distribution as well as the concentration (volume fraction) play a basic role in the modification of Al-MMC.

A typical way to improve the mechanical properties of Al-MMC is by additional reinforcement or interfacial modification. Recently, the spinel $MgAl_2O_4$ has been applied for this purpose. In the last decade, $MgAl_2O_4$ has been produced by several methods such as solid-state reaction [30], high-energy ball milling [31], polymerization method [32], pyrolysis [33], sol-gel processes [34, 35], and coprecipitation techniques [36–38]. The shape and size of the spinel particles vary depending on the preparation techniques [39, 40]. This spinel can be prepared in the form of ceramic foams [41, 42], whiskers [43–45], thin films [46, 47], nanoparticles [37, 48, 49], and also layer of particle bonding in composite materials [50]. For the Al/SiC composites, in which Al acts as the matrix and SiC acts as the filler reinforcement, Al_4C_3 phase can be formed at the interface between matrix and fillers. This phase weakens the mechanical strength of the composite. Up to now, several experimental techniques have been proposed to prevent this problem including adding excessive Si into the matrix and introducing a thin coating layer on the SiC particle surface. The formation of interlayer phase is believed to be a stable interfacial bonding that improves the mechanical properties of the composite. However, the detail of this mechanism remains unclear.

Particulate-reinforced composites with micron-sized filler of various materials are the most commonly used composites in daily life. Particles are typically introduced to enhance the matrix elastic modulus and yield strength. It has been shown that the novel properties of composites are improved by adding nano-scaled particles as the fillers [51]. The key role of mechanical strength in composites is the interface-bonding quality between matrix and filler materials. The nano-sized particles have much greater total area compared to particles with larger size at the same volume. The interface interaction can therefore be extended by reducing particle size of fillers down to nanometer scale.

Meanwhile, synthesis of particles in nanometer size of various materials is still an active area of research. Because of their very small size, nanoparticles tend to be aggregate to each other to form larger size particles [52]. As nanometer-scaled filler, the particles have to be kept separate in the matrix. In this context, an effort to maintain nanoparticles as fillers from agglomeration was achieved by the introduction of a surfactant. Tetra-methyl-ammonium-hydroxide (TMAH) could be used as surfactant in the process of filler-matrix-mixing media. Aluminum was chosen as matrix, while MgO and $MgAl_2O_4$ were used as nano-fillers. In the case of Al/SiC composites, a thin-layer $MgAl_2O_4$ is coated onto the SiC filler to enhance matrix-filler interfacial bonding leading to increased mechanical strength. The SiC fillers were coated by spinel $MgAl_2O_4$ by electroless plating, soaking SiC particulate powder in solutions containing Mg and Al ions before heating process to grow spinel phase on the surface

5. Preparation of Fe_3O_4 nanoparticles from iron sands

The detail of synthesis method for the preparation of Fe_3O_4 nanoparticles was described in our previous works [53–57]. The iron sand was extracted by a magnetic separator to obtain dominantly Fe_3O_4 powders. Using a magnetic stirrer, the hydrochloric acid (HCl) was used to dissolve the Fe_3O_4 powder to produce $FeCl_3$ and $FeCl_2$ solutions at room temperature as written in Eq. (1).

$$Fe_3O_4 + 8HCl \rightarrow 2FeCl_3 + FeCl_2 + 4H_2O \tag{1}$$

The solution of $FeCl_3$ and $FeCl_2$ was then reacted by dropping slowly ammonium hydroxide (NH_4OH) to produce black precipitate at room temperature. The reaction mechanism for producing Fe_3O_4 particles is described as Eq. (2) [54]

$$2FeCl_3 + FeCl_2 + 8NH_4OH \rightarrow Fe_3O_4 + 8NH_4Cl + 4H_2O \tag{2}$$

In order to synthesize $Mn_xFe_{3-x}O_4$ nanoparticles discussed later and investigate its magnetic properties, the variation of Mn concentration from $MnCl_2$ solution was also prepared to form $Mn_xFe_{3-x}O_4$ ($0<x\leq 1$). The dissolved $MnCl_2$ solution was mixed using a magnetic stirrer followed by a drop-wise addition of NH_4OH to obtain the precipitate of Fe_3O_4 and $Mn_xFe_{3-x}O_4$. The magnetic moment of Mn^{2+} ion is 25% higher than that of Fe^{2+} ion. Therefore, introducing Mn^{2+} into Fe_3O_4 to form $Mn_xFe_{3-x}O_4$ structure will theoretically enhance the magnetization of Fe_3O_4 [58]. A washing process of the precipitate was then carried out for several times using distilled water until achieving normal pH condition. Finally, the Fe_3O_4 powders were prepared by drying process at 100°C for 1 h. For the preparation of Fe_3O_4 magnetic fluid, tetra-methyl-

ammonium-hydroxide (TMAH), as the stabilizing layers of the particles, was added in the precipitate. In the present investigation, Fe_3O_4 ferrogels were also synthesized using freezing-thawing method as described in Refs. [57, 59].

To study the crystal structure, the particle size and its distribution as well as form and structural factor, and the magnetic properties of the sample, the Fe_3O_4 powders were then characterized by means of X-ray diffractometry (XRD), small-angle neutron scattering (SANS) spectrometer and superconducting quantum interference device (SQUID) magnetometer, respectively. The XRD measurements were carried out in the 2θ range from 20° to 70° using Cu-Kα radiation. A 36-m SANS spectrometer was used to investigate the primary and secondary particles as well as the fractal structure of Fe_3O_4 nanoparticles. The detailed SANS measurements have been given elsewhere [54, 60]. The two-lognormal analysis was also performed to investigate the clustering effect of the magnetic fluid and hydrogel [55, 57].

6. Preparation of spinel $MgAl_2O_4$ as filler reinforcers

Magnesium oxide (MgO) and magnesium-aluminum oxide ($MgAl_2O_4$) as fillers were synthesized by employing coprecipitation process. Mg (Aldrich, 99.9%) and Al (Aldrich, 99.9%) were dissolved in HCl (12.63 molar) with stoichiometric molar fraction. The NH_4OH (6.5 molar) was used as precipitating agent to produce $Mg(OH)_2$ and $MgAl_2O_4$ after filtering, washing process with distilled water for several times, and finally drying at 100°C. The resulting powders were then heated at 500°C for 1 h to convert them into MgO and $MgAl_2O_4$ phases. The powders were checked with XRD and transmission electron microscope (TEM) to explore the phase purity, crystallite size, particle shape, and size.

The MgO and $MgAl_2O_4$ powders were mixed with TMAH as surfactant to form a colloid system, consisting of individual MgO or $MgAl_2O_4$ particles which were homogeneously dispersed. To fabricate composites, Al powders as matrix were added into the colloid system and followed by thorough stirring and grinding to achieve homogeneity. The mixed powders were dried at 100°C for 3 h and then pressed into pellets having a diameter of 1 cm. During pelletization, the powders were compacted using a force of 1.5 kN applied for 15 min in an isostatic die with zinc stearate on the inner disc. Before sintering process at 500°C for 1 h, pellets were pre-sintered at, respectively, 200 and 400°C for 20 min each. All heat treatments were conducted in a furnace with a controlled atmosphere of low vacuum ($\sim 10^{-3}$ atm).

The samples produced consist of Al/MgO and Al/$MgAl_2O_4$ composites with various volume fractions (% vol) of fillers covering 10, 20, and 30%. We have also prepared the corresponding samples fabricated without TMAH for comparison. To study the effects of surfactant addition during the process on the fabricated composites, measurements of densities, porosity, elastic modulus by compressive test, and microhardness (Vicker's hardness number, VHN) of the samples were carried out.

Furthermore, to prepare the Al/SiC composites, the aluminum (PA) powders and particles of SiC ceramics (220 mesh) were employed as starting materials. The SiC reinforcement particles were cleaned by ultrasonic cleaner in alcohol (90%), and then dried in oven at 100°C. The SiC particles were then soaked in an electrolyte media of 40 ml HNO_3 containing Al and Mg ions as the part of the electroless-plating mechanism. The process of metal oxide coating was done using a magnetic stirrer at 125°C. The oxidation of SiC particles to grow the $MgAl_2O_4$ layer was performed in a furnace at 200°C for 1 h, and continued at 400°C for another 1 h. The electrolyte media for electroless plating can be controllably adjusted to grow various metal oxides on the SiC surface, such as MgO, CuO, Al_2O_3, and so forth, besides $MgAl_2O_4$. The metal oxide-coated SiC particles were mixed with Al powders in n-butanol (wet mixing) using magnetic stirrer at 100°C, having SiC volume fraction (%vol) of 10, 20, and 30%. The heating process of the green pellets was carried out in a vacuum (10^{-3} Torr, rotary pump) by applying a pre-sintering at 200°C for 20 min and followed by sintering at 600°C for 1 h.

7. Results and discussion

7.1. Fe_3O_4 nanoparticles: magnetic properties and applications

The XRD pattern of Fe_3O_4 nano-powders is shown in **Figure 1**. All peaks in the pattern show a single phase of spinel structure corresponding to the crystal structure of Fe_3O_4 with PDF No. 19-0629 without any impurity. Based on the Rietveld analysis, the sample has lattice parameter $a = b = c$ of approximately 8.377 Å.

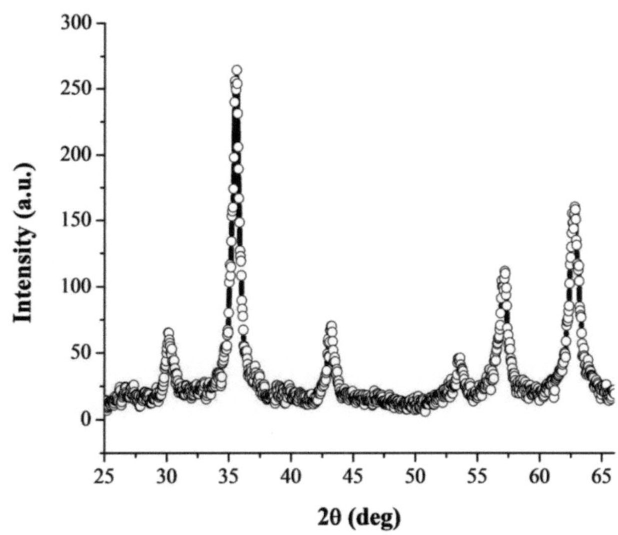

Figure 1. The XRD pattern of Fe_3O_4 nano-powders (using Cu-Kα radiation).

The SANS pattern of the Fe_3O_4 nano-powders is presented in **Figure 2**. The SANS data were analyzed using a lognormal spherical model as a form factor $P(R)$ and mass fractal model as a structure factor $S(q)$, following Eqs. (3) and (4), respectively,

$$P(R) = \frac{1}{\sigma R\sqrt{2\pi}} \exp\left(\frac{\ln^2(R/R_0)}{2\sigma^2}\right) \tag{3}$$

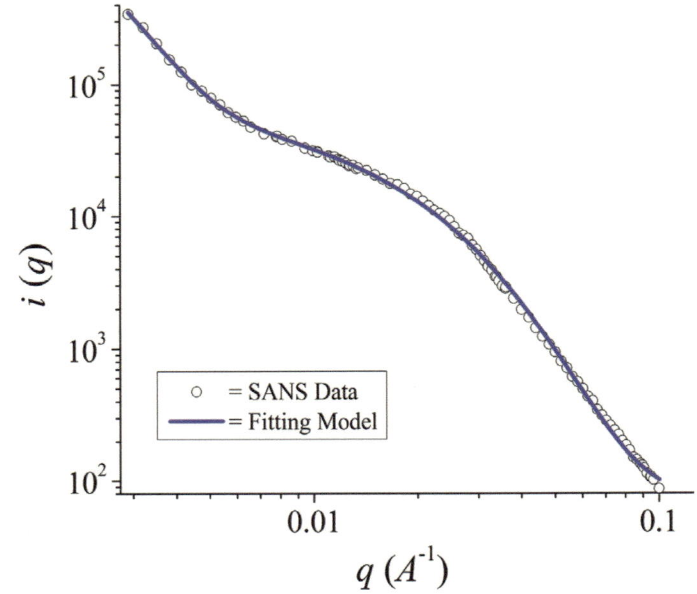

Figure 2. SANS data of Fe_3O_4 nano-powders.

with R_0 and σ representing the radius of the distribution and standard deviation, respectively [61],

$$S(q) = 1 + \frac{D\Gamma(D-1)}{(qR)^D\left[1+1/(q^2\xi)^2\right]^{(D-1)/2}} \sin\left[(D-1)\tan^{-1}(q\xi)\right] \tag{4}$$

where q, Γ, D, and ξ represent the scattering vector, gamma function, fractal dimension, and cut-off distance, respectively [62]. The SANS curve was fitted globally using two lognormal spherical model as form factor combining with mass fractal model as structure factor regarding Eq. (5) [54]

$$I(q) \approx \int_0^\infty N_1(R_1) F_N^2(q,R_1) dR_1 + \int_0^\infty N_2(R_2) F_N^2(q,R_2) dR_2 S(q,\xi,D,R_2) \tag{5}$$

Here, I is the density, N is the number density of particles, R_1 is the primary particles, R_2 is the secondary particles or clusters, and F is the scattering amplitude.

http://dx.doi.org/10.5772/66293

Based on the analysis using lognormal and mass fractal models, the Fe_3O_4 nano-powders have hierarchical nanostructure with the primary particles of 3.8 nm as a building block constructing secondary particles as clusters of 9.3 nm. The clusters of Fe_3O_4 nanoparticles have a fractal structure in three-dimension with fractal dimension of 2.9. The SANS data analysis of magnetic nanoparticles coincides with the image produced by high-resolution transmission electron microscopy (HRTEM) as shown in the previous work [54]. The details of the analysis of SANS data of the Fe_3O_4 nanoparticles were presented in the previous work [54].

The magnetic properties of the Fe_3O_4 nanoparticles as well as M versus H variations and zero-field-cooling (ZFC) curves are presented in **Figures 3** and **4**. The M versus H curve of the Fe_3O_4 nanoparticles was collected at room temperature by sweeping the magnetic field from -5 to 5 T. The Fe_3O_4 nanoparticles have saturated magnetization of 37.1 emu/g. Regarding **Figure 3**, it is clear that the M-H curve of sample has S-shape with nearly zero coercivity field, indicating the superparamagnetic behavior at room temperature. This result is consistent with the recent paper by Abboud et al. [63] where the magnetization was almost zero in the absence of external magnetic field. The superparamagnetic phenomenon in the magnetic nanoparticles has also been observed in the Fe_3O_4@SiO_2 core-shell composites [64], the dispersed Fe_3O_4 in the polymer matrix [65], and Fe_3O_4-$LiMo_3Se_3$ [66]. Moreover, the zero-field-cooling (ZFC) measurement was also carried out to investigate the superparamagnetic phenomenon of the sample as shown in **Figure 4**. Based on the results in **Figure 4**, it is evident that the sample has a maximum peak of magnetic blocking temperature, T_B, at 243 K. Theoretically, T_B depends on the particle size and shape so the increase of nanoparticle's volume will increase the T_B value [67].

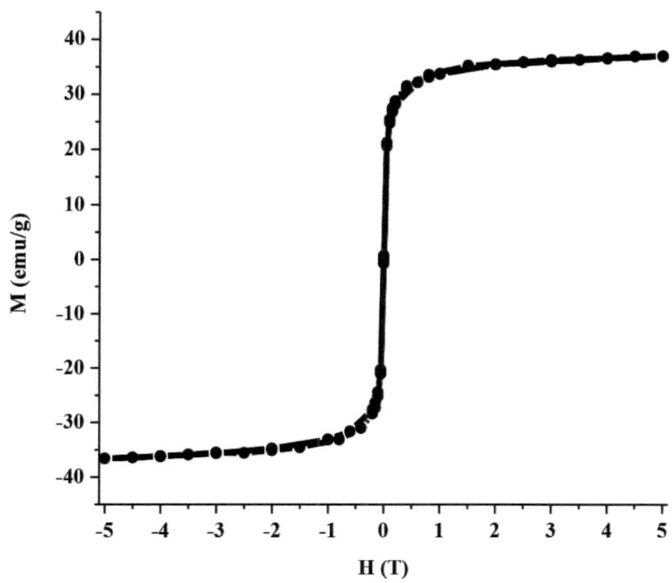

Figure 3. M-H curve of Fe_3O_4 nanoparticles at room temperature.

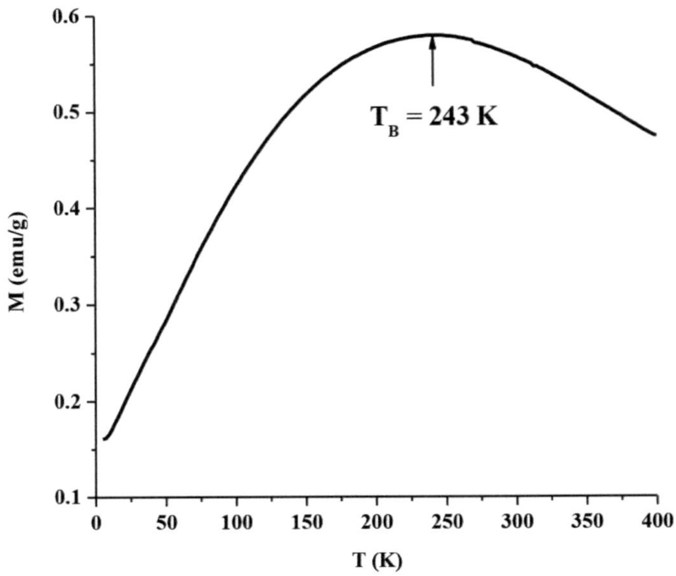

Figure 4. Zero-field-cooling (ZFC) curve of Fe_3O_4 nanoparticles (T_B = 243 K).

Dutta et al. [68] have reported that Fe_3O_4 nanoparticles with the particles size ranging from 4 to 12 nm have blocking temperature lower than 100 K, which is not similar to the result of this work. The difference of these results can be explained in a sense of the presence of clusters or aggregations phenomena in the present work. In Ref. [68], the samples were constructed with primary particles in relatively homogeneous samples without any clusters or aggregations. On the other hand, the sample in this work consists of hierarchical nanostructures of primary particles forming secondary particles or clusters with fractal dimension in three-dimensions. Despite the particles size and anisotropy constant, the clusters of magnetic nanoparticles give an effect on the blocking temperature. Theoretically, at magnetic-blocking temperature, the thermal energy of particles is comparable with the anisotropy energy barrier. In bulk, the Fe_3O_4 is ferrimagnetic generated by net magnetic moments at tetrahedral and octahedral sites. The effect of clusters on magnetic properties was also documented by other researchers [69, 70].

In the magnetic nanoparticles, there is the so-called blocking temperature or the energy barrier that can be obtained from ZFC and field-cooling (FC) magnetization curves. With increasing temperature, a curve peak should appear in the ZFC measurement. This peak temperature can be considered as the average T_B in the magnetic material. On the other hand, the increase of magnetic field should decrease the barrier energy, and results in a shift of the blocking temperature to lower temperatures. A typical of nanocrystalline Fe_3O_4 has shown superparamagnetic behavior (**Figure 3**). The superparamagnetism is related to the random fluctuation of the magnetization of a single domain when the thermal energy overcomes the anisotropy energy barrier. Below the T_B, the magnetization of each domain of nanoparticles is oriented parallel to a certain crystallographic direction or the easy axis with minimum energy. Consequently, the magnetization is blocked in the nanoparticles. Above T_B, the thermal energy can overcome the anisotropy energy (barrier energy) and the magnetization starts fluctuating and the magnetic susceptibility follows modified Curie-law behavior at higher temperatures [71].

The effect of Mn^{2+} substitution on the magnetic properties of $Mn_xFe_{3-x}O_4$ nanoparticles has been intensively studied recently [54, 72]. It has been found that the lattice parameters and crystal volume of the $Mn_xFe_{3-x}O_4$ increase with the increase of Mn content. This is due to the fact that the ionic radius of Fe^{2+} is smaller than that of Mn^{2+} substituted in the spinel structure. Fe_3O_4 without any substitution has a cubic spinel structure, where the Fe^{2+} and Fe^{3+} ions occupy the tetrahedral (A) and octahedral (B) sites represented by $(Fe^{3+})_A(Fe^{3+}Fe^{2+})_BO^{2-}{}_4$. The Fe_3O_4 is one example of the cubic inverse spinels, in which there is a mixed valence of Fe ions on the octahedral sublattice. In the $Mn_xFe_{3-x}O_4$, the cationic distribution can be written as follows [54]:

$$\left(\mathrm{Mn}_x^{2+}\mathrm{Fe}_{1-x}^{3+}\right)_A\left[\mathrm{Fe}_{1-x}^{2+}\mathrm{Fe}_{1+x}^{3+}\right]_B \tag{6}$$

The Mn^{2+} ions occupy the A sites of the spinel structure. A further analysis of the SANS data of the $Mn_xFe_{3-x}O_4$ nanoparticles from $x = 0$ to 1 using the lognormal function has found that the particle size of the $Mn_xFe_{3-x}O_4$ depends on the Mn content. In this study, the increase of Mn content results in the smaller size of the nanoparticles. Moreover, the nanoparticles of $Mn_xFe_{3-x}O_4$ tend to become larger aggregates or clusters. The Mn content, particle size, and particle clustering are important factors influencing the magnetic properties of nanomaterials.

Taufiq et al. [55] have successfully synthesized Fe_3O_4 magnetic fluid with a chain-like spinel structure using coprecipitation method. These results show that the Fe_3O_4 ferrofluid has a primary particle size of about 7.6 nm and the size of fractal aggregates of about 45 nm constructing the chain-like structure. Even though there were chain aggregates, a homogeneous particle distribution with low polydispersity value of about 0.4 was found, which is lower than that in the former paper [55]. It is reported that both nanoparticles of Fe_3O_4 and $Mn_xFe_{3-x}O_4$ (x up to 1) exhibit superparamagnetic behavior and their saturation magnetization decreases with increasing Mn content [54, 72].

In the research of magnetic hydrogels, Fe_3O_4 ferrogels have been investigated by Sunaryono et al. [57]. It has been reported that the magneto-elastic properties of the ferrogels were strongly affected by the preparation technique, the ratio of the magnetic particle and the polymer, and magnetic particle content as well as the particle clustering and distribution. It has also been shown that the magneto-elasticity of the ferrogel tends to decrease with the increasing Fe_3O_4 content from 2.5 to 15% [56, 57]. Further analysis of the SAXS data using the two-lognormal distribution function showed that Fe_3O_4 in the hydrogel has primary particles of about 3 nm with an average particle distance of about 18 nm. Higher Fe_3O_4 concentration in the hydrogel leads to the increase of Fe_3O_4 cluster size. Due to the surface effect, it influences the magnetization value of the ferrogel. Consequently, the saturation magnetization drops with decreasing particle size and the cluster of Fe_3O_4 nanoparticles in the hydrogel. It has been confirmed that Fe_3O_4 ferrogel exhibits a superparamagnetic behavior at room temperature which is a crucial parameter for biomedical applications.

7.2. Improved mechanical properties of aluminum matrix composites with surfactant-coated $MgAl_2O_4$ nanofillers and $MgAl_2O_4$-coated SiC particulates

Figure 5 presents the XRD spectra (using Cu-Kα radiation, $\lambda = 1.54$ angstrom) of MgO and $MgAl_2O_4$ particles as results of coprecipitation process. The powders seem to be single phase as can be seen from the spectra, where no other additional diffraction peak is observed, except those associated with Miller indexes belonging to the MgO (periclase phase) and $MgAl_2O_4$ (spinel phase) as specified in **Figure 5**. Besides, the diffraction peaks are much broadened, featuring that the crystal is in nanometer size. If we take the full-width at half maximum (FWHM) of diffraction peaks, by using Scherrer formula [73] and taking the apparatus-broadening correction into account, we arrive at crystallite sizes of 17.5 ± 2.3 and 6.3 ± 0.7 nm for MgO and $MgAl_2O_4$ powders, respectively.

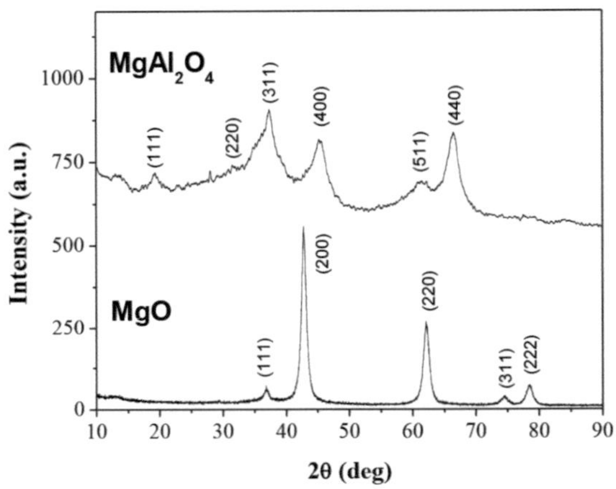

Figure 5. XRD spectra of MgO and $MgAl_2O_4$.

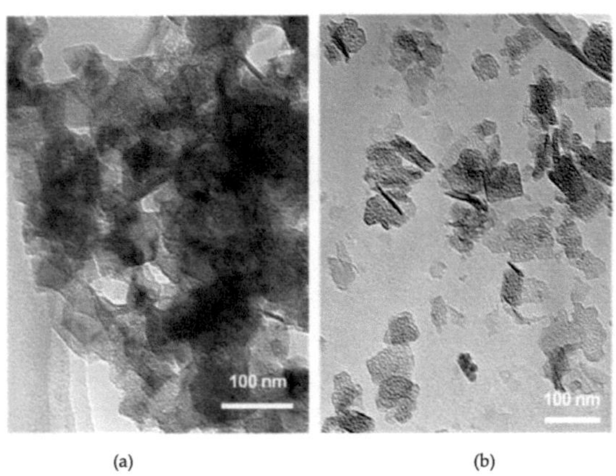

Figure 6. TEM images of (a) MgO and (b) $MgAl_2O_4$.

It is interesting to correlate the above crystallite size to the particle size of powders from TEM images depicted in **Figure 6**. The particle size for both powders, MgO in **Figure 6a** and $MgAl_2O_4$ in **Figure 6b**, is seen to be between 50 and less than 100 nm, confirming that the powders can be classified as nanoparticles. The shape of particles is observed to be varied from spherical, square, square with rounded edge, to semi-rod. Comparing the crystal and the particle sizes, one can consider that both MgO and $MgAl_2O_4$ powders contain secondary particles, where each particle is constituted by several crystals or grains.

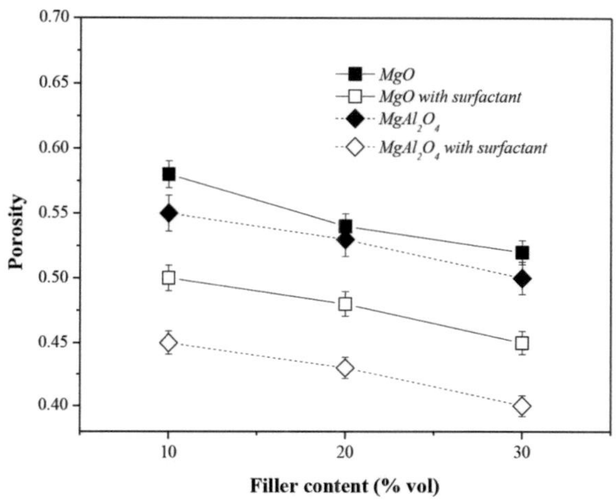

Figure 7. Porosity of sintered Al/MgO and Al/$MgAl_2O_4$ composites with fillers specified.

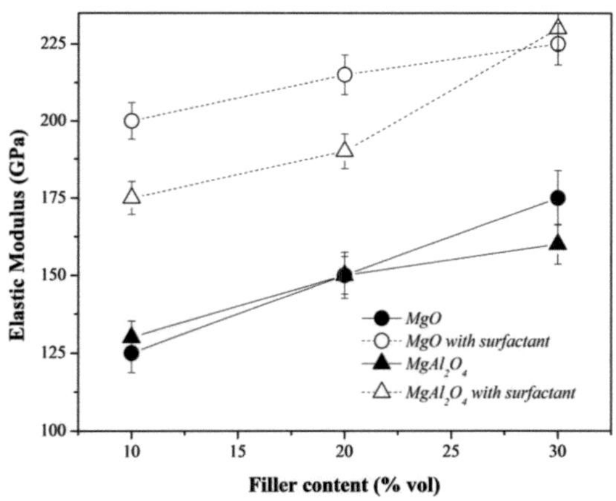

Figure 8. Elastic moduli of sintered Al/MgO and Al/$MgAl_2O_4$ composites with fillers specified.

To discuss further, the physical properties of composites, the density of pellets in the green state (after compaction) and after sintering significantly increases with increasing volume fraction of fillers and the use of surfactant (coated fillers by surfactant). The enhanced density has led to the lowering porosity in the pellets. By applying the formula to define porosity [52]: $\rho = 1 - \rho_s/\rho_t$, where ρ_s and ρ_t stand for, respectively, sintered and theoretical densities, the porosity of samples is exhibited in **Figure 7**. The increase of pellet densification has directly affected the mechanical properties, as reflected by the results of compressive tests (yielding elastic modulus) and hardness (VHN) measurements (see **Figures 8** and **9**).

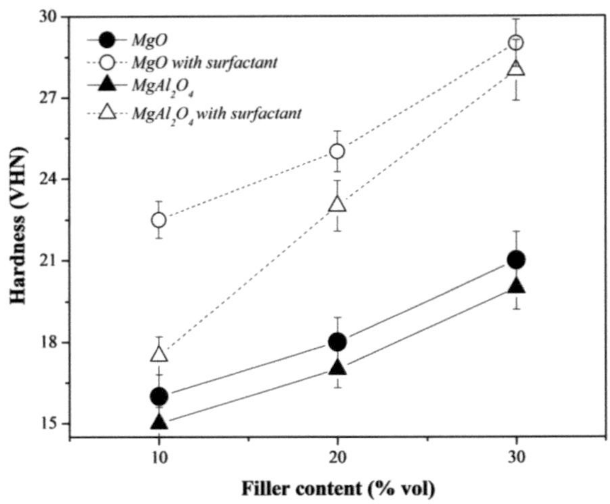

Figure 9. Micro-hardness of sintered Al/MgO and Al/$MgAl_2O_4$ composites with fillers specified.

It is worth noting that the role of surfactant is to enhance the physical and therefore mechanical properties of these Al/MgO and Al/$MgAl_2O_4$ nanocomposites. The TMAH as a mixing media of matrix and filler is an ionic molecule which will attach on the surface of nanoparticles. A metal-oxide nanoparticle is usually deficient of oxygen ion [74], hence its surface is positively charged. The negatively charged end of TMAH molecule is therefore attracted by metal-oxide surface, while the opposite end is dangled. It may further create the situation where one may consider that MgO or $MgAl_2O_4$ nanoparticle is "coated" by TMAH. This in turn avoids nanoparticles from aggregation to each other in the matrix. So, the filler is still in nanometer scale to maintain its larger surface area, especially compared to the agglomerated fillers having the same volume fraction in the composites fabricated without TMAH. The larger filler surface will lead to stronger interface bonding between matrix and filler.

The next attractive point to address is that the surfactant-coated MgO filler has led to considerably enhanced elastic modulus (~60%) and microhardness (~40%) of the Al-matrix nanocomposites (**Figures 8** and **9**). The periclase phase of MgO has simpler crystal structure than the spinel $MgAl_2O_4$ one [75, 76]. The former phase is transformed from $Mg(OH)_2$ at 500°C, whereas the latter one is already stable from 100°C as a result of coprecipitation process. Besides particles consolidation during the sintering at 500°C, the MgO filler will be possibly able to react further in a limited amount with Al matrix to form spinel phase on the matrix-filler interface, creating a strong bonding. Such kind of process, except consolidation, may not occur in the Al/$MgAl_2O_4$ composite, because of stable $MgAl_2O_4$.

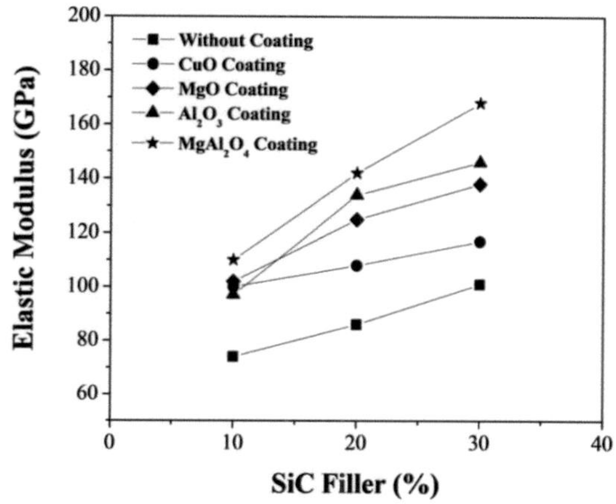

Figure 10. Elastic modulus of Al/SiC composites depending on filler content, using SiC coated by metal oxides as specified.

Going to further study on Al/SiC composites, it can be seen in **Figure 10** that a very clear distinction exists between SiC particles without coating and with coated metal oxides on the surface based on the modulus elasticity. The entire filler volume fraction (%vol) of SiC in Al-SiC composites using coatings has higher value than that without coating. The experimental results show that the modulus of elasticity of the composites using metal-oxide coating approaches the theoretical prediction [77], estimating a limit of lower and upper bounds. Based on the stress analysis, the composite having filler without metal oxide coating, for all low-volume fraction of reinforcement, the modulus value lies in the outside of the upper and lower bounds, which indicates weak interaction between matrix and filler. This is much different from composites with filler using metal oxides, where for the entire volume fraction of reinforcement, the elastic modulus lies inside the region of the upper and lower bounds, implying that the compactibility between matrix and filler, and reinforcements are therefore going well. One can see in **Figure 10** that coating with $MgAl_2O_4$ onto the SiC surface has produced the superior Al/SiC composites among the prepared samples, followed by Al_2O_3 and MgO. The three metal oxides coating materials mentioned seem to be compatible in facilitating and therefore enhancing interfacial bonding between Al matrix and SiC filler in the composites.

8. Conclusions

1. Fe_3O_4 and the $Mn_xFe_{3-x}O_4$ ($0 < x \leq 1$) nanoparticles have been successfully prepared by coprecipitation method utilizing a local natural source (viz. natural iron sand) as the main raw material.

2. The structures of Fe_3O_4 and $Mn_xFe_{3-x}O_4$ ($0 < x \leq 1$) nanoparticles, magnetic fluid, and magnetic hydrogel have intensively been studied by XRD, TEM, SANS, and SAXS data. These studies show that the magnetic materials have inverse spinel structure with a primary particle size of around 3–4 nm.

3. The magnetic properties have also been investigated by observing the hysteresis (M-H) behavior. It is found that Fe_3O_4 and $Mn_xFe_{3-x}O_4$ ($0 < x < 1$) nanoparticles exhibit superparamagnetic behavior at room temperature which depends on the particle size, doping content, and particle clustering.

4. The physical and mechanical properties of Al/MgO and Al/$MgAl_2O_4$ nanocomposites have been enhanced by the introduction of TMAH as surfactant during the mixing process of matrix and filler. The surfactant-coated MgO seems to be a more efficient filler compared to the surfactant-coated $MgAl_2O_4$ for the lower content (<30%). The experiment results show that the elastic modulus of the samples with volume fraction of fillers up to 30% was enhanced by more than 60% and 40%, respectively, for Al/MgO and Al/$MgAl_2O_4$. The use of other kind of surfactant needs to be explored to further enhance the mechanical properties.

5. A layer of the spinel $MgAl_2O_4$ micro-sized grown on the surface of SiC filler has induced the enhancement of mechanical properties of the Al/SiC composites.

Acknowledgements

This chapter is based on research partially supported by LPPM ITS, Ministry of Research and Technology, and Ministry of Education and Culture, Republic of Indonesia, 2006–2015.

Author details

Malik Anjelh Baqiya[1], Ahmad Taufiq[2], Sunaryono[2], Khuroti Ayun[1], Mochamad Zainuri[1], Suminar Pratapa[1], Triwikantoro[1] and Darminto[1*]

*Address all correspondence to: darminto@physics.its.ac.id

1 Advanced Materials Research Group, Department of Physics, Sepuluh Nopember Institute of Technology (ITS), Surabaya, Indonesia

2 Department of Physics, State University of Malang (UM), Malang, Indonesia

References

[1] D. Ghanbari, M. Salavati-Niasari, M. Ghasemi-Kooch, J. Ind. Eng. Chem. 20 (2014) 3970–3974.

[2] L.F. Cótica, V.F. Freitas, G.S. Dias, I.A. Santos, S.C. Vendrame, N.M. Khalil, R.M. Mainardes, M. Staruch, and M. Jain, J. Magn. Magn. Mater. 324 (2012) 559–563.

[3] F. Chen, S. Xie, X. Huang and X. Qiu, J. Hazard. Mater. 322 Part A (2017) 152–162.

[4] J. Sato, M. Kobayashi, H. Kato, T. Miyazaki, and M. Kakihana, J. Asian Ceram. Soc. 2 (2014) 258–262.

[5] T. Lu, J. Wang, J. Yin, A. Wang, X. Wang, and T. Zhang, Colloids Surf. A 436 (2013) 675–683.

[6] X. Zhang, D. Han, Z. Hua, and S. Yang, J. Alloys Compd. 684 (2016) 120–124.

[7] X. Cui, S. Belo, D. Krüger, Y. Yan, R.T.M. de Rosales, M. Jauregui-Osoro, H. Ye, S. Su, D. Mathe, N. Kovács, I. Horváth, M. Semjeni, K. Sunassee, K. Szigeti, M.A. Green, and P.J. Blower, Biomater. 35 (2014) 5840–5846.

[8] X. Zeng, B. Yang, X. Li, R. Li, and R. Yu, Mater. Des. 101 (2016) 35–43.

[9] C. Xingxing, Z. Xuebin, S. Hao, and F. Yi, Rare Met. Mater. Eng. 43 (2014) 2330–2334.

[10] C.-C. Lin and J.-M. Ho, Ceram. Int. 40 (2014) 10275–10282.

[11] I. Martínez-Mera, M.E. Espinosa-Pesqueira, R. Pérez-Hernández, and J. Arenas-Alatorre, Mater. Lett. 61 (2007) 4447–4451.

[12] A. Amirabadizadeh, H. Farsi, M. Dehghani, and H. Arabi, J. Supercond. Novel Magn. 25 (2012) 2763–2765.

[13] J.M. Niu and Z.G. Zheng, Adv. Mater. Res. 900 (2014) 172–176.

[14] H. Iida, K. Takayanagi, T. Nakanishi, and T. Osaka, J. Colloid Interface Sci. 314 (2007) 274–280.

[15] G. Gnanaprakash, S. Mahadevan, T. Jayakumar, P. Kalyanasundaram, J. Philip, and B. Raj, Mater. Chem. Phys. 103 (2007) 168–175.

[16] S. Lian, Z. Kang, E. Wang, M. Jiang, C. Hu, and L. Xu, Solid State Commun. 127 (2003) 605–608.

[17] D. Jin, G. Yongqian, C. Lanlan, Z. Guifu, L. Yue, and Q. Yitai, Nanotechnology 17 (2006) 4923.

[18] K. He, C.-Y. Xu, L. Zhen, and W.-Z. Shao, Mater. Lett. 61 (2007) 3159–3162.

[19] C. Scherer and A.M.F. Neto, Brazilian J. Phys. 35 (2005) 718–727.

[20] W. Yuan, C. Yin, P. Xiao, X. Wang, J. Sun, S. Huang, X. Chen, and Z. Cao, Microfluid. Nanofluid. 11 (2011) 781–785.

[21] C.L. Sansom, P. Jones, R.A. Dorey, C. Beck, A. Stanhope-Bosumpim, and J. Peterson, J. Magn. Magn. Mater. 335 (2013) 159–162.

[22] A. Labarta, J. Župan, and M.M. Renjo, 20th Int. Conf. Magn. ICM 2015, Physics Procedia, 75 (2015) 1458–1467.

[23] H. Ruican, L. Huagang, G. Wen, Z. Na, and H. Ruixiao, J. Magn. Magn. Mater. 416 (2016) 231–235.

[24] H. Ke, W. Huang, and X. Wang, Tribol. Int. 93, Part A (2016) 318–323.

[25] I. Sharifi, H. Shokrollahi, and S. Amiri, J. Magn. Magn. Mater. 324 (2012) 903–915.

[26] R.V. Ramanujan and L.L. Lao, Smart Mater. Struct. 15 (2006) 952.

[27] Y. Li, Z. Hu, and Y. Chen, J. Appl. Polym. Sci. 63 (1997) 1173–1178.

[28] M. Zrínyi, L. Barsi, and A. Büki, Polym. Gels and Networks, 5 (1997) 415–427.

[29] I. Ganesh, Int. Mater. Rev. 58 (2013) 63–112.

[30] I. Ganesh, B. Srinivas, R. Johnson, B.P. Saha, and Y.R. Mahajan, J. Eur. Ceram. Soc. 24 (2004) 201–207.

[31] H.B. Bafrooei and T. Ebadzadeh, Ceram. Int. 39 (2013) 8933–8940.

[32] P.Y. Lee, H. Suematsu, T. Yano, and K. Yatsui, J. Nanopart. Res. 8 (2006) 911–917.

[33] S. Tripathy and D. Bhattacharya, J. Asian Ceram. Soc. 1 (2013) 328–332.

[34] L.-Z. Pei, W.-Y. Yin, J.-F. Wang, J. Chen, C.-G. Fan, and Q.-F. Zhang, Mater. Res. 13 (2010) 339–343.

[35] S. Sanjabi and A. Obeydavi, J. Alloys Compd. 645 (2015) 535–540.

[36] M.F. Zawrah, H. Hamaad, and S. Meky, Ceram. Int. 33 (2007) 969–978.

[37] J. Rufner, D. Anderson, K. van Benthem, and R.H.R. Castro, J. Am. Ceram. Soc. 96 (2013) 2077–2085.

[38] H. Li, Y.Q. Liu, H. Liu, and Z.J. Yang, Mater. Res. Innovations, 19 (2015) S9-20-S29-23.

[39] A. Goldstein, J. Eur. Ceram. Soc. 32 (2012) 2869–2886.

[40] I. Ganesh, G. Jaganatha Reddy, G. Sundararajan, S.M. Olhero, P.M.C. Torres, and J.M.F. Ferreira, Ceram. Int. 36 (2010) 473–482.

[41] B. Ma, Y. Li, G. Liu, and D. Liang, Ceram. Int. 41 (2015) 3237–3244.

[42] S. Tripathy, D.S. Saini, and D. Bhattacharya, J. Asian Ceram. Soc. 4 (2016) 149–154.

[43] Z. Yu, N. Zhao, E. Liu, C. Shi, X. Du, and J. Wang, Composites Part A 43 (2012) 631–634.

[44] L. Xing, Y. Zhang, C. Shi, Y. Zhou, N. Zhao, E. Liu, and C. He, Mater. Sci. Eng., A 617 (2014) 235–242.

[45] Y. Zhou, Z. Yu, N. Zhao, C. Shi, E. Liu, X. Du, and C. He, Mater. Des. 46 (2013) 724–730.

[46] J.H. Boo, S.B. Lee, S.J. Ku, W. Koh, C. Kim, K.S. Yu, and Y. Kim, Appl. Surf. Sci. 169–170 (2001) 581–586.

[47] H. Jiang, Z. Cao, R. Yang, L. Yuan, H. Xiao, and J. Dong, Thin Solid Films, 539 (2013) 81–87.

[48] M.Y. Nassar, I.S. Ahmed, and I. Samir, Spectrochim. Acta, Part A 131 (2014) 329–334.

[49] P. Fu, W. Lu, W. Lei, K. Wu, Y. Xu, and J. Wu, Mater. Res. 16 (2013) 844–849.

[50] M.O. Bodunrin, K.K. Alaneme, and L.H. Chown, J. Mater. Res. Technol. 4 (2015) 434–445.

[51] E.T. Thostenson, C. Li, and T.-W. Chou, Compos. Sci. Technol. 65 (2005) 491–516.

[52] R.M. German, Liquid Phase Sintering, 1 ed., Springer US, New York, 1985.

[53] Sunaryono, A. Taufiq, Mashuri, S. Pratapa, M. Zainuri, Triwikantoro, and Darminto, Mater. Sci. Forum 827 (2015) 229-234.

[54] A. Taufiq, Sunaryono, E.G. Rachman Putra, A. Okazawa, I. Watanabe, N. Kojima, S. Pratapa, and Darminto, J. Supercond. Novel Magn. 28 (2015) 2855–2863.

[55] A. Taufiq, Sunaryono, E.G. Rachman Putra, S. Pratapa, and Darminto, Mater. Sci. Forum 827 (2015) 213–218.

[56] Sunaryono, A. Taufiq, Munaji, B. Indarto, Triwikantoro, M. Zainuri, and Darminto, AIP Conf. Proc. 1555 (2013) 53–56.

[57] Sunaryono, A. Taufiq, E.G.R. Putra, A. Okazawa, I. Watanabe, N. Kojima, S. Rugmai, S. Soontaranon, M. Zainuri, Triwikantoro, S. Pratapa, and Darminto, Nano, 11 (2016) 1650027.

[58] K.M. Wang, D.S. Lee, L. Horng, G. Chern, J. Magn. Magn. Mater. 282 (2004) 73–77.

[59] P.J. Reséndiz-Hernández, O.S. Rodríguez-Fernández, and L.A. García-Cerda, J. Magn. Magn. Mater. 320 (2008) e373–e376.

[60] E.G.R. Putra, Bharoto, E. Santoso, and A. Ikram, Nucl. Instrum. Methods Phys. Res., Sect. A 600 (2009) 198–202.

[61] F.L.O. Paula, R. Aquino, G.J. da Silva, J. Depeyrot, F.A. Tourinho, J.O. Fossum, and K.D. Knudsen, J. Appl. Crystallogr. 40 (2007) s269–s273.

[62] J. Teixeira, J. Appl. Crystallogr. 21 (1988) 781–785.

[63] M. Abboud, S. Youssef, J. Podlecki, R. Habchi, G. Germanos, and A. Foucaran, Mater. Sci. Semicond. Process. 39 (2015) 641–648.

[64] M. Gao, W. Li, J. Dong, Z. Zhang, and B. Yang, World J. Condens. Matter Phys. 1 (2011) 49–54.

[65] I. Volkov, S. Gudoshnikov, N. Usov, A. Volkov, M. Moskvina, A. Maresov, O. Snigirev, and S. Tanaka, J. Magn. Magn. Mater. 300 (2006) e294–e297.

[66] F.E. Osterloh, H. Hiramatsu, R.K. Dumas, and K. Liu, Langmuir, 21 (2005) 9709–9713.

[67] C. Daniela, C. Gabriel, and J.O.C. Charles, J. Phys. D: Appl. Phys. 40 (2007) 5801.

[68] P. Dutta, S. Pal, M.S. Seehra, N. Shah, and G.P. Huffman, J. Appl. Phys. 105 (2009) 07B501.

[69] M. Li, H. Gu, and C. Zhang, Nanoscale Res. Lett. 7 (2012) 1–11.

[70] S. Shen, S. Wang, R. Zheng, X. Zhu, X. Jiang, D. Fu, and W. Yang, Biomater. 39 (2015) 67–74.

[71] M. S. Seehra and A. Punnoose, Phys. Rev. B 64 (2001) 132410.

[72] J. Amighian, E. Karimzadeh, and M. Mozaffari, J. Magn. Magn. Mater. 332 (2013) 157–162.

[73] A.L. Patterson, Phys. Rev. 56 (1939) 978–982.

[74] Shipra, A. Gomathi, A. Sundaresan, and C.N.R. Rao, Solid State Commun. 142 (2007) 685–688.

[75] T.S. Duffy, C. Meade, Y. Fei, H.-K. Mao, and R.J. Hemley, Am. Mineral. 80 (1995) 222–230.

[76] L. Rooi Ping, A.-M. Azad, T. Wan Dung, Mater. Res. Bull. 36 (2001) 1417–1430.

[77] M. Zainuri, E.S. Siradj, D. Priadi, A. Zulfia, Darminto, MAKARA Sains, 12 (2008) 126–133.

Chapter 12

Photothermal Conversion Applications of the Transition Metal (Cu, Mn, Co, Cr, and Fe) Oxides with Spinel Structure

Pengjun Ma, Qingfen Geng and Gang Liu

Additional information is available at the end of the chapter

http://dx.doi.org/10.5772/67210

Abstract

The transition metal (Cu, Mn, Co, Cr, and Fe) oxides with spinel structure can be used as solar absorber materials due to their unique properties. Copper-based spinel ceramic pigments have been successfully prepared by sol-gel combustion method at low temperatures. Subsequently, spinel ceramic pigments have been employed to fabricate selective absorber paint coating by spraying-coating. The paint coating showed good spectral selectivity and thermal stability at low-to-mid temperature region. Spinel ceramic films have also been deposited on metal substrates by one dipping/annealing cycle. Spinel phase for single-layer ceramic film could be achieved at low annealing temperatures, and the single-layer ceramic film showed good spectral selectivity and benign thermal stability. Results presented here show that spinel compounds based on transition metal (Cu, Mn, Co, Cr, and Fe) oxides are promising materials for photothermal conversion applications.

Keywords: solar absorber coating, sol-gel combustion, spinel ceramic pigment, sol-gel dip-coating, spinel ceramic film, spectral selectivity

1. Introduction

Spectrally selective absorber coatings have attracted more attention because of the ability of its absorber to convert solar radiation into heat. The high photothermal conversion efficiency is usually dependent on selectivity absorption of the coating, which is required to have a high solar absorbance (α_s) in the solar wavelength range (0.3–2.5 μm) and low thermal emittance (ε_T) in infrared wavelength ranges (2.5–20 μm) [1]. During the last few decades, the transition metal (Cu, Mn, Co, Cr, and Fe) oxides with spinel structure have attracted significant attention due to their unique properties such as chemical inertness, high corrosion resistance, high mechanical strength, good thermal-shock resistance, and excellent optical and catalytic properties. This makes them very suitable for potential applications ranging from optics, electronics, magnetism, and catalysis to energy conversion and storage [2–4]. Furthermore, the existence of numerous spin-allowed electron transitions between partially filled d-orbitals contributes to their high absorption of radiation across the whole solar radiation spectrum [5]. These features endow these materials with promising application as solar selective absorber in solar-thermal energy conversion systems.

In recent years, several different approaches to the preparation of the spinel ceramic powders have been developed. For example, to prepare the spinel ceramic pigment, techniques of high temperature solid-state reaction, coprecipitation, sol-gel, hydrothermal synthesis, and thermal decomposition have been employed [6–8]. However, it is the major drawback of the above methods that the presence of heterogeneous products and requirement of high temperature and long durations result in tremendous wastage of energy. The sol-gel self-combustion method contributes to synthesize the better chemical homogeneity, small grain size, and high purity powder, which requires relatively low crystallization temperature [9]. In addition, crystallinity, size, and morphology of the particle, surface area, as well as other particular properties of the particles could be directed by tuning reaction parameters such as kinds of fuel, the mole ratio of fuel and oxidizer ratio, kinds and contents of the solvent, and the annealing temperature [10]. The as-prepared spinel ceramic powders are then utilized as solar-absorbing pigments to fabricate thickness-sensitive spectrally selective (TSSS) paint coatings by a convenient and economical spray-coating technology. Spray-coating technique is quick, easily adaptable to different coating solutions, and complex shapes can be coated. This makes it adoptable for an in-line process with minimum of material waste. These advantages with the spray-coating method suggest that this is the technique to prefer when scaling up the process.

Sol-gel techniques are promising synthesis methods for these spectral selectivity absorber coatings. The optical properties and durability of the spectral selectivity absorber coating can be easily controlled by fine-tuning relevant design parameters such as heating temperature or precursor concentrations in the synthesis process. In light of this, there are many knowledge gaps that need to be filled in the context of technicalities regarding the sol-gel processes and the optical and morphological characteristics of these coatings. The sol-gel processes are a soft chemistry technique where the precursors are generally in the form of a colloidal-based solution that eventually "transforms" into a widespread network of either discrete or continuously linked molecules [11]. Sol-gel techniques facilitate control of coating parameters such as absorber particle size, particle size distribution, homogeneity, chemical composition, and film thickness. The techniques also show good potential for scaling up to an industrial scale [12]. The most significant advantage of sol-gel over other established coating methods is its ability to tailor the microstructure of the deposited film at relatively low temperatures [13].

http://dx.doi.org/10.5772/67210

2. Experimental procedures

2.1. Synthesis of spinel ceramic pigments and spectrally selective paint coatings

Spinel ceramic pigments were synthesized by sol-gel self-combustion technique. Metal nitrate was first dissolved in an adequate amount of ultrapure water with the appropriate molar ratio. An appropriate amount of citric acid was then added into the prepared aqueous solution to chelate metal ion. After stirring for some time, the polyethylene glycol was added to the solution as an esterifying agent, which took part in chelation reaction. The mixture solution was adjusted to pH = 6.0–4.5 by slowly dropping ammonia and successively stirred to obtain a homogeneous solution. The prepared solution was subsequently heated for the adequate period of time to form the xerogel. Then, the xerogel was ignited in the atmosphere and burned in a self-combustion manner with rapid evolution of a large quantity of fume, yielding voluminous powders. Finally, the as-burned powders were annealed at different temperatures to obtain spinel ceramic pigments. Pigment dispersion was first prepared by mixing the pigments with the commercially corresponding binders and solvent in specific proportions and ground in a ball mill to form paint. Ultimately, the paint was sprayed on metal substrate to obtain paint coatings. A diagram for the sample preparation procedures is shown in **Figure 1**.

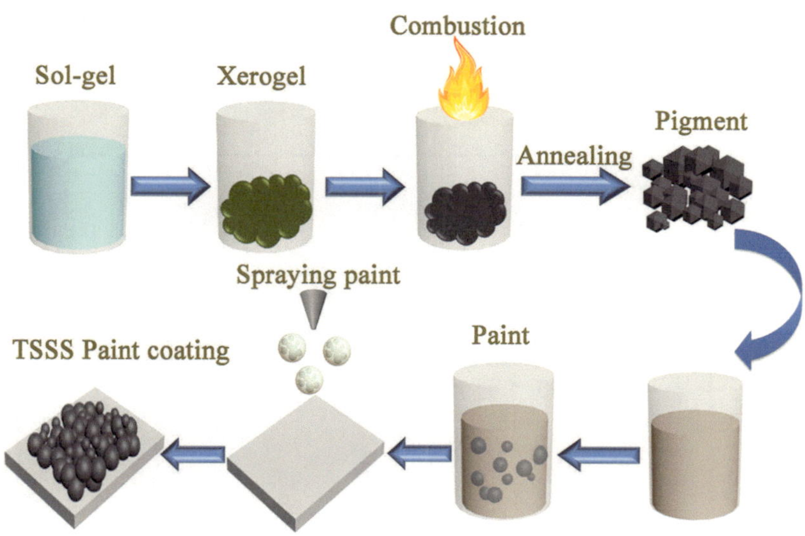

Figure 1. Spectrally selective paint coatings based on spinel ceramic pigments are fabricated [14].

2.2. Synthesis of spinel ceramic film for spectrally selective coatings

Metallic precursor sol was prepared by dissolving metal nitrates in absolute ethanol with suitable mole ratio of 1:1. This was followed by the addition of the citric acid as chelating agents. After stirring for a period of time, appropriate amount of polyethylene glycol was added under magnetic stirring. The resulting solution was then used for coating deposited on metallic substrates by soakage method and subsequently heated to obtain xerogel films. The films were then air annealed in an oven at different temperatures.

3. Results and discussion

3.1. Spectral selectivity paint coating based on $Cu_{1.5}Mn_{1.5}O_4$ spinel ceramic pigments

$Cu_{1.5}Mn_{1.5}O_4$ spinel ceramic pigments have been prepared by a facile and cost-effective sol-gel self-combustion method and annealed at the temperature ranging from 500 to 900°C for 1 h [14]. Ceramic pigments are utilized to fabricate the TSSS paint coatings by means of the convenient and practical spray-coating technique, and TSSS paint coatings based on pigments annealed at 700°C show solar absorbance of α_s = 0.914–0.923 and thermal emittance of ε_T = 0.244–0.357, which are calculated from the reflectance spectral shown in **Figure 2**. Furthermore, it is seen from reflectance spectral that the absorption edge is shifted toward longer wavelength, which means that the coating becomes thick and both of solar absorptance and thermal emittance are increased. As can be seen from **Figure 3**, the reflectance spectra show the marginal changes after the accelerated thermal test is carried out. Therefore, TSSS paint coating shows the thermal stability at the temperature of 227°C. Meanwhile, TSSS paint coatings exhibit no observable visual changes, and the performance criterion (PC) values reach the qualified requirement.

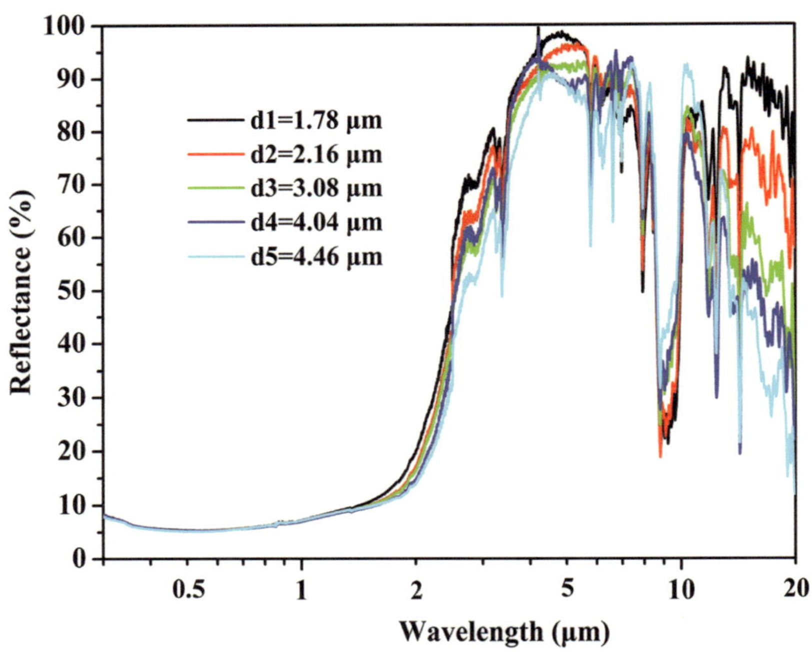

Figure 2. Reflectance spectra of paint coatings with different thickness based on spinel ceramic pigments [14].

http://dx.doi.org/10.5772/67210

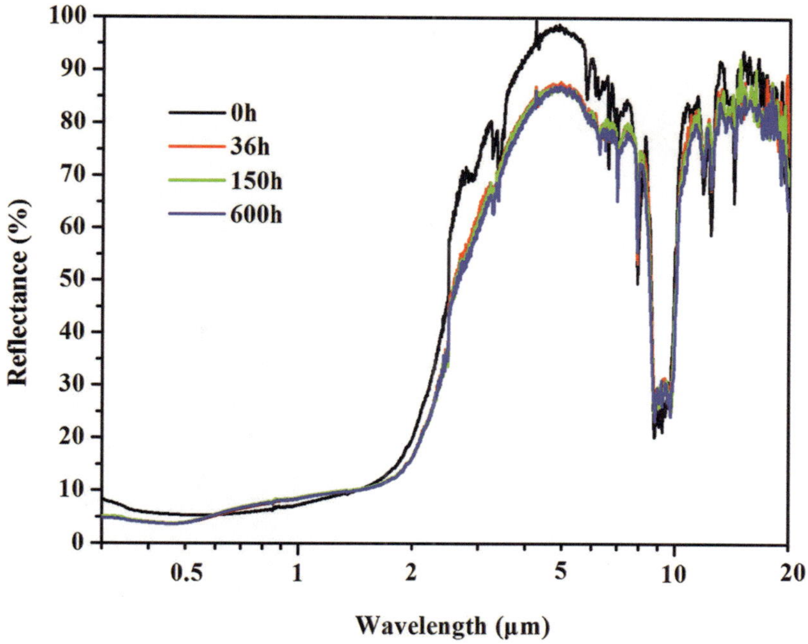

Figure 3. Reflectance spectra of paint coating subjected to thermal stability test [14].

3.2. Spectral selectivity paint coating based on $CuCr_2O_4$ spinel ceramic pigments

Single-phase $CuCr_2O_4$ spinel crystals are obtained after heat treatment of the as-burned powder at a low temperature (600°C) for 1 h, and the average crystallite size, morphology, and crystallinity of the $CuCr_2O_4$ are greatly influenced by the annealing temperature It can be seen from SEM images (**Figure 4**) that the as-burned powder has numerous voids and pores embedding into lamella. Increasing the annealing temperature, there are obvious flat face and clear edge appearing on the particles. Particles take on regular octahedron-shaped morphology and perfection of crystals at the annealing temperature of 800°C [15]. Comparison of TSSS paint coating based on metal oxide powder, the as-burned powder, and $CuCr_2O_4$ spinel ceramic pigments as solar absorber pigments shows that TSSS paint coatings based on the spinel ceramic pigment exhibit the relative high solar selective absorption. The SEM representative morphologies of the similar thickness TSSS paint coatings ((a) sample A3 based on CuO and Cr_2O_3, (b) sample B4 based on as-burned pigment, (c) sample C4 based on pigment annealed at 600°C, and (d) sample D2 based on pigments annealed at 800°C) are shown in **Figure 5**. It can be observed that the surface morphologies of all samples exhibit microscale bumps and protrusions, which are brought about particles agglomeration. For the TSSS paint coating, pigment particle agglomeration in the resin can cause uneven distribution and formation of clusters. The corresponding 3D surface profile images of samples are coincident with the surface morphologies shown by the SEM images. Water contact angles exhibited on sample surface can testify the different surface roughness (**Figure 6**). Furthermore, the TSSS paint coatings based on the spinel ceramic pigment show low surface roughness value and good hydrophobicity.

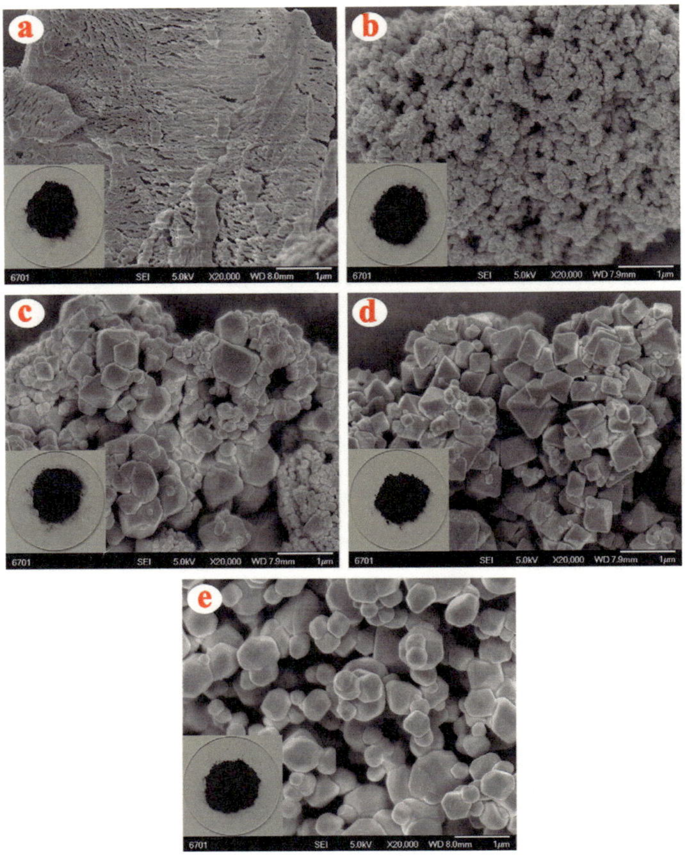

Figure 4. FE-SEM images and corresponding (inset) photographs of the powders: (a) the as-burned powder and powders annealed at (b) 600°C, (c) 700°C, (d) 800°C, and (e) 900°C for 1 h [15].

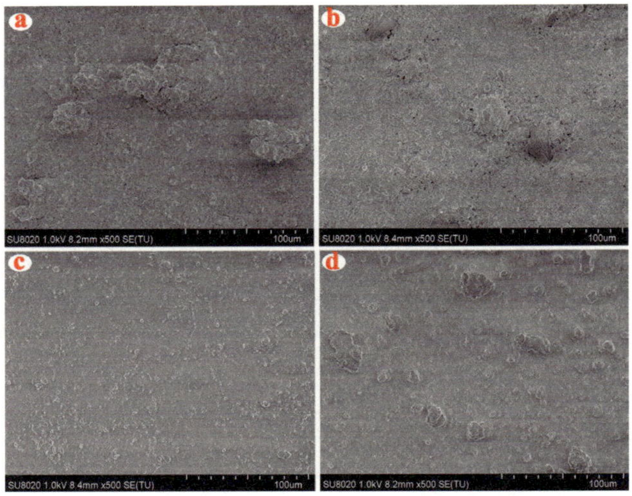

Figure 5. SEM images of paint coatings for (a) the sample A3, (b) the sample B4, (c) the sample C4, and (d) the sample D2 [15].

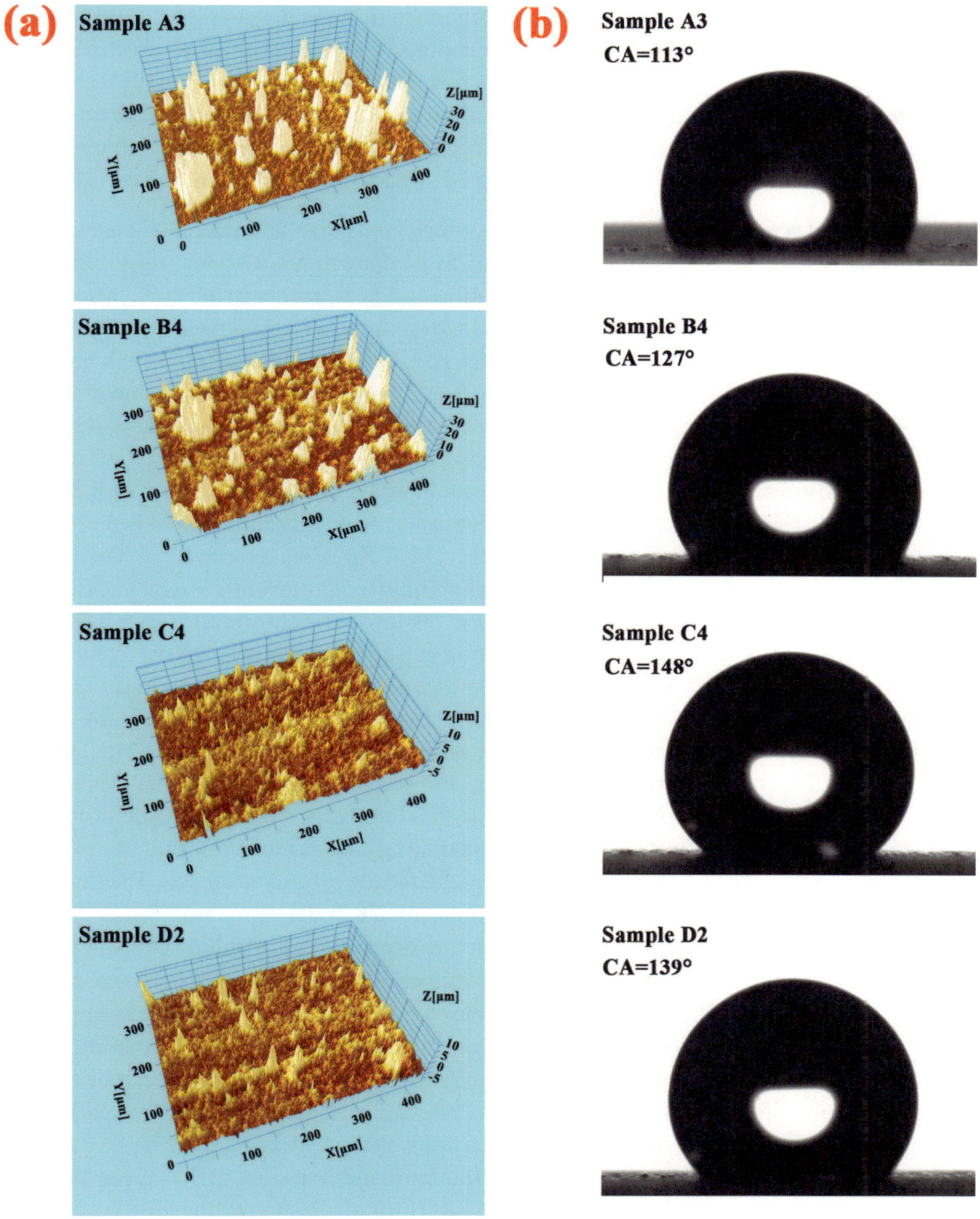

Figure 6. (a) 3D surface roughness profiles of samples and (b) images of water droplet on samples [15].

3.3. Spectral selectivity absorber coating based on $CuMnO_x$ spinel ceramic film

Spray-coating technique is quick, low-cost, easily adaptable to different coating solutions, and suitable for the establishment of a large-scale process, and there is a minimum of material waste. But as paint coatings are comparatively thicker and the organic binders also absorb in the thermal

IR range, these coatings usually suffer from the higher thermal emittance ($\varepsilon_{100} > 0.2$) [16]. Hence, the preparation and investigation of the spinel thin films by sol-gel route have attracted considerable attention. There is a great demand for low-cost and environmentally friendly techniques for synthesizing high-quality spectral selectivity absorber coatings. Such coatings are capable of absorbing most of the incoming solar radiation (high solar absorbance) without losing much of the thermal energy through reradiation from heated surface (low thermal emittance) [17].

The term spinel refers to a group of minerals, which crystallize in a cubic (isometric) crystal structure. Kaluza et al. [18] have succeeded in synthesizing $CuFeMnO_4$ black film spinel solar absorber coating using sol-gel dip-coating and heat treatment at 500°C. Mn-acetate tetrahydrate, Cu-chloride, and Fe-chloride hexahydrate precursors are used in a molar ratio of 3:3:1, respectively. To protect the spinel from corrosion, a 3-aminopropyltriethoxysilane (3-APTES) silica precursor is added to the Cu, Mn, and Fe sol precursors with a molar ratio of (Mn-Cu-Fe): silica = 1:1. Analytical results show that the films consist of two layers: the lower is amorphous SiO_2, and the upper is a spinel having the composition of $Cu_{1.4}Mn_{1.6}O_4$. The films exhibit solar absorbance values of around $\alpha_s = 0.6$ and thermal emittance values of $\varepsilon = 0.29–0.39$. Copper-cobalt oxide thin coatings have been deposited on highly infrared-reflecting aluminum substrate via a four-dipping/annealing-cycle sol-gel dip-coating route [19]. Nevertheless, the high annealing temperature and long annealing time would severely wreck the mechanical strength of aluminum substrate. Furthermore, the low solar absorptance ($\alpha_s = 0.834$) was merely obtained. He and Chen [5] added a complexing agent and an esterifying agent to fabricate the precursor sol, and thus $CuCoMnO_x$ coatings were deposited on aluminum substrate with a α_s value up to 0.93. Mahallawy et al. [17] also successfully synthesized $CuCoMnO_x$ coatings on aluminum and copper substrates by sol-gel dip-coating method. $CuMnO_x$ monolayer coating, $CuMnO_x/SiO_2$ two-layer coating, and $CuMnSiO_x/CuMnSiO_x/SiO_2$ three-layer coating have been fabricated [20], which provide good design strategies for ceramic spectral selectivity (CSS) coatings. $Cu_{1.5}Mn_{1.5}O_4$-based CSS coating is deposited on aluminum substrate using sol-gel dip-coating method from a stable metal nitrate precursor sol [21]. The Chelating agent citric acid, acting as a reducing agent simultaneously in the exothermic redox reaction, lowered the annealing temperature required by the formation of crystalline $Cu_{1.5}Mn_{1.5}O_4$. By optimizing the withdrawal rates and annealing temperatures, coating with optical parameter values as good as $\alpha_s = 0.876$ and $\varepsilon_{100} = 0.057$ is achieved after only one dipping/annealing cycle. Furthermore, the recycling experiment should be implemented to certify the reproducibility and stability of the metallic precursor sol. After reserved for 20 days, the precursor sol was deposited on aluminum substrate to obtain the CSS coating. **Figure 7** shows the typical x-ray diffraction (XRD) patterns and field emission scanning electron microscopy (FE-SEM) images of the CSS coating, which is deposited at 120 mm/min and annealed at 500°C for 1 h after recycling experiment. As can been seen from the XRD diffraction spectra, the diffraction peaks of the sample at 2θ values of 30.46°, 35.85°, 37.52°, and 57.82° correspond to (2 2 0), (3 1 1), (2 2 2), and (3 3 3) crystal planes of cubic spinel structure $Cu_{1.5}Mn_{1.5}O_4$. The morphology of $Cu_{1.5}Mn_{1.5}O_4$ coating indicates the presence of jagged and uneven pores, which can be attributed to the liberation of abundant H_2O, CO_2, O_2, NO_2, and other NO_x during the heat treating process. One interesting thing worthy to discuss is that the pores are conducive to enhance solar absorber of the CSS coatings. The pores look like light traps where the reflected light can be refracted consecutively and enters into the absorber layer again.

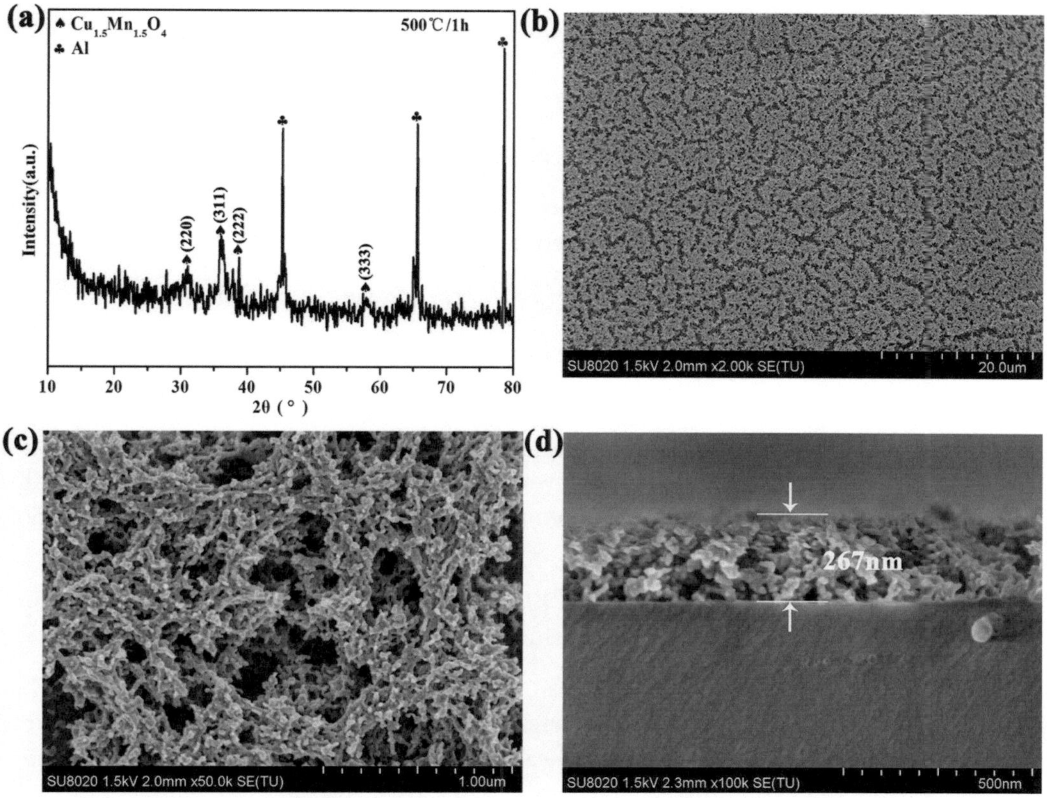

Figure 7. The XRD patterns and FE-SEM images of the solar selective absorber coating deposited on aluminum substrate after recycling experiment [21].

4. Conclusions

Black-colored transition metal oxides with spinel structure are easy to synthesize via sol-gel methods, and most of them show the high spectral selectivity and thermal stability. However, most of the coatings based on those materials are relatively lower spectral selectivity than the commercial absorber surfaces. More explorations on precursor's combinations, absorber stack configuration and compositions, as well as the application of superior antireflection layer are needed to improve their spectral selectivity. Furthermore, there are still problems associated with the reproducibility of the sol, and few studies have been done on this type of absorber coatings using sol-gel methods. Some problems such as the border effect and the heterogeneity of the film, which have a negative effect on the practical application of spinel films, arise when sol-gel dip-coating method is used for large-scale deposition. The use of highly soluble raw materials and the avoidance of compounds that easily settle in precursor's preparation are the robust ways to solve the reproducibility problem. Additionally, other pertinent factors such as the thickness of silica (especially if used as a matrix), abrasion, corrosion resistance, and the durability of the spinel absorber film should also be examined more extensively in future research.

Author details

Pengjun Ma[1], Qingfen Geng[2] and Gang Liu[1]*

*Address all correspondence to: gangliu@licp.cas.cn

1 Research & Development Center for Eco-material and Eco-chemistry, Lanzhou Institute of Chemical Physics, Chinese Academy of Sciences, Lanzhou, China

2 CEPREI Certification Body, China CEPREI Laboratory, Guangzhou, China

References

[1] N. Selvakumar, H.C. Barshilia, Review of physical vapor deposited (PVD) spectrally selective coatings for mid- and high-temperature solar thermal, Sol. Energy Mater. Sol. Cells, 2012, 98, 1–23.

[2] M. Rubat du Merac, I.E. Reimanis, H.J. Kleebe, Electrochemical impedance spectroscopy of transparent polycrystalline magnesium aluminate ($MgAl_2O_4$) spinel, J. Am. Ceram. Soc., 2015, 98, 2130–2138.

[3] H. Kim, D.H. Seo, H. Kim, I. Park, J. Hong, K.Y. Park, K. Kang, Multicomponent effects on the crystal structures and electrochemical properties of spinel-structured M_3O_4 (M = Fe, Mn, Co) anodes in lithium rechargeable batteries, Chem. Mater., 2012, 24,720–725.

[4] A.B. Rinkevich, A.V. Korolev, M.I. Samoylovich, S.M. Klescheva, D.V. Perov, Magnetic properties of nanocomposites based on opal matrices with embedded ferrite-spinel nanoparticles, J. Magn. Magn. Mater., 2016, 399, 216–220.

[5] M.Y. He, R.Y. Chen, Structural and optical properties of $CuMnCoO_x$ spinel thin films prepared by a citric acid-based sol-gel dip coating route for solar absorber applications, J. Sol-Gel Sci. Technol., 2015, 74, 528–536.

[6] R. Prasad, P. Singh, Applications and preparation methods of copper chromite catalysts: a review, Bull. Chem. React. Eng. Catal., 2011, 6, 63–113

[7] M. Edrissi, S.A. Hosseini, M. Soleymani, Synthesis and characterisation of copper chromite nanoparticles using coprecipitation method, Micro Nano Lett., 2011, 6, 836–839.

[8] M.H. Habibi, F. Fakhri, Sol-gel combustion synthesis and characterization of nanostructure copper chromite spinel, J. Therm. Anal. Calorim., 2014, 115, 1329–1333.

[9] O. Padmaraj, M. Venkateswarlu, N. Satyanarayana, Structural, electrical and dielectric properties of spinel type $MgAl_2O_4$ nanocrystalline ceramic particles synthesized by the gel-combustion method, Ceram. Int., 2015, 41, 3178–3185.

[10] M.Th. Makhlouf, B.M. Abu-Zied, T.H. Mansoure, Effect of fuel/oxidizer ratio and the calcination temperature on the preparation of microporous-nanostructured tricobalt tetraoxide, Adv. Powder Technol., 2014, 25, 560–566.

[11] T. Boström, G. Westin, E. Wäckelgård, Optimization of a solution-chemically derived solar absorbing spectrally selective surface, Sol. Energy Mater. Sol. Cells, 2007, 91, 38–43.

[12] N. Selvakumar, H.C. Barshilia, Developments in the synthesis of flat plate solar selective absorber materials via sol-gel methods: a review, Renew. Sust. Energ. Rev., 2014, 36, 316–328.

[13] C.J. Brinker, G.C. Frye, A.J. Hurd, C.S. Ashley, Fundamentals of sol-gel dip coating, Thin Solid Films, 1991, 201, 97–108.

[14] P.J. Ma, Q.F. Geng, X.H. Gao, S.R. Yang, G. Liu, Spectrally selective $Cu_{1.5}Mn_{1.5}O_4$ spinel ceramic pigments for solar thermal applications, RSC Adv., 2016, 39, 32947–32955.

[15] P.J. Ma, Q.F. Geng, X.H. Gao, S.R. Yang, G. Liu, $CuCr_2O_4$ spinel ceramic pigments synthesized by sol-gel self-combustion method for solar absorber coatings, J. Mater. Eng. Perform., 2016, 25, 2814–2823

[16] M. Lundh, T. Blom, E. Wäckelgård, Antireflection treatment of Thickness Sensitive Spectrally Selective (TSSS) paints for thermal solar absorbers, Sol. Energy, 2010, 84, 124–129.

[17] N.E. Mahallawy, M. Shoeib, Y. Ali, Application of $CuCoMnO_x$ coat by sol gel technique on aluminum and copper substrates for solar absorber application, J. Coat. Technol. Res., 2014, 11, 979–991.

[18] L. Kaluza, A. Šurca-Vuk, B. Orel, G. Dražič, P. Pelicon, Structural and IR spectroscopic analysis of sol-gel processed $CuFeMnO_4$ spinel and $CuFeMnO_4$/silica films for solar absorbers, J. Sol-Gel Sci. Technol., 2001, 20, 61–83.

[19] A. Amri, Z.T. Jiang, T. Pryor, C.Y. Yin, Z.H. Xie, N. Mondinos, Optical and mechanical characterization of novel cobalt-based metal oxide thin films synthesized using sol-gel dip-coating method, Surf. Coat. Technol., 2012, 207, 367–374.

[20] R. Bayón, G.S. Vicente, Á. Morales, Durability tests and up-scaling of selective absorbers based on copper-manganese oxide deposited by dip-coating, Sol. Energy Mater. Sol. Cells, 2010, 94, 998–1004.

[21] P.J. Ma, Q.F. Geng, X.H. Gao, S.R. Yang, G. Liu, $Cu_{1.5}Mn_{1.5}O_4$-based ceramic spectrally selective coatings for efficient solar absorber applications, J. Alloy. Compd., 2016, 675, 423–432.

Chapter 13

Anti-Corrosive Properties and Physical Resistance of Alkyd Resin–Based Coatings Containing Mg-Zn-Fe Spinels

Kateřina Nechvílová, Andrea Kalendová and Miroslav Kohl

Additional information is available at the end of the chapter

http://dx.doi.org/10.5772/65545

Abstract

The anti-corrosive and physical properties of organic coatings containing spinel pigments $Mg_{0.2}Zn_{0.8}Fe_2O_4$ wer studied. Pigments exhibiting different particle morphologies were synthesized by high-temperature solid phase reactions. Core-shell pigments containing Fe-Mg-Zn ferrite shells deposited on non-isometric particles of mineral cores consisting of layered silicates were also prepared. The pigments were used in paints, the pigment volume concentrations in the binder being 5, 10 and 15%. Anti-corrosive efficiency was investigated for paint films containing one of three ferro-spinel ($Mg_{0.2}Zn_{0.8}Fe_2O_4$)-based pigments or one of two core-shell pigments consisting of Fe-Mg-Zn shell and lamellar silicate-based cores. The paint properties were examined by accelerated corrosion tests and by physico-mechanical tests. The relationships between the pigment particle shape and the paint properties were examined. The effect of the pigment particle morphology on the mechanical properties of the paint films was also investigated. The dependence of the paint film properties on the pigment volume concentration was studied and the optimum concentrations providing the most efficient anti-corrosive protection were determined for each pigment.

Keywords: anti-corrosive pigment, organic coating, spinel, ferrite, core-shell pigment, accelerated corrosion test

1. Introduction

Metals are routinely protected against atmospheric corrosion by coating with pigmented paints possessing anti-corrosive properties [1]. Development of non-toxic anti-corrosive pigments for paints protecting the substrate efficiently against corrosion is a major trend in the science and technology of organic coating materials. The goal of research in this area is to develop non-toxic pigments possessing a high anti-corrosive efficiency with a view to replacing lead and chromium (VI)-based pigments that used to be applied in the past [2–4]. One direction in this area focuses on chemical or physical modification of known pigments, typically of the phosphate type [3, 5]. New inorganic pigments or organic corrosion inhibitor are also developed [6]. Among rather new anti-corrosive pigments are spinel-type pigments, finding application as thermally stable colour pigments and also as anti-corrosive pigments to protect the substrate against corrosion [7, 8]. Owing to their stable structure, spinel pigments exhibit a high resistance against high temperatures and chemical effects [9]. Mixed oxides in the spinel form, particularly ferrites containing Ca, Zn, Mg and their solid solutions, applied in paints already assume a firm position in anti-corrosion protection of metals [10, 11]. More efficient than the spinel-structure ferrite-type pigments with a combination of two cations in the lattice structure ($ZnFe_2O_4$, $MgFe_2O_4$) are second-generation spinel pigments with three cations ($Mg_{1-x}Zn_xFe_2O_4$ and $Ca_{1-x}Zn_xFe_2O_4$) [12]. Ferrite pigments are reactive particles on the surface in which there occur chemical reactions leading to the formation of less aggressive substances and a protective film during the diffusion of the corrosion substances through the polymeric film [13].

The protective properties of organic coatings are dependent, among other factors, on their impermeability to liquids and gases [14–16]. Protective film impermeability to liquids and gases can be improved by using non-isometric fillers [17, 18], e.g. micaceous iron oxide (specularite) [19–21]. **Figure 1** shows schematically the effect of non-isometric pigment particles on the physical properties of the paint films, consisting in an improvement of the mechanical properties of the films, reduction of the rate of diffusion through the films and protection of the binder against UV radiation [22, 23].

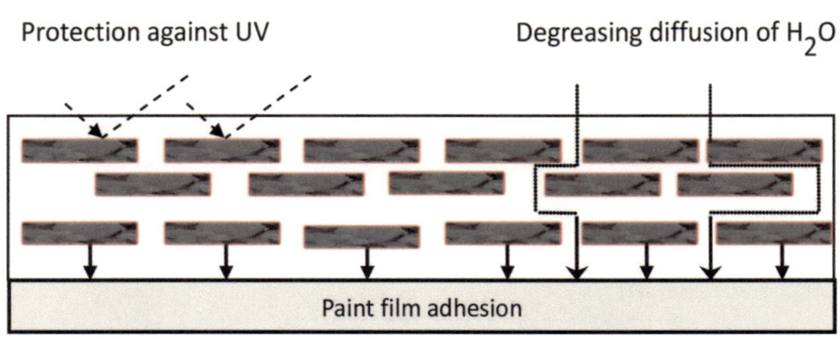

Figure 1. Pigment particles with a lamellar morphology in the paint film [29, 30].

Among pigments exhibiting non-isometric particle morphology are also some core-shell pigments [24]. The composition of both the core and the shell can be very diverse. Inorganic substances or metal powders are frequently used [25, 26]. Among assets of core-shell pigments is their economy, because raw materials from natural sources can often be used as the cores [27]. Control of the particle morphology, owing to which the product particles retain the shape of the initial core particles, is another reason why such pigments are used [28].

An ideal anti-corrosive pigment will not only possess good corrosion-inhibition properties but it will also have a favourable effect on the paint film's mechanical resistance [31, 32]. Among candidates that are studied in this context are pigments based on mixed oxides, specifically ferrites. The development of pigments whose primary particles possess a non-isometric, lamellar shape represents another direction in the development of anti-corrosion pigments. Such pigments, with their alkaline properties, can enhance both the barrier and inhibiting mechanisms of the protective paint system [33]. Examples include Zn, Ca, Mg ferrites with a lamellar, non-isometric particle shape which, apart from their anti-corrosion properties, improve the paint film's mechanical properties such as bending strength, elasticity and adhesion to the substrate [34]. The resulting ferrite spinels were found to retain particle morphology of the initial iron oxides [35]. The most efficient anti-corrosive ferrites include ferro-spinels containing zinc and magnesium in specific ratios [9, 36]. It has been found that ferrites with Mg^{2+} and Zn^{2+} as the divalent cations exhibit a very good anti-corrosive efficiency, especially if their molar ratio is 0.2:0.8 [37]. The studies are continued in this work, where one of the phenomena investigated was the effect of the particle size on the properties of the protective paint films based on a modified alkyd resin.

The aim of this study was to synthesize and investigate the properties of Mg-Zn-Fe ferro-spinel-based pigment possessing different particle morphologies with the goal to prepare such paints pigmented with them as they exhibit very good physical and anticorrosive properties. Core-shell pigments containing silicate cores and Mg-Zn-Fe shell were also prepared and studied with focus on their properties in protective coatings in a simulated corrosive environment. The particle morphology of such pigments is dictated by the core particle's shape and size, affecting the physical properties and barrier properties of the organic coating in corrosive environments. In order to get a deeper insight into the role of the pigment in the paint, the effect of the core material structure and chemical composition as well as the effect of the pigment concentration on the barrier capacity and the physical and anti-corrosive properties of the paint in the corrosive environment should be investigated [38, 39]. Layered silicates, specifically talc and calcined kaolin, exhibited differences in the composition, physical properties of their surface and particle texture, which may affect the pigment properties in the paints. Talc, $Mg_3(OH)_2$ (Si_4O_{10}), is a soft hydrophobic mineral (hardness 1 on the Mosh scale) with lamellar particles. Calcined kaolin obtained by calcination of the mineral kaolinite $Al_4(OH)_8$ (Si_4O_{10}) at 600°C contains morphologically diverse particles, largely lamellar, is harder than talc and contains more structural phases (largely mullite and quartz) [40–42]. All that may play a role in the overall anti-corrosive properties of the core-shell pigments [43, 44]. Core-shell pigments represent an economical solution for the synthesis of anti-corrosive pigments [27] obtained by the high-temperature reaction. Moreover, talc and kaolin are natural materials which are reasonably well available and are environmentally and toxicologically acceptable.

2. Experimental methods and procedures

2.1. Pigment synthesis

The pigments were formulated as the oxides $Mg_{0.2}Zn_{0.8}Fe_2O_4$ possessing a spinel structure. Efforts were made during the synthesis of the ferrite $Mg_{0.2}Zn_{0.8}Fe_2O_4$ ("pigment ***A***") to achieve (create) the spinel structure and isometric particles. Therefore, α-Fe_2O_3 (hematite), whose particles are regularly nodular, was used as the starting material. The aim of the synthesis of the next $Mg_{0.2}Zn_{0.8}Fe_2O_4$ ferrite ("pigment ***B***") was to obtain a with a non-isometric—specifically needle-shaped—particle morphology. Goethite α-FeO (OH), with needle-shaped particles, was used as the ferric oxide source. The third $Mg_{0.2}Zn_{0.8}Fe_2O_4$ ferrite ("pigment ***C***") was prepared by an identical procedure with the aim to prepare a non-isometric spinel pigment with a lamellar particle shape. Specularite, a ferric oxide Fe_2O_3 with a lamellar particle structure, was used as the starting material. The aim of the synthesis of the core-shell pigments with Mg-Zn-Fe shell was to coat the mineral core with a functional layer of mixed Mg-Zn-Fe oxides in the 0.2:0.8:2.0 ratios. Calcined Kaolin and talc $Mg_3(OH)_2(Si_4O_{10})$, i.e. materials exhibiting a lamellar particle shape, were used as the mineral carriers (cores). The core-shell pigments were synthesized by applying the matrix core (kaolin, talc)-to-mixed oxide (Mg-Zn-Fe) shell weight ratio 1:1.

The incoming materials were homogenized for 45 minutes, not only to obtain intimate mixtures but also to achieve mechanical-chemical activation of the materials, enhance contact of the particles in the powder mixture and increase the contact area for the reaction [4]. The high-temperature reaction was achieved by reaction mixture calcination in an electric furnace. The process was conducted as a two-stage procedure: the pigments were first calcined at 1000°C for 2 hours and then at 1180°C. The calcination temperatures were selected based on previous X-ray analysis of the products of tentative experiments. The resulting calcined material was multiply rinsed with distilled water. In order to obtain the right pigment particle size for use in paints, the calcinate was subjected to wet milling in a planetary ball mill (Pulverisette 6, Nietzsche, Germany). Accommodated in a milling vessel made of a zirconium silicate ceramics, the material was milled at 480 rpm for 420 minutes. The milling bodies were rollers made of the same ceramic material. The finely ground pigments were rinsed with water again and dried at 105°C in a laboratory-scale electric furnace. The core-shell pigments (or the mixed oxides forming the functional layer on the core) were also synthesized by high temperature solid-phase reaction, viz. by two-stage calcination like the non-core ferrites [43, 44].

Structural purity of the products was checked and their X-ray diffraction spectra were measured on a D8 Advance Diffractometer (Bruker AXS). The pigment particle surface and shape were examined on a JEOL-JSM 5600 LV scanning electron microscope (JEOL, Japan) [6].

2.2. Measuring the anticorrosion efficiency of the pigments in paints

In order to assess their potential anti-corrosion efficiency, the pigments were added to a solution of an alkyd resin modified with soy oil (density 1.1 g/cm^3, dry matter fraction 58.9%). The pigment volume concentrations (PVC) selected for the paints were 5, 10 and 15%. The

PVC/CPVC ratio (CPVC = critical PVC) was invariably adjusted to 0.35 by using limestone ($CaCO_3$) as an anti-corrosive-neutral filler. Cobalt octoate at a fraction of 0.3 wt.% was used as the siccative. The paints were prepared by dispersing the powdered pigments in the liquid binder in a Dispermat CV pearl mill. Test samples were prepared by coating test steel panels (deep-drawn cold-rolled steel manufactured by Q-panel, UK) 150 mm × 100 mm × 0.9 mm size with the paints by means of a box-type application ruler 200 μm slot width as per ISO 1514. The dry film thickness (DFT) was measured as per ISO 2808. Ten panels were prepared for each paint. A thin line 7 cm long and penetrating as deep as the substrate was cut into the paint film by using a sharp blade. Paint films were also prepared on polyethylene sheets, allowed to dry and peeled off and cut to pieces, 1 mm × 1 mm size. Such unsupported films were used to prepare 10% aqueous suspensions of the paint films in redistilled water [45].

The cyclic corrosion test in an environment with condensing water and SO_2 was performed as per CSN EN ISO 3231. This test consisted of 24-hour cycles comprising condensation of distilled water with SO_2 (0.2 mg/l) at 36°C for 8 hours followed by drying at 23 ± 2°C for 16 hours. The samples were evaluated after 1392 test hours. The corrosion tests were evaluated in accordance with the standards ASTM D 714-87, CSN ISO 2409, ASTM D 610-85 and ASTM D 1654-92. The following corrosion effects were assessed: formation (= size and frequency of occurrence) of blisters on the film surface and near the test cut made in the film; extent of substrate metal surface corrosion (corrosion-affected surface area fraction in %); and propagation (in mm) of corrosion in the vicinity of the test cut. Anti-corrosion efficiency on the 100–0 scale (100 = excellent anti-corrosion efficiency, 0 = poor anti-corrosion efficiency) was assigned to the corrosion effects in an environment with SO_2 [46].

2.2.1. *Determination of corrosion-induced steel mass loss in aqueous extracts of the pigments*

In this test, steel panels of defined size and known weight were submerged for 10 days in filtrates of 10 wt.% aqueous suspensions of the powdered pigments tested. The observed corrosion weight losses were converted to relative data (%) with respect to the values in water. The relative data obtained from the suspensions of the powdered pigments are denoted X_{corrp}. Relative corrosion losses (X_{corrf}) of steel panels exposed to 10% aqueous extracts of the loose pigmented paint films during 10 days were also measured [47, 48].

2.2.2. *Effect of the pigments on the physico-mechanical properties of the organic coatings*

The following tests were performed on the paint films: resistance to bending over a cylindrical mandrel; resistance to the cupping test (Erichsen test); resistance to a weight dropping on the reverse side of the test panels and adhesion (cross-cut test, or lattice test, knife spacing 1 mm). Degree of adhesion of the paints (ISO 2409) was performed by the cross-cut test. Impact resistance (ISO 6272) measured the maximum height of free drop of a weight (1000 g) at which the paint film still resisted damage. Resistance of the paint film against cupping (ISO 1520) was made in an Erichsen cupping tester. Resistance of the coating during bending over a cylindrical mandrel (ISO 1519) provides the largest diameter of the mandrel (in mm) causing disturbance of the paint film when the test panel is bent over it [49]. The results of the tests were used to calculate the overall physical-mechanical efficiency, i.e. overall paint film resistance to mechanical effects. A resistance score on the 100–0 scale (100 = excellent resistance, 0 = poor resistance) was assigned to each test result. The overall physical-mechanical resistance of the paints (M_E) was calculated as the arithmetic mean of the cohesion score from the bending test on the cylindrical mandrel, the resilience score from the impact test, the degree of adhesion score and the strength score from the cupping [50].

The starting hematite was also examined (pigment ***F***). The non-pigmented organic coating material was subjected to the mechanical resistance tests and corrosion tests as well.

3. Results and discussion

3.1. Pigment morphology and structure

Morphology of the pigment particles synthesized is shown in the SEM photographs in **Figure 2**. The aim of the synthesis of the pigments ***A***, ***B*** and ***C*** was to obtain pigments possessing the spinel structure whose chemical formula is $Mg_{0.2}Zn_{0.8}Fe_2O_4$. Mixed spinel ferrites containing zinc and magnesium cations constituted the majority phases. The magnesium atoms can be iso-morphically substituted by zinc atoms, owing to which the mixed spinels can be formed across an unlimited range of concentrations. The magnesium-to-zinc cation ratio could not be determined precisely from the analysis because the lattice parameters in $MgFe_2O_4$ and $ZnFe_2O_4$ approach each other and the diffraction lines may overlap. The X-ray diffraction analysis of the pigments ***A***, ***B*** and ***C*** did not exhibit any crystal phases of the starting materials; in other words, the starting materials had completely reacted to the products. These results are consistent with [37]. Furthermore, the three spinel types were synthesized with a view to obtaining pigments $Mg_{0.2}Zn_{0.8}Fe_2O_4$ possessing different particle shapes, making use of the different particle shapes of the starting ferric oxide types: the particles of pigment ***A*** were isometric, the particles of pigment ***B*** were needle-shaped (acicular) and the particles of pigment ***C*** were lamellar. The three pigments are also referred to as isometric $Mg_{0.2}Zn_{0.8}Fe_2O$, acicular $Mg_{0.2}Zn_{0.8}Fe_2O_4$ and lamellar $Mg_{0.2}Zn_{0.8}Fe_2O_4$, respectively, throughout this text.

The core-shell pigment ***D*** ($Mg_{0.2}Zn_{0.8}Fe_2O_4$/*kaolin*) was prepared as a pigment whose non-isometric (lamellar) core particles of calcined kaolin are coated with a functional shell of the mixed oxide $Mg_{0.2}Zn_{0.8}Fe_2O_4$. For this, the starting materials for the shell were mixed in proportions providing the Mg-Zn-Fe cation ratio 0.2:0.8:2. Mixed oxides of aluminium, iron and magnesium were identified (the mixed oxide hercynite). The analysis gave evidence that chemical reactions had occurred between the kaolin surface and the remaining starting materials: indeed, the mixed oxide hercynite ($FeMg_3Al_2O_4$) contains the aluminium cation, which initially was present in the starting kaolin. In other words, a chemically bonded functional layer of mixed oxides had formed on the kaolin particles. The X-ray diffraction analysis showed that the pigment also included silicates ($MgSiO_3$) and crystalline silicon oxide (SiO_2) phases (cristobalite) from the starting kaolin. The resulting pigment ***D*** (simplified referred to as $Mg_{0.2}Zn_{0.8}Fe_2O_4$/*kaolin* or $Mg_{0.2}Zn_{0.8}Fe_2O_4/Al_6Si_2O_{13}$) exhibited a lamellar particle shape.

Figure 2. Morphology of the particles of tested pigments (SEM): (a) $Mg_{0.2}Zn_{0.8}Fe_2O_4$ isometric; (b) $Mg_{0.2}Zn_{0.8}Fe_2O_4$ acicular; (c) $Mg_{0.2}Zn_{0.8}Fe_2O_4$ lamellar; (d) mixed oxide Mg-Zn-Fe/kaoline; (e) mixed oxide Mg-Zn-Fe/talc.

Pigment *E* was synthesized with a view to obtaining a core-shell pigment where the mixed Mg-Zn-Fe oxide shell covers a mineral core possessing a lamellar particle structure, specifically talc $(Mg_3(OH)_2Si_4O_{10})$. The highest diffraction line belonged to the crystalline ferrite phase. As in the remaining pigments, isomorphic replacement of the zinc cation by a magnesium cation had occurred in this ferrite layer and so their precise ratio could not be determined. Components of minor importance in this pigment included magnetite (Fe_3O_4) from the starting hematite and the silicon oxide cristobalite from talc in the core. The resulting pigment *E* (simply referred to as $Mg_{0.2}Zn_{0.8}Fe_2O_4/talc$ or $Mg_{0.2}Zn_{0.8}Fe_2O_4/Mg_3(Si_4O_{10})(OH)_2$) exhibited a lamellar particle shape.

3.2. Steel panel mass loss in aqueous extracts of the powdered pigments and of the loose pigmented films, and pH and conductivity of the pigmented aqueous extracts

Table 1 contains the calculated relative corrosion losses of the steel panels in aqueous extracts of the powdered pigments (X_{corrp}) and in aqueous extracts of the paint films (X_{corrf}), and pH values (pH_f) and specific electric conductivities (χ_f) of aqueous extracts of the loose paint films.

Pigment	Corrosion losses*		pH[a] pH_f (-)	Specific conductivity[b] χ (μS/cm)
	X_{corrf} (%)	$X_{corrp.}$ (%)	pH_{28}	χ_{28}
A ($Mg_{0.2}Zn_{0.8}Fe_2O_4$) isometric	19.66	78.82	6.2 ± 0.01	190
B ($Mg_{0.2}Zn_{0.8}Fe_2O_4$) acicular	15.99	76.76	6.90 ± 0.01	195
C ($Mg_{0.2}Zn_{0.8}Fe_2O_4$) lamellar	4.29	33.31	7.90 ± 0.01	199
D (Mg-Zn-Fe) mixed oxide/kaolin	9.40	36.09	6.10 ± 0.01	200
E (Mg-Zn-Fe) mixed oxide/talc	9.98	43.93	6.70 ± 0.01	180
F (hematite) Fe_2O_3	28.15	89.87	5.10 ± 0.01	170
Water	-	100	-	-
Non-pigmented paint film	100	-	3.04 ± 0.01	216

*Parameters are given as arithmetic averages within three measured values.
[a]pH was measured with an accuracy of ±0.01.
[b]Conductivity was measured with an accuracy of ±0.5%.

Table 1. Relative steel mass losses due to corrosion in aqueous extracts of the powdered pigments (X_{corrp}) and in aqueous extracts of the loose paint films ($X_{corrf,}$) and pH and specific electric conductivity values (pH_f, χ_f) of the aqueous extracts of the loose paint films (PVC = 10%).

The corrosion loss data (X_{corp}) characterize the pigment's ability to affect the resistance of the metal to corrosion in the pigment extract, where ions passivating the metal surface are present to a larger or lesser extent. The data demonstrate that the corrosion phenomena are partly inhibited by the presence of the pigments [18]. Low steel losses by corrosion were observed with the core-shell pigments ***D*** (*mixed Mg-Zn-Fe oxide/kaolin*), and ***E*** (mixed *Mg-Zn-Fe oxide/talc*) and with the lamellar ferrite ***C*** (lamellar $Mg_{0.2}Zn_{0.8}Fe_2O_4$), viz. 36, 39 and 44%, respectively. Those (comparable) corrosion losses were related to the extracts' specific conductivities or pH levels [29]. Lower corrosion losses were observed with the core-shell pigments, where the silicate core and the silicate phases contributed favourably to corrosion protection of the steel panels [44].

The pigments also affected the pH and specific conductivity levels of the paint films pigmented with them. The alkaline nature of the pigments induced changes in the pH of the pigmented films compared to the non-pigmented film. The pigments shifted the pH values upwards due to the neutralization of the acid components of the binder [5, 51]. This effect was most pronounced for pigment ***C*** (lamellar $Mg_{0.2}Zn_{0.8}Fe_2O_4$), where pH_f was 7.9 (**Table 1**). The core-shell pigment ***E*** (mixed *Mg-Zn-Fe oxide/talc*) was a next pigment imparting an alkaline nature to the paint. The effects were reasonable, the pigments themselves inducing alkalinity in their aqueous extracts. Most acid ($pH_f = 3.0$) was the aqueous extract of the non-pigmented film. The conductivities of the aqueous extracts of the loose paint films were also affected by the pigments. The conductivity of the extract of the non-pigmented film was $\chi_f = 216$ μS/cm. The conductivity was lower if a pigment was present. This was due to neutralization of the binder's acid components (R-COOH) giving rise to metallic soaps, where dissociation was suppressed [52]. This effect was marked, once again, with pigment ***C*** (lamellar $Mg_{0.2}Zn_{0.8}Fe_2O_4$) and, to a lesser extent, with the core-shell pigment ***E*** (*mixed Mg-Zn-Fe oxide/talc*) [43, 44].

The anti-corrosive efficiency of a paint film depends on the pigment's ability to release inhibiting components that are involved in the reactions inside the film and affect the diffusing environment. The corrosion losses in the suspensions of the organic films (X_{corrf}) containing anti-corrosive pigments provide information on the potential reactions occurring inside the paint in the liquid state, during the film formation and, to some extent, during the ageing of the cured film [48, 53]. The binder was an alkyd resin, a polyester modified by a fatty acid, which is well suited to the indirect examination of the behaviour of the ferrite and of the paint film. The carboxy groups in the binder were responsible for the acid nature of the aqueous extract of the film: the observed value was pH 3.1 when no pigment was present. However, if a pigment was present, the pH levels were largely as high as pH 6–8, due to the formation of metallic soaps inside the film [54]. Alkaline anti-corrosive pigments neutralized the acid groups in the binder [55]. This effect is basically similar to the inhibiting effect of minimum. Metallic soaps may also exhibit inhibiting properties at the protected metal–organic coating interface [6]. The alkyd binder contains both acid -COOH groups and an alkaline $Zn_{0.8}Mg_{0.2}Fe_2O_4$ pigment, as shown schematically in **Figure 3** [56]. Neutralization of the carboxy groups inside the film results in more or less neutral products: neutral metallic soaps M^{n+} ($^-OOCR)n$ acidic metallic soaps M^{n+} (OOCR)n/RCOOH, and alkaline metallic soaps M^{n+} $(O^{2-})x(OH^-)y(^-OOCR)z$ $(x/2 + y + z = n)$. In the formulas, R is a hydrocarbon and M is a metal (e.g. Zn, Mg) whose oxidation state [36]. The acid components of the binder that were released into the aqueous medium may also be neutralized by the alkaline extract of the anti-corrosive pigment [54, 57]. The low corrosion losses in the extracts of the loose paint films corroborated the results obtained with the extracts of the powdered pigments. The most marked active chemical protection was observed with the core-shell pigments ***D*** (*mixed Mg-Zn-Fe oxide/kaolin*) (X_{corrf} = 36.09%) and **C** (*lamellar* $Mg_{0.2}Zn_{0.8}Fe_2O_4$) (X_{corrf} = 39.31%).

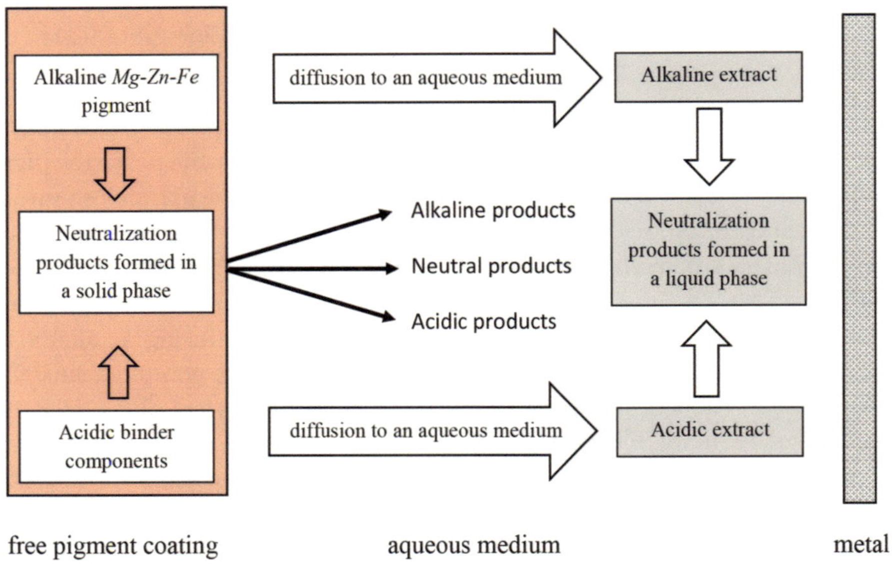

Figure 3. Schematic illustration of the processes running on extraction of alkyd coatings containing Mg-Zn-Fe pigment in an aqueous medium [54].

3.3. Mechanical resistance tests of the paint films containing the pigments

Outstanding mechanical properties constitute a precondition for good anti-corrosion efficiency of a paint film. **Table 2** presents the results of mechanical tests of the paints' films containing the pigments. The non-pigmented binder and a paint containing the starting ferric pigment ***F*** (hematite Fe_2O_3) were also measured as reference materials. Resistance in the dropping weight test was 100 cm for all the paint films. Similarly, resistance to the bending test was invariably less than 4 cm. Since the values were identical for all the pigments, they are not included in **Table 2**. The pigmented paints films exhibited very good resistance against mechanical stress at any PVC. The paint film was never disturbed when bent over the mandrel 4 mm in diameter. No defects were observed on the films after the test hammer was dropped from the largest height, i.e. 100 cm. Good results were also obtained for all of the pigments in the cross-cut test, any defects were negligible and nearly all of the films scored 1 in this test. The only pigment for which defects were observed on the lattice (particularly at higher PVC levels) was the core-shell pigment ***D*** (*mixed Mg-Zn-Fe oxide/kaolin*), and so the films scored 2. The films gave very good results also in the cupping test, no film disturbance was observed with a magnifying

http://dx.doi.org/10.5772/65545

glass after the body penetration to 8 mm distance. The paint film resistance in this test was PVC-dependent, viz. so that the resistance against penetration decreased slightly with increasing PVC. Pigment ***A*** (*isometric* $Mg_{0.2}Zn_{0.8}Fe_2O_4$,) was an exception in this respect. **Table 3** also demonstrates that the highest resistance against the test body penetration was obtained with pigment ***B*** having needle-shaped particles (*acicular* $Mg_{0.2}Zn_{0.8}Fe_2O_4$), viz. 9.5 mm at PVC = 5%. Good results were also obtained with the core-shell pigment ***E*** (*mixed Mg-Zn-Fe oxide/talc*), where the film was disturbed at 9.4 mm.

Pigment	**Cupping test* ISO 1520 (mm)**	**Cross-cut test* ISO 2409 (-)**	**Overall physico-mechanical resistance M_E**
A	8.9	1	95
$Mg_{0.2}Zn_{0.8}Fe_2O_4$ isometric	8.8	1	95
	9.0	1	95
B	9.5	1	95
$Mg_{0.2}Zn_{0.8}Fe_2O_4$ acicular	9.2	1	95
	8.8	1	95
C	9.0	1	95
$Mg_{0.2}Zn_{0.8}Fe_2O_4$ lamellar	8.8	1	95
	8.5	1	95
D	9.3	1	95
Mg-Zn-Fe mixed oxide/kaolin	9.2	2	91
	8.8	2	91
E	9.5	1	95
Mg-Zn-Fe mixed oxide/talc	9.0	1	95
	8.8	1	95
F	8.7	1	95
hematite Fe_2O_3	8.9	1	95
	8.8	1	95
Non-pigmented paint film	8.8	1	95

*Parameters are given as arithmetic averages within three measured values.

Table 2. Results of mechanical tests of the paints containing the pigments synthesized and the reference pigment and of the reference paint (dry film thickness DFT = (60 ± 10) μm.

Nearly all of the paint films exhibited very good overall physico-mechanical resistance, their score was 95. Only the paints with the core-shell pigment ***D*** (*mixed Mg-Zn-Fe oxide/kaolin*) provided poorer cross-cut test at high PVCs and scored 91 only. The results give evidence that owing to their morphological properties, the pigments tested do not detract from the mechanical resistance of the alkyd resin–based paints. These results are very important for practical applications of these pigments in paints.

Pigment	PVC (%)	Paint film: Degree of blistering ASTM D 714-87 – Around the cut	Paint film: Degree of blistering ASTM D 714-87 – On the film surface	Steel panel: Corrosion in the cut (mm) ASTMD 1654-92	Steel panel: Metal surface corrosion (%) ASTMD 610-85	Calculated anti-corrosive efficiency E_{SO_2}
A	5	2F	-	2.0–3.0	3	73
$Mg_{0.2}Zn_{0.8}Fe_2O_4$ isometric	10	4M	6F	2.0–3.0	33	49
	15	4MD	6F	1.0–2.0	50	42
B	5	4F	-	0.5–1.0	3	79
$Mg_{0.2}Zn_{0.8}Fe_2O_4$ acicular	10	4MD	-	0.5–1.0	1	71
	15	4MD	6F	0.5–1.0	16	54
C	5	6M	-	0.5–1.0	16	68
$Mg_{0.2}Zn_{0.8}Fe_2O_4$ lamellar	10	6F	-	0.5–1.0	16	68
	15	2F	2F	1.0–2.0	33	45
D	5	6D	8M	0.5–1.0	33	48
Mg-Zn-Fe mixed oxide/kaolin	10	6D	8M	0.5–1.0	>50	36
	15	4D	8MD	0.5–1.0	>50	30
E	5	8D	8M	1.0–2.0	>50	35
Mg-Zn-Fe mixed oxide/talc	10	8MD	8M	0.5–1.0	>50	43
	15	6MD	8M	1.0–2.0	>50	39
F	5	4M	8M	1.0–2.0	10	58
hematite Fe_2O_3	10	6MD	8D	1.0–2.0	>50	29
	15	6D	8D	0.5–1.0	>50	26
Non-pigmented paint film		4MD	2F	1.0–2.0	33	45

Table 3. Corrosion resistance of the paints with the pigments synthesized and with the reference pigment and of the reference paint, measured after 912 hours exposure of the steel panels coated with the paints in an atmosphere with SO_2, DFT = (85 ± 10) μm.

3.4. Corrosion tests of the pigmented paints

Table 3 lists the results of corrosion resistance measurements of the paint films in the atmosphere with condensed moisture and SO_2, obtained after 912 hours of exposure. Nearly all of the paint films had osmotic blisters on their surface after the exposure. The paint films with pigment ***C*** (*lamellar* $Mg_{0.2}Zn_{0.8}Fe_2O_4$) were most resistant in this respect, especially at PVC =10%, where no blisters were found on the film surface and a few small blisters only (6F) were observed around the test cut. The results observed when the pigment concentration was PVC

http://dx.doi.org/10.5772/65545

= 5% were also very good (6M). Equally good results were obtained with the paint containing pigment **B** (*acicular* $Mg_{0.2}Zn_{0.8}Fe_2O_4$). This was true particularly at PVC = 5%, whereas a higher occurrence of blisters was observed at higher concentrations. So, the blisters observed around the test cut at PVC = 10% were categorized as 4MD, as against the 4F score at PVC = 5%.

Similar behaviour was observed for the paint containing pigment **A** (*isometric* $Mg_{0.2}Zn_{0.8}$). Here, too, the resistance against blistering in the cut was poorer at higher concentrations: the scores at PVC = 5, 10 and 15% were 2F, 4M and 4MD, respectively.

The occurrence of blisters was more pronounced for the remaining paints. So, when the paint with the core-shell pigment **E** (*mixed Mg-Zn-Fe oxide/talc*) was used, the paint film surface was speckled with very small blisters (frequency of occurrence 8M). This applies to the entire PVC range. The effects were very similar with the core-shell pigment **D** (*mixed Mg-Zn-Fe oxide/ kaolin*), where, in addition, more abundant (score 6D) blisters were found near the test cut. The poorest results were obtained by using the paint pigmented with the reference pigment **F** (hematite Fe_2O_3) at PVC =10% and at PVC = 15%, where the paint film surface was densely covered with very small blisters (score 8D). The differences in the extent of blistering on the various paint films are apparent from **Figure 4**.

a) b) c)

Figure 4. Photographs of the steel panels coated with paints containing: (a) pigment **F** (hematite Fe_2O_3) at PVC = 10%, (left); (b) pigment **B** (*acicular* $Mg_{0.2}Zn_{0.8}Fe_2O_4$) at PVC = 5% (middle); and (c) pigment **C** (lamellar $Mg_{0.2}Zn_{0.8}Fe_2O_4$) at PVC = 10% (right).

The paint films were removed from the steel panels in order to examine corrosion on the metal surface (beneath the initially present films) and corrosion propagation from the test cut. Corrosion in the test cut was similar for all of the paints. Appreciable corrosion propagation from the cut, reaching a 3 mm distance, was only observed for the paint with the isometric pigment **A** (*isometric* $Mg_{0.2}Zn_{0.8}Fe_2O_4$) at PVC = 5% and PVC = 10%, followed by the paint with the isometric reference pigment **F** (1.0–2.0 mm). The degree of corrosion was higher in this atmosphere than in the atmosphere with condensed moisture for nearly all of the paints. The best protection was provided by the paint with pigment **B** (*acicular* $Mg_{0.2}Zn_{0.8}Fe_2O_4$) at PVC = 10% and at PVC= 5%, the affected surface fraction being 1 and 3%, respectively. A comparable protection was provided by the paint with pigment *A* (*isometric* $Mg_{0.2}Zn_{0.8}Fe_2O_4$) at PVC = 5% (corroded surface fraction 3%). This pigment exhibited a sharp drop of its protective capacity against corrosion beneath the film with increasing PVC: the affected fraction was as high as 50% at PVC=10%.

A comparison of the steel panel corrosion beneath selected paint films is presented in **Figure 5**. The photograph on the right shows the steel panel initially protected by the paint with pigment ***B*** (*acicular* $Mg_{0.2}Zn_{0.8}Fe_2O_4$) at PVC = 5%. The other two photographs show panels with appreciable degree of corrosion beneath the paint films.

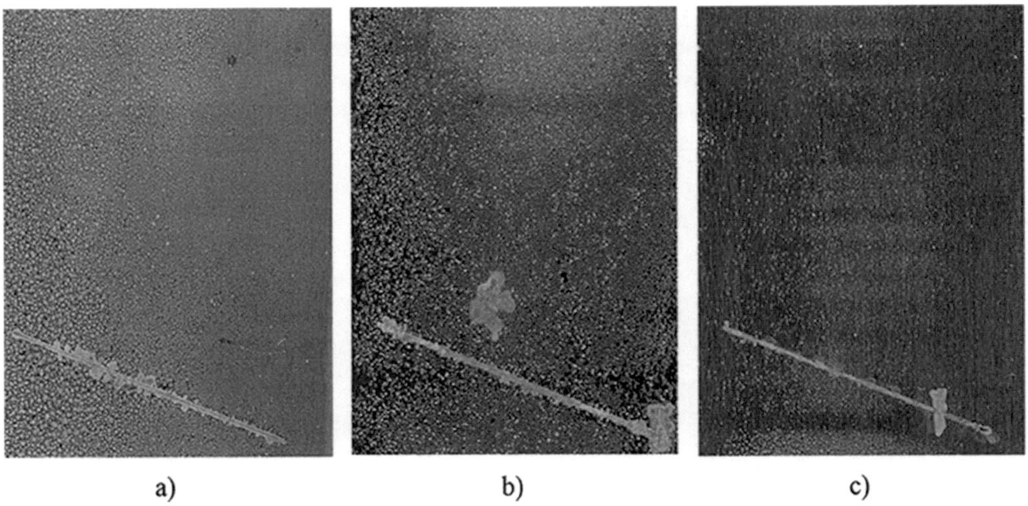

Figure 5. Photographs of the steel panels after removing the paint films: (a) pigment ***E*** (mixed *Mg-Zn-Fe oxide/talc*) at PVC = 15% (left); (b) pigment ***D*** (mixed *Mg-Zn-Fe oxide/kaolin*) at PVC = 15% (middle); (c) pigment ***B*** (acicular $Mg_{0.2}Zn_{0.8}Fe_2O_4$) at PVC = 5% (right).

The paint that was found most efficient in protection against corrosion in the environment with condensed moisture and SO_2 was that with the non-isometric pigment ***B*** (*acicular* $Mg_{0.2}Zn_{0.8}Fe_2O_4$). Paint films with this pigment were well resistant to blistering and efficiently protected the metal against corrosion. Favourably, this pigment is alkaline and can neutralize the strong acid aqueous medium acting on the paint film in the environment of the SO_2 chamber whose acidity is about pH 2. In the optimum arrangement, the needle-shaped pigment particles enhance cohesion of the paint and slow down diffusion of substances dissolved in the environment with condensing water [55]. The pigment is capable of forming soaps by reacting with the fatty acids (RCOOH) in the binder. Good resistance against blistering was also provided by the paint with pigment ***C*** (*lamellar* $Mg_{0.2}Zn_{0.8}Fe_2O_4$). At PVC = 15%, however, the protective capacity of this paint was as poor as that of the non-pigmented film—the affected metal surface fraction was 33%. In the optimum arrangement, particularly at low PVCs, the lamellar pigment particles enhance paint adhesion to the metal and it is more difficult for the corrosive medium to diffuse through the paint film. Similar results were obtained with the paint containing pigment ***A*** (*isometric* $Mg_{0.2}Zn_{0.8}Fe_2O_4$).

The following conclusion can be drawn from the observations in this corrosive environment:

- The non-isometric pigments exhibited the highest degree of protection against the formation of osmotic blisters on the paint film surface—higher than the isometric pigments.
- The isometric pigments (***A*** and ***F***) exhibit a higher ability to prevent corrosion of the metal at low PVCs. This may be associated with the nature of the corrosive environment: a paint film with a higher PVC is more porous and the SO_2 molecules penetrate faster and more readily through the film. As a result, the efficiency of the paints containing the isometric pigment particles at high concentrations decreases to the level of that of the non-pigmented film.
- Unlike the non-core-shell pigments, the core-shell pigments provide very low protection against metal surface corrosion.
- The chemically inactive pigment ***F***, acting by the barrier effect only, as well as the large particles of the core-shell pigments, is unable to completely prevent diffusion of the SO_2 molecules through the protective coating.
- Active protection against corrosion in the environment with SO_2 can be achieved by using a non-isometric pigment with small particles. The efficiency of such a pigment is higher than that of a core-shell pigment, which exhibits a larger particle size distribution.

The overall anti-corrosive efficiency was calculated for all the paint films tested. The most efficient in this respect was the paint with pigment ***B*** (*acicular* $Mg_{0.2}Zn_{0.8}Fe_2O_4$) at PVC = 5% (E_{SO_2} = 79) and at PVC = 10% (E_{SO_2} =71), followed by the paint with pigment ***C*** (*lamellar* $Mg_{0.2}Zn_{0.8}Fe_2O_4$) at PVC= 5% and 10% (E_{SO_2} = 68). Both pigment types exhibit a fairly uniform anti-corrosive efficiency. The corrosive environment contains mobile SO_2 ions [58] creating a strongly acid medium containing sulphite ions $\left(SO_3^{2-}\right)$ and sulphate ions $\left(SO_4^{2-}\right)$, and so a pigment with small particles exerting an alkaline effect is needed to attain a high corrosion resistance. Such a pigment will hinder diffusion through the paint film and neutralize the action of acid substances [52]. The paint with pigment ***A*** (*isometric* $Mg_{0.2}Zn_{0.8}Fe_2O_4$), was found efficient, its concentration, however, must be as low as PVC = 5% (E_{SO_2} =73).

The paint properties are not so favourable, in other words, the paint is not so efficient, if a core-shell pigment with lamellar particles is used. Such a paint is even less efficient than the coating material containing no pigment at all. Ideal for a paint exposed to the environment in question is a pigment with needle-shaped particles exerting an alkaline effect, which will also strengthen and reinforce the film.

4. Conclusions

Ferro-spinels (Zn-Mg-Fe) differing in the primary particle shapes were synthesized and added as pigments to a paint formula, and the paints containing them at various volume concentrations were investigated with respect to their anti-corrosive properties. Every property of a pigmented paint is at its optimum at a specific pigment concentration; this applies particularly to the physical and anti-corrosive properties [37]. So it is possible to identify a pigment concentration at which a specific property is at its best or at which the overall anti-corrosive efficiency is at its maximum [59]. It is convenient that such pigment concentrations are usually not very high.

The spinels fall in the class of chemically acting pigments that help slow down corrosion processes on the metal surface beneath the paint film through their alkaline nature and by neutralization of the carboxy groups. The lamellar shape of the pigment particles enhances paint film adhesion to the substrate and its cohesion and reduces the formation of blisters on the paint film surface (and around the test cut). The pigmented paint films exhibited very good physico-mechanical properties, commensurable with those of the alkyd resin alone.

The best anti-corrosive efficiency in the accelerated corrosion test in the atmosphere with SO_2 was found for the paint containing the non-isometric pigment ***B*** (*acicular* $Mg_{0.2}Zn_{0.8}Fe_2O_4$) at PVC = 5%.

It follows from the results that the morphology the pigment particles plays a major role in the paints' anti-corrosive properties. Paints with non-isometric pigments gave better results than paints with isometric pigments during nearly all measurements. This is primarily due to the barrier effect to substances penetrating through the protective film. Furthermore, the aggressive substances in the penetrating medium come in closer contact with the pigment particles, which may result in neutralization of the former. The best combined anti-corrosive effect was found for pigment ***C*** with lamellar particles (lamellar $Mg_{0.2}Zn_{0.8}Fe_2O_4$). This pigment acted very favourably by the chemical inhibition mechanism; induced saponification in the alkyd resin owing to by its high alkalinity; and acted by the barrier effect. The core-shell pigments ***D*** (mixed *Mg-Zn-Fe oxide/kaolin*) and ***E*** (mixed *Mg-Zn-Fe oxide/talc*), which also exhibit very good anti-corrosive efficiency in many parameters, may also be used as convenient variants: they are less expensive than pigment ***C*** and contain less zinc because the mineral core represents a considerable fraction of the pigment particle weight.

Paints with the pigments tested are also advantageous in that they protect the metal against corrosion. And they are environmentally friendly and might be favourably used to replace the toxic chromium (VI)-based pigments.

Author details

Kateřina Nechvílová, Andrea Kalendová* and Miroslav Kohl

*Address all correspondence to: andrea.kalendova@upce.cz

Faculty of Chemical Technology, University of Pardubice, Pardubice, Czech Republic

http://dx.doi.org/10.5772/65545

References

[1] A. Kalendova, D. Vesely, M. Kohl and J. Stejskal, Prog. Org. Coat. 78, 1 (2015).

[2] P. Benda and A. Kalendova, Phys. Proced. 44, 185 (2013).

[3] R. Naredi, M. Mahdavian and A. Darvish, Prog. Org. Coat. 76, 302 (2016).

[4] A. Darvish, R. Naredi and M. M. Attar, Prog. Org. Coat. 77, 1768 (2014).

[5] E. Langer, H. Kuczyńska, E. K. Tarnawska and J. Lukastcyk, Prog. Org. Coat. 71, 162 (2011).

[6] M. A. Deyab, Prog. Org. Coat. 85 146 (2015).

[7] X.-Z. Gao, H.-J. Liu and F. Cheng, Chem. Eng. J. 283, 682 (2016).

[8] M. Conradi, A. Koujan, M. Zorko and I. Jerman. Prog. Org. Coat. 74, 392 (2012).

[9] M. E. H. Sadek, R. M. Khattab, A. A. Gaber and M. F. Zawrah, Spectrochim. Acta Part A . 125, 353 (2014).

[10] A. Kalendová, Pigm. Resin Technol. 27, 225 (1998).

[11] A. Kalendová, Anti-Corros. Methods Mater. 45, 344 (1998).

[12] A. Kalendová, Pigm. Resin Technol. 29, 215 (2000).

[13] A. Kalendová, Pigm. Resin Technol. 29, 164 (2000).

[14] L. Anez, S. Calas-Etienne, J. Primera and T. Woignier, Microporous and Mesoporous Mater. 200, 79 (2014).

[15] T. Marzi, Constr. Build. Mater. 97, 119 (2015).

[16] J. Šadauskiene, V. Stankevičius, R. Bliudžuus and A. Gailius, Constr. Build. Mater. 23, 2788 (2009).

[17] M. Ulbrich and A. Kalendova, Phys. Proced. 44, 247 (2013).

[18] J. Du, S. Liang, X. Wang and Z. Fan. Acta Metall. Sinica (English lett.). 22, 263 (2009).

[19] S. Y. Arman, B. Ramezanzadeh, S. Farghadani, M. Mehdijour and A. Rajabi. Corros. Sci. 77, 118 (2013).

[20] B. Nikravesh, B. Ramezanzedeh, A. A. Sarabi and S. M. Kasiriha, Corros. Sci. 53, 1592 (2011).

[21] J.-M. Hu, J.-T. Zhang, J.-Q. Zhang and C.-N. Cao, Corros. Sci. 47, 2607 (2005).

[22] I. Kos, I. G. Schwarz and K. Suton, Procedia Engineer. 69, 881 (2014).

[23] M. Kohl and A. Kalendova. Prog. Org. Coat. 86, 96 (2015).

[24] S. Singh, N. Kumar, R. Bhargava, M. Sahni, K. Sung and J. H. Jung, J. Alloys and Compd. 587, 437 (2014).

[25] L. Lu, I. Swonkaev, A. Kumar and D. V. Goia, Powder Technol. 261, 87 (2014).

[26] S. Roy, S. Kar, B. Bagchi and S. Das, Appl. Clay Sci.107, 205 (2015).

[27] C. Liu, M. Ye, A. Han and J. Li, Ceram. Int. 41, 5537 (2015).

[28] D. Vesely, A. Kalendova and P. Nemec, Sur. Coat. Technol. 204, 2032 (2010).

[29] G. Toussaint, M. Brisbois, J. Grandjean, R. Cloots and C. Menrist, J. Colloid and Interface Sci. 329, 120 (2009).

[30] M. Ren, H. Yin, Y. Hang, C. Ge, Y. Zhang, A. Wang, L. Yu and Z. Wu, Powder Technol. 214, 31 (2011).

[31] J. Li, L. Ecco, M. Fedel, V. Ermini, G. Delmas and J. Pan, Prog. Org. Coat. 87, 179 (2015).

[32] A. Gergely, I. Bertoti, T. Totok, E. Pfeifer and E. Kalman, Prog. Org. Coat. 76, 17 (2013).

[33] A. Kalendová and D. Veselý, Pigm. Resin Technol. 36, 195 (2007).

[34] A. Kalendová and D. Veselý, Anti-Corros. Methods Mater. 54, 3 (2007).

[35] P. Porta, F. S. Stone and R. G. Turner, J. Solid State Chem. 11, 135 (1974).

[36] K. Sun, G. Wu, B. Wang, Q. Zhong, Y. Yang, Z. Yu, C. Wei, X. Jiang and Z. Lan, J. Alloys Compd. 650, 363 (2015).

[37] A. Kalendová, P. Rysanek and K. Nechvilova, Prog. Org. Coat. 86, 147 (2015).

[38] A. L. Fernandez and L. de Pablo, Pigm. Resin Technol. 31, 350 (2002).

[39] M. F. Zawrah, H. Mamaad and S. Meky, Ceram. Int. 33, 969 (2007).

[40] J. B. R. Neto and R. Moreno, Appl. Clay Sci. 37, 157 (2007).

[41] Z. Li, T. Katsumi, T. Inui and A. Takai, Soils Found. 53, 680 (2013).

[42] D. Veselý, A. Kalendova and M. V. Manso, Prog. Org. Coat. 74, 82 (2012).

[43] K. J. Jotti and K. Palanivelu, Appl. Surf. Sci. 288, 60 (2014).

[44] A. Kalendová and D. Veselý, Prog. Org. Coat. 64, 5 (2009).

[45] A. Kalendová, D. Veselý and P. Kalenda, Applied Clay Science 48, 581 (2010).

[46] A. Kalendová, P. Kalenda and D. Veselý, Prog. Org. Coat. 57, 1 (2006).

[47] A. Kalendova, Prog. Org. Coat. 44, 201 (2002).

[48] Y. Hao, F. Liu and E. Man, J. Electrochem. Soc. 159, 403 (2012).

[49] M. Kohl, A. Kalendová, R. Boidin and P. Němec, Prog. Org. Coat. 77, 1369 (2014).

[50] A. Kalendová, D. Veselý, M. Kohl and J. Stejskal, Prog. Org. Coat. 77, 1465 (2014).

[51] A. M. Atta, R. A. El-Ghazawy and A. M. El-Saeed, Int. J. Electrochem. Sci. 8, 5136 (2013).

[52] M. Conradi, A. Kocijan, D. Ker-Merl, M. Zorko and I. Verpoest, Appl. Surf. Sci. 292, 432 (2014).

[53] M. F. Montemor, Surf. Coat. Technol. 258, 17 (2014).

[54] A. Kalendová, Prog. Org. Coat. 38, 199 (2000).

[55] G. Wang and J. Yang, Surf. Coat. Technol. 204, 1186 (2010).

[56] D. Sun, M. X. Wang, Z. H. Zhang, H. L. Tao, M. He, B. Song and Q. Li, Solid State Commun. 223, 12 (2015).

[57] N. M. Ahmed, M. G. Mohamed, M. R. Mabrouk and A. A. ElShami, Construction and Building Mater. 98, 399 (2015).

[58] H. S. Emira and F.F. Abdel-Mohsen, Pigm. Resin Technol. 32, 259 (2003).

[59] A. M. Banerjee, M. R. Pai, S. S. Meena, A. K. Tripathi and S. R. Bharadwaj, Int. J. Hydrogen Energy 36, 4768 (2011).